Learn PostgreSQL

Build and manage high-performance database solutions
using PostgreSQL 12 and 13

Luca Ferrari
Enrico Pirozzi

BIRMINGHAM - MUMBAI

Learn PostgreSQL

Commissioning Editor: Amey Varangaonkar
Acquisition Editor: Siddarth Mandal
Content Development Editor: Joseph Sunil
Senior Editor: David Sugarman
Technical Editor: Sonam Pandey
Copy Editor: Safis Editing
Project Coordinator: Aishwarya Mohan
Proofreader: Safis Editing
Indexer: Manju Arasan
Production Designer: Alishon Mendonca

First published: October 2020

Production reference: 1081020

Published by Packt Publishing Ltd.
Livery Place
35 Livery Street
Birmingham
B3 2PB, UK.

ISBN 978-1-83898-528-8

www.packt.com

To my beautiful wife, Emanuela; I love her like Santa loves his reindeer.
To my great son, Diego, who has changed our lives on 1283788200.
To my parents, Miriam and Anselmo; my greatest fans since day one.

– Luca Ferrari

In loving memory of my father, Ilario.

- Enrico Pirozzi

About Packt

Subscribe to our online digital library for full access to over 7,000 books and videos, as well as industry leading tools to help you plan your personal development and advance your career. For more information, please visit our website.

Why subscribe?

- Spend less time learning and more time coding with practical eBooks and Videos from over 4,000 industry professionals

- Improve your learning with Skill Plans built especially for you

- Get a free eBook or video every month

- Fully searchable for easy access to vital information

- Copy and paste, print, and bookmark content

Did you know that Packt offers eBook versions of every book published, with PDF and ePub files available? You can upgrade to the eBook version at `www.packt.com` and as a print book customer, you are entitled to a discount on the eBook copy. Get in touch with us at `customercare@packtpub.com` for more details.

At `www.packt.com`, you can also read a collection of free technical articles, sign up for a range of free newsletters, and receive exclusive discounts and offers on Packt books and eBooks.

Contributors

About the authors

Luca Ferrari has been passionate about computer science since the Commodore 64 era, and today holds a master's degree (with honors) and a Ph.D. from the University of Modena and Reggio Emilia. He has written several research papers, technical articles, and book chapters. In 2011, he was named an Adjunct Professor by Nipissing University. An avid Unix user, he is a strong advocate of open source, and in his free time, he collaborates with a few projects. He met PostgreSQL back in release 7.3; he was a founder and former president of the **Italian PostgreSQL Community** (**ITPUG**). He also talks regularly at technical conferences and events and delivers professional training.

Enrico Pirozzi has been passionate about computer science since he was a 13-year-old, his first computer was a Commodore 64, and today he holds a master's degree from the University of Bologna. He has participated as a speaker at national and international conferences on PostgreSQL. He met PostgreSQL back in release 7.2, he was a co-founder of the first PostgreSQL Italian mailing list and the first Italian PostgreSQL website, and he talks regularly at technical conferences and events and delivers professional training. Right now, he is employed as a PostgreSQL database administrator at Nexteam (Zucchetti Group S.p.a.).

About the reviewers

Marcelo Diaz is a software engineer with more than 15 years of experience, and with a special focus on PostgreSQL. He is passionate about open source and has promoted its application in critical and high-demand environments where he has worked as a software developer and consultant for both private and public companies. He currently works very happily at Cybertec and as a technical reviewer for Packt Publishing. He enjoys spending his leisure time with his daughter, Malvina, and his wife, Romina. He also likes playing football.

Ilja Everilä is a software developer and consultant with over a decade of experience in various projects. He has done both frontend and backend work, along with database administration, in PostgreSQL. All in all, he is very much into database work and found this book an interesting read on an important subject.

Packt is searching for authors like you

If you're interested in becoming an author for Packt, please visit `authors.packtpub.com` and apply today. We have worked with thousands of developers and tech professionals, just like you, to help them share their insight with the global tech community. You can make a general application, apply for a specific hot topic that we are recruiting an author for, or submit your own idea.

Table of Contents

Section 3: Administering the Cluster

Preface

PostgreSQL is one of the fastest-growing open source object-relational **Database Management Systems** (**DBMS**) in the world. As well as being easy to use, it's scalable and highly efficient. In this book, you'll explore PostgreSQL 12 and 13 and learn how to build database solutions using it. Complete with hands-on tutorials, this guide will teach you how to achieve the right database design required for a reliable environment.

You'll learn how to install and configure a PostgreSQL server and even manage users and connections. The book then progresses to key concepts of relational databases, before taking you through the **Data Definition Language** (**DDL**) and commonly used DDL commands. To build on your skills, you'll understand how to interact with the live cluster, create database objects, and use tools to connect to the live cluster. You'll then get to grips with creating tables, building indexes, and designing your database schema. Later, you'll explore the **Data Manipulation Language** (**DML**) and server-side programming capabilities of PostgreSQL using PL/pgSQL, before learning how to monitor, test, and troubleshoot your database application to ensure high-performance and reliability.

By the end of this book, you'll be well-versed in the Postgres database and be able to set up your own PostgreSQL instance and use it to build robust solutions.

Who this book is for

This Postgres book is for anyone interested in learning about the PostgreSQL database from scratch. Anyone looking to build robust data warehousing applications and scale the database for high-availability and performance using the latest features of PostgreSQL will also find this book useful. Although prior knowledge of PostgreSQL is not required, familiarity with databases is expected.

What this book covers

Chapter 1, *Introduction to PostgreSQL*, explains what the PostgreSQL database is, the community and development behind this great and robust relational database, as well as how to get help and recognize different PostgreSQL versions and dependencies. You will also learn how to get and install PostgreSQL through either binary packages or by compiling it from sources. A glance at how to manage the cluster with your operating system tools (systemd and rc scripts) will be taken.

Chapter 2, *Getting to Know Your Cluster,* shows you the anatomy of a PostgreSQL cluster by specifying what is on the file system, where the main configuration files are, and how they are used. The `psql` command-line utility will be described in order to make you connect to the database cluster and check it's working.

Chapter 3, *Managing Users and Connections*, provides a complete description of how users and connections are managed by a running instance and how you can prevent or limit user connections. The architecture and terminology of the database will be detailed. The concept of "role" will be described, and you will learn how to create single-user accounts, as well as groups.

Chapter 4, *Basic Statements*, shows how to create and destroy main database objects, such as databases, tables, and schemas.
The chapter also takes a glance at basic statements, such as SELECT, INSERT, UPDATE, and DELETE.

Chapter 5, *Advanced Statements,* introduces the advanced statements PostgreSQL provides, such as common table expressions, UPSERTs, and queries with RETURNING rows. This chapter will provide practical examples of when and how to use them.

Chapter 6, *Window Functions*, introduces a powerful set of functions that provide aggregation without having to collapse the result in a single row. In other words, thanks to window functions, you can perform aggregation on multiple rows (windows) and still present all the tuples in the output. Window functions allow the implementation of business intelligence and make reporting easy.

Chapter 7, *Server-Side Programming*, tackles the fact that while SQL is fine for doing most of the day-to-day work with a database, you could end up with a particular problem that requires an imperative approach. This chapter shows you how to implement your own code within the database, how to write functions and procedures in different languages, and how to make them interact with transaction boundaries.
Chapter 8, *Triggers and Rules*, presents both triggers and rules with practical examples, showing advantages and drawbacks.

Chapter 9, *Partitioning,* explores partitioning – the capability to split a table into smaller pieces. PostgreSQL has supported partitioning for a long time, but with version 10 it introduced so-called "declarative partitioning." After having a quick lock at old-school inheritance-based partitioning, the chapter focuses on all the features related to declarative partitioning and its tuning parameters.

Chapter 10, *Users, Roles, and Database Security*, first glances at user management: roles, groups, and passwords.
You will learn how to constrain users to access only particular databases and from particular machines, as well as how to constrain the usage of database objects such as tables. You then will see how row-level security can harden your table contents and prevent users from modifying tuples that do not belong to them.

Chapter 11, *Transactions, MVCC, WALs, and Checkpoints*, presents a very fundamental concept in PostgreSQL: the Write-Ahead Log. You will learn why such a log is so important, how it deals with transactions, and how you can interact with transactions from a SQL point of view. The chapter also presents you with the concept of transaction isolation, ACID rules, and how the database can implement them. Then you will discover how the WAL can speed up database work and, at the very same time, can protect it against crashes. You will understand what MVCC is and why it is important. Lastly, the chapter provides insight into checkpoints and related tunables.

Chapter 12, *Extending the Database*, introduces a handy way to plug new functionalities into your cluster – extensions. This chapter will show you what an extension is, how to get and install an extension, and how to search for already available extensions in the PostgreSQL ecosystem.

Chapter 13, *Indexes and Performance Optimization*, addresses the fact that optimizing for performance is an important task for every database administrator. Indexes are fast ways to let the database access the most commonly used data, but they cannot be built on top of everything because of their maintenance costs. The chapter presents the available index types, then it explains how to recognize tables and queries that could benefit from indexes and how to deploy them. Thanks to tools such as explain and autoexplain, you will keep your queries under control.

Chapter 14, *Logging and Auditing*, tackles questions such as *What is happening in the database cluster? What happened yesterday?*
Having a good logging and auditing ruleset is a key point in the administration of a database cluster. The chapter presents you with the main options for logging, how to inspect logs with external utilities such as pgFouine, and how to audit your cluster (in a way that can help you make it compliant with GDPR).

Chapter 15, *Backup and Restore*, broaches the fact that things can go wrong, and in such cases, you need a good backup to promptly restore in order for your database to always be available. The chapter presents the basic and most common ways to back up a single database or a whole cluster, as well as how to do archiving and point-in-time recovery. External tools such as Barman and pgBackRest will be introduced.

Chapter 16, *Configuration and Monitoring*, presents the cluster catalog, the way in which PostgreSQL exports its own internal status. It does not matter how finely you tuned your cluster, you need to monitor it to understand and promptly adjust it to incoming needs. Knowing the catalog is fundamental for a database administrator, in order to be able to see what is going on in the live system. Thanks to special extensions, such as pg_stat_activity, you will be able to monitor in real time what your users are doing against the database.

Chapter 17, *Physical Replication*, covers built-in replication, a mechanism that allows you to keep several instances up and in sync with a single master node, which PostgreSQL has supported since version 9. Replication allows scalability and redundancy, as well as many other scenarios such as testing and comparing databases. This chapter presents so-called "physical replication," a way to fully replicate a whole cluster over another instance that will continuously follow its leader. Both asynchronous and synchronous replication, as well as replication slots, will be presented.

Chapter 18, *Logical Replication*, covers logical replication, which allows very fine-grained replication specifying which tables have to be replicated and which don't – supported by PostgreSQL since version 10. This, of course, allows a very new and rich scenario of data sharing across different database instances. The chapter presents how logical replication works, how to set it up, and how to monitor the replication.

Chapter 19, *Useful Tools and Useful Extensions*, is to be considered as an appendix to the book. In this chapter, we will talk about some tools and some extensions that allow the DBA to maximize the work done while minimizing the effort.

Chapter 20, *Toward PostgreSQL 13*, looks at the latest version of the database – PostgreSQL 13, which at the time of writing is in the beta-2 state. This chapter presents the main changes and highlights the differences between PostgreSQL 12 and version 13, and looks at how to upgrade to the new production-ready version once it is available.

To get the most out of this book

For this book to be useful, basic knowledge of the Linux operating system in any distribution or knowledge of the FreeBSD operating system is required. All the SQL examples can be run using the `psql` program or using the GUI tool pdAdmin. This makes them applicable to most platforms. Some scripts will be executed using the bash scripting language.

Software/Hardware covered in the book	OS Requirements
PostgreSQL 12 - 13	Linux OS / FreeBSD

If you are using the digital version of this book, we advise you to type the code yourself or access the code via the GitHub repository (link available in the next section). Doing so will help you avoid any potential errors related to the copying and pasting of code.

Download the example code files

You can download the example code files for this book from your account at `www.packt.com`. If you purchased this book elsewhere, you can visit `www.packtpub.com/support` and register to have the files emailed directly to you.

You can download the code files by following these steps:

1. Log in or register at `www.packt.com`.
2. Select the **Support** tab.
3. Click on **Code Downloads**.
4. Enter the name of the book in the **Search** box and follow the onscreen instructions.

Once the file is downloaded, please make sure that you unzip or extract the folder using the latest version of:

- WinRAR/7-Zip for Windows
- Zipeg/iZip/UnRarX for Mac
- 7-Zip/PeaZip for Linux

The code bundle for the book is also hosted on GitHub at `https://github.com/PacktPublishing/Learn-PostgreSQL`. In case there's an update to the code, it will be updated on the existing GitHub repository.

We also have other code bundles from our rich catalog of books and videos available at `https://github.com/PacktPublishing/`. Check them out!

Download the color images

We also provide a PDF file that has color images of the screenshots/diagrams used in this book. You can download it here: `https://static.packt-cdn.com/downloads/9781838985288_ColorImages.pdf`.

Conventions used

There are a number of text conventions used throughout this book.

`CodeInText`: Indicates code words in text, database table names, folder names, filenames, file extensions, pathnames, dummy URLs, user input, and Twitter handles. Here is an example: "`psql` is a powerful environment in which to manage our data and our databases."

A block of code is set as follows:

```
CREATE DATABASE databasename
```

When we wish to draw your attention to a particular part of a code block, the relevant lines or items are set in bold:

```
postgres=# \c forumdb
You are now connected to database "forumdb" as user "postgres".
forumdb=#
```

Bold: Indicates a new term, an important word, or words that you see onscreen. For example, words in menus or dialog boxes appear in the text like this. Here is an example: "The first field is an **object identifier** (**OID**), which is a number that uniquely identifies the database called `forumdb`."

Warnings or important notes appear like this.

 Tips and tricks appear like this.

Get in touch

Feedback from our readers is always welcome.

General feedback: If you have questions about any aspect of this book, mention the book title in the subject of your message and email us at customercare@packtpub.com.

Errata: Although we have taken every care to ensure the accuracy of our content, mistakes do happen. If you have found a mistake in this book, we would be grateful if you would report this to us. Please visit www.packtpub.com/support/errata, selecting your book, clicking on the Errata Submission Form link, and entering the details.

Piracy: If you come across any illegal copies of our works in any form on the Internet, we would be grateful if you would provide us with the location address or website name. Please contact us at copyright@packt.com with a link to the material.

If you are interested in becoming an author: If there is a topic that you have expertise in and you are interested in either writing or contributing to a book, please visit authors.packtpub.com.

Reviews

Please leave a review. Once you have read and used this book, why not leave a review on the site that you purchased it from? Potential readers can then see and use your unbiased opinion to make purchase decisions, we at Packt can understand what you think about our products, and our authors can see your feedback on their book. Thank you!

For more information about Packt, please visit packt.com.

Section 1: Getting Started

In this section, you will learn what PostgreSQL is, what is new in version 12 (and version 13), and how to install and run this great open source database.

This section contains the following chapters:

- *Chapter 1, Introduction to PostgreSQL*
- *Chapter 2, Getting to Know Your Cluster*
- *Chapter 3, Managing Users and Connections*

Introduction to PostgreSQL

1

PostgreSQL is a well-known open-source relational database, and its motto states what the project intends to be: *the most advanced open-source database in the world*.

The main qualities that attract masses of new users every year and keep current users enthusiastic about their projects are its rock-solid stability, scalability, and safeness, as well as the features that an enterprise-level database management system provides.

But PostgreSQL is not just a database; it has grown to be a whole ecosystem of extensions, tools, and languages tied together by communities spread around the world.

PostgreSQL is an open-source project and is fully developed in the open-source world. That means that there is no single entity in charge of the project and the result is that PostgreSQL is not a commercial product. In other words, PostgreSQL belongs to everyone, and anyone can contribute to it. Thanks to a very permissive BSD-style license, PostgreSQL can be used in any project or scenario, either open or closed source.

Of course, contributing to a project of that size requires experience in software development, database concepts, and, of course, a positive attitude to open source and collaborative efforts. But it does also mean that PostgreSQL will continue to live pretty much forever without the risk of a single company going out of business and sinking with the database.

PostgreSQL 12 is the latest release of this great database, and at the time of writing, efforts for PostgreSQL 13 have already begun. This book will focus on PostgreSQL, starting from the basics and moving toward the most exciting and complex tasks (such as replicating your datasets to prevent disasters). Of course, given that PostgreSQL is a project of this size with so many features, a single book cannot cover it all in detail, so our aim is to introduce the whole set of qualities PostgreSQL provides to you, giving practical use cases and examples, as well as external resources to help you learn more about particular aspects.

This book covers PostgreSQL 12 and 13, but the concepts explained in this book can apply also to later versions (as well as to previous ones when the same features are present). At the time of writing, PostgreSQL 12 is the stable release, while PostgreSQL 13 is in its second beta public release and is expected, therefore, to be stable enough for testing its features.

This chapter will introduce you to this great open source database starting from the project history and goals, which is very important to help you decide whether you want to use PostgreSQL in the first place. You will learn basic PostgreSQL terminology, which is very important to help you search the documentation and understand the main error messages, in case you need to. Finally, you will see how to install PostgreSQL in different ways so that you will get a basic knowledge of how to install it on different platforms and contexts.

The following topics are covered in this chapter:

- PostgreSQL at a glance
- Exploring PostgreSQL terminology
- Installing PostgreSQL 12 or higher

Technical requirements

You can find the code for this chapter at the following GitHub repository: `https://github.com/PacktPublishing/Learn-PostgreSQL`.

PostgreSQL at a glance

As a relational database, PostgreSQL provides a lot of features, and it is quite difficult to "scare" a PostgreSQL instance. In fact, a single instance can contain more than 4 billion individual databases, each with unlimited total size and capacity for more than 1 billion tables, each containing 32 TB of data. Moreover, if there's any concern that those upper limits won't suffice, please consider that a single table can have 1,600 columns, each 1 GB in size, with an unlimited number of multi-column (up to 32 columns) indexes. In short, PostgreSQL can store much more data than you can possibly think of!

Therefore, there is no amount of data that PostgreSQL cannot handle, but of course, in order to perform well with certain big databases, you need to understand PostgreSQL and its features.

PostgreSQL is fully ACID-compliant and has a very strong foundation in data integrity and concurrency. It ships with a procedural language, named PL/pgSQL, which can be used to write reusable pieces of code, such as functions and routines, and it supports before and after triggers, views, materialized views, and partitioned tables.

ACID is an acronym of the properties used to indicate that the database engine provides **atomicity**, **consistency**, **isolation**, and **durability**. Atomicity means that a complex database operation is processed as a single instruction even when it is made up of different operations. Consistency means that the data within the database is always kept consistent and that is it is not corrupted due to partially performed operations. Isolation allows the database to handle concurrency in the "right way"—that is, without having corrupted data from interleaved changes. Lastly, durability means that the database engine is supposed to protect the data it contains, even in the case of software and hardware failures, as much as it can.

PostgreSQL can be extended with other embedded languages, such as Perl, Python, Java, and even Bash! And if you think the database does not provide you with enough features, you can plug in extensions to obtain different behaviors and enhancements—for instance, **geospatial references** (**GIS**), scheduled jobs, esoteric data types, and utilities in general.

PostgreSQL runs on pretty much every operating system out there, including Linux, Unix, Mac OS X, and Microsoft Windows, and can even run on commodity hardware such as Raspberry Pi boards. There are also several cloud computing providers that list PostgreSQL in their software catalog.

Thanks to its extensive tuning mechanism, it can be adapted very well to the hosting platform. The community is responsible for keeping the database and documentation at a very high-quality level, and also the mailing lists and IRC channels are very responsive and a valuable source for solutions and ideas.

In the experience of the authors, there has never been a case where PostgreSQL has not been able to adapt to an application scenario.

The PostgreSQL project has a very rich and extensive set of a mailing lists that range from general topics to very specific details. It is a good habit to search for problems and solutions on the mailing list archives; see the web page at `https://www.postgresql.org/list/` to get a better idea.

A brief history of PostgreSQL

PostgreSQL takes its name from its ancestor: Ingres. Ingres was a relational database developed by professor Michael Stonebraker. In 1986, Professor Stonebraker started a post-Ingres project to develop new cool features in the database landscape and named this project **POSTGRES** (**POST-Ingres**). The project aimed to develop an object-relational database, where "object" means the user would have the capability to extend the database with their own objects, such as data types, functions, and so on.

In 1994, POSTGRES was released with version 4.2 and an MIT license, which opened up collaboration from other developers around the world. At that time, POSTGRES was using an internal query language named **QUEL**.

Two Berkeley students, Andrew Yu and Jolly Chen, replaced the QUEL query language with the hot and cool SQL language, and the feature was so innovative that the project changed its name to Postgre95 to emphasize the difference compared to other preceding versions.

Eventually, in 1996, the project gained a public server to host the code, and five developers, including Marc G. Fournier, Tom Lane, and Bruce Momjan, started the development of the new branded project named **PostgreSQL**. Since then, the project has been kept in good shape and up to date.

This also means that PostgreSQL has been developed for over 30 years, again emphasizing the solidity and openness of the project itself. If you are curious, it is also possible to dig into the source code down to the initial commit in the open source world:

```
$ git log `git rev-list --max-parents=0 HEAD`
commit d31084e9d1118b25fd16580d9d8c2924b5740dff
 Author: Marc G. Fournier <scrappy@hub.org>
 Date:   Tue Jul 9 06:22:35 1996 +0000

 Postgres95 1.01 Distribution - Virgin Sources
```

What's new in PostgreSQL 12?

PostgreSQL 12 was released on October 3, 2019. It includes a rich set of new features with regard to its predecessor versions, including the following:

- Several performance optimizations, ranging from inlining Common Table Expressions to huge table partition management and an improved user-defined statistic hint for multi-column selections

- A few administrative optimizations, including the concurrent rebuilding of indexes, off-line check-summing, and, most notably, reporting about maintenance processes' progress
- Security features including multi-factor authentication and TCP/IP encryption via GSSAPI
- Support for the SQL JSON path language
- Stored generated columns

PostgreSQL 12 also contains a set of changes aimed to make the **database administrator** (**DBA**)'s life easier—for instance, removing conflicting options and obsolete SQL terms and types. This emphasizes the fact that PostgreSQL developers do always take care of the database and its adherence to the current SQL standard.

What's new in PostgreSQL 13?

PostgreSQL 13 will contain a very rich set of optimizations under the hood, with particular regard to the following:

- **Partitioning**, which now includes the ability to execute *before* triggers on partitioned tables, the capability to prune partitions in particular edge cases to speed up query execution, and a better way to join partitions in queries (referred to as **partition-wise joins**).
- **Replication**, which can now work at the logical level even on partitioned tables, automatically publishing all the partitions. Also, there is now no automatic promotion of a server if it does not reach the specific target to recovery and a slave server can be promoted without cancelling any pending pause requests. It is worth noting that it is possible to change the settings of a streaming replication without having to restart the cluster, therefore having a no-downtime impact.
- **Indexes**, which are now more efficient in general for storing data and accepting operators with parameters.
- **Statistics**, with particular regard to improvements in the extended statistics, the data collected and used by the optimizer and a few changes in the monitoring catalogs.

There are a lot more changes that will be discussed in an appropriate chapter at the end of the book, but as usual, a new release of PostgreSQL contains performance improvements as well as security improvements and, as always, a better configuration system.

PostgreSQL release policy, version numbers, and life cycle

PostgreSQL developers release a new major release once per year, usually near October. A *major release* is a stable version that introduces new features and possible incompatibilities with previous versions. During its life cycle, a major release is constantly improved by means of *minor releases*, which are usually bug-fixing and maintenance releases.

The PostgreSQL version number identifies the major and minor release. Since PostgreSQL 10 (released in 2011), the version number is specified as `major.minor`; so, for instance, `12.0` indicates the first major release, `12`, while `12.1` indicates the minor release, `1`, of major release `12`. In short, the greater the number, the more recent the version you are managing.

However, before PostgreSQL 10, the version number was made by three different groups of digits—`brand.year.minor`—where the *brand* is the main development topic (for instance, "replication"), the *year* represents the year of development of that brand, and the *minor* is the minor version. What is important to keep in mind is that the *brand* and *year* pair made a major number in PostgreSQL versions prior to 10. So, for instance, PostgreSQL `9.6.16` is the 16th minor release done on brand *9* during the 6th year of development, and therefore could be incompatible with 9.5.20 because the two major versions are 9.6 and 9.5.

But what does it mean, in a practical sense, that two major versions are possibly incompatible?

PostgreSQL stores its own data on the storage system, often the hard disk. This data is stored in binary format, for optimization of performances and space consumption, and this format could possibly change between major versions. This means that, while you are able to upgrade PostgreSQL between minor versions on the fly, you probably will have to dump and restore your database content between major version upgrades. As you will see in this book, PostgreSQL provides ad hoc tools to support you even in the worst case of a major upgrade with a lot of incompatibilities, but keep in mind that while a minor version upgrade is something you usually do without any ahead planning, a major version upgrade could imply downtime.

The recommendation, as for much other software, is to run the most recent version of PostgreSQL available to you: PostgreSQL developers put in a lot of effort in order to provide bug-free products, but new features could introduce new bugs, and regardless of the very extensive testing platform PostgreSQL has, it is software after all, and software could have bugs. Despite internal bugs, new releases also include fixes for security exploits and performance improvements, so it is a very good habit to keep up to date with your running PostgreSQL server.

Last but not least, not all PostgreSQL versions will live forever. PostgreSQL provides support and upgrades for 5 years after a new release is issued; after this length of time, a major release will reach its **end of life** (**EOL**) and PostgreSQL developers will no longer maintain it. This does not mean you cannot run an ancient version of PostgreSQL, it simply means this version will not get any upgrades from the official project and, therefore, will be out of date. As an example, since PostgreSQL 12 was released in 2019, it will reach its EOL in 2024.

With that in mind, we'll now introduce the main PostgreSQL terminology, as well as further useful-to-understand concepts.

Exploring PostgreSQL terminology

A PostgreSQL instance is called a **cluster** because a single instance can serve and handle multiple databases. Every database is an isolated space where users and applications can store data.

A database is accessed by allowed users, but users connected to a database cannot cross the database boundaries and interact with data contained in another database, unless they explicitly connect to the latter database too.

A database can be organized into namespaces, called *schemas*. A schema is a mnemonic name that the user can assign to organize database objects, such as tables, into a more structured collection. Schemas cannot be nested, so they represent a flat namespace.

Database objects are represented by everything the user can create and manage within the database—for instance, tables, functions, triggers, and data types. Every object belongs to one and only one schema that, if not specified, is the default *public* schema.

Users are defined at a cluster-wide level, which means they are not tied to a particular database in the cluster. A user can connect with and manage any database in the cluster they have been allowed to.

PostgreSQL splits users into two main categories:

- **Normal users**: These users are the ones who can connect to and handle databases and objects depending on their privilege set.
- **Superusers**: These users can do anything with any database object.

PostgreSQL allows the configuration of as many superusers as you need, and every superuser has the very same permissions: they can do everything with every database and object and, most notably, can also control the life cycle of the cluster (for instance, they can terminate normal user connections, reload the configuration, stop the whole cluster, and so on).

PostgreSQL internal data, such as users, databases, namespaces, configuration, and database runtime status, is provided by means of catalogs: special tables that present information in a SQL-interactive way. Many catalogs are trimmed depending on the user who is inspecting them, with the exception that superusers usually see the whole set of available information.

PostgreSQL stores the user data (for example, tables) and its internal status on the local filesystem. This is an important point to keep in mind: PostgreSQL relies on the underlying filesystem to implement persistence, and therefore tuning the filesystem is an important task in order to make PostgreSQL perform well. In particular, PostgreSQL stores all of its content (user data and internal status) in a single filesystem directory known as PGDATA. The PGDATA directory represents what the cluster is serving as databases, so it is possible for you to have a single installation of PostgreSQL and make it switch to different PGDATA directories to deliver different content. In effect, this is a possible way to implement quick upgrades between major versions. As you will see in the next sections, the PGDATA directory needs to be initialized before it can be used by PostgreSQL; the initialization is the creation of the directory structure within PGDATA itself and is, of course, a one-time operation.

The detailed content of PGDATA will be explained later in the next chapter, but for now, it will suffice for you to remember that the PGDATA directory is where PostgreSQL expects to find data and configuration files. In particular, the PGDATA directory is made by at least the **write-ahead logs** (**WALs**) and the data storage. Without either of those two parts, the cluster is unable to guarantee data consistency and, in some critical circumstances, even start.

WALs are a technology that many database systems use, and even some transaction filesystems (such as ZFS, ReiserFS, UFS with Soft Updates, and so on) provide. The idea is that, before applying any change to a chunk of data, an intent log will be made persistent. In this case, if the cluster crashes, it can always rely on the already-written intent log to understand what operations have been completed and what must be recovered (more details on this in later chapters). Please note that with the term "crash," we refer to any possible disaster that can hit your cluster, including a software bug, but more likely the lack of electrical power, hard disk failures, and so on. PostgreSQL does commit to providing to you the best data consistency it can, and therefore, it makes a great effort to ensure that the intent log (WAL) is as secure as possible.

Internally, PostgreSQL keeps track of the tables structures, indexes, functions, and all the stuff needed to manage the cluster in dedicated storage named the catalog. The PostgreSQL catalog is fundamental for the life cycle of the cluster and reflects pretty much every action the database does on the user's structures and data. PostgreSQL provides access to the catalog from database superusers by means of an SQL interface, which means the catalog is totally explorable and, to some extent, manipulable, via SQL statements.

The SQL standard defines a so-called **information schema**, a collection of tables common to all standard database implementations, including PostgreSQL, that the DBA can use to inspect the internal status of the database itself. For instance, the information schema defines a table that collects information about all the user-defined tables so that it is possible to query the information schema to see whether a specific table exists or not.

The PostgreSQL catalog is what some call an "information schema on steroids": the catalog is much more accurate and PostgreSQL-specific that the general information schema, and the DBA can extract a lot more information about the PostgreSQL status from the catalog. Of course, PostgreSQL does support the information schema, but throughout the whole book, you will see references to the catalogs because they provide much more detailed information.

When the cluster is started, PostgreSQL launches a single process called the *postmaster*. The aim of the postmaster is to wait for incoming client connections, often made over a TCP/IP connection, and fork another process named the *backend process*, which in turn is in charge of serving one and only one connection.

This means that every time a new connection against the cluster is opened, the cluster reacts by launching a new backend process to serve it until the connection ends and the process is, consequently, destroyed. The postmaster usually starts also some utility processes that are responsible to keep PostgreSQL in good shape while it is running; these processes will be discussed later in this process.

To summarize, PostgreSQL provides you with executables that can be installed wherever you want on your system and can serve a single cluster. The cluster, in turn, serves data out of a single PGDATA directory that contains, among other stuff, the user data, the cluster internal status, and the WALs. Every time a client connects to the server, the postmaster process forks a new backend process that is the minion in charge of serving the connection.

This is a quick recap of the main terms used within PostgreSQL:

- **Cluster**: Cluster refers to the whole PostgreSQL service.
- **Postmaster**: This is the first process the cluster executes, and this process is responsible for keeping track of the activities of the whole cluster. The postmaster forks itself into a backend process every time a new connection is established.
- **Database**: The database is an isolated data container to which users (or applications) can connect to. A cluster can handle multiple databases. A database can be made by different objects, including schemas (namespaces), tables, triggers, and other objects you will see as the book progresses.
- PGDATA: PGDATA is the name of the directory that, on persistent storage, is fully dedicated to PostgreSQL and its data. PostgreSQL stores the data within such a directory.
- **WALs**: WALs contains the intent log of database changes, used to recover data from a critical crash.

Now that we've discussed the basic terminology related to PostgreSQL, it is time to get it installed on your machine.

Installing PostgreSQL 12 or higher

PostgreSQL can run on several Unix and Unix-like operating systems, such as Linux, as well as on Microsoft Windows. So far, the most supported platform remains Linux because most PostgreSQL developers work on this platform, and so it is the one with the most tested use cases. However, deploying on other platforms should not present any problems and, most importantly, is not going to put your data at any risk.

This section will focus on installing PostgreSQL 12, since it is the latest stable version available worldwide. You will learn, however, how to build your own version of PostgreSQL, and this may also be the way to install PostgreSQL 13 on your system.

Before installing PostgreSQL 12, you need to choose, or at least evaluate, how to install it. There are two main ways to get PostgreSQL 12 up and running, as follows:

- Compiling from sources
- Using a binary package

Binary packages are provided by the PostgreSQL community or the operating system, and using them has the advantage that it can provide you with a PostgreSQL installation very quickly. Moreover, binary packages do not require a compilation toolchain, and therefore are much easier to adopt. Lastly, a binary package adheres to the operating system conventions it has been built for (for instance, on where to place configuration files) and upgrades can be managed by the operating system as well. Since binary packages need to be pre-built from vendors, they could possibly not be the very latest released version.

On the other hand, installing from sources requires a compilation toolchain, as well as much more time and CPU consumption to build the PostgreSQL executables. You have full control over which components will be available in the final product, and can trim and optimize your instance for very high performances and shrink resource consumption to a minimum. In the long term, however, you will be responsible for maintaining the installation and upgrading it in a similar manner.

What to install

PostgreSQL is split across several components to install:

- The PostgreSQL **server** is the part that can serve your databases to applications and users and is required to store your data.
- The PostgreSQL **client** is the library and client tool to connect to the database server. It is not required if you don't need to connect to the database on the very same machine, while it is required on client machines.
- The PostgreSQL `contrib` package is a set of well-known extensions and utilities that can enhance your PostgreSQL experience.
- The PostgreSQL **docs** is the documentation related to the server and the client.
- PostgreSQL PL/Perl, PL/Python, and PL/Tcl are three components to allow the usage of programming languages— Perl, Python, and Tcl, respectively—directly within the PostgreSQL server.

The recommended set of components is the server, the client, and the `contrib` modules; these modules will be used across the book. You are free to decide whether to install the other components as you wish.

Installing PostgreSQL 12 from binary packages

In the following sections, you will see how to install PostgreSQL 12 on a few popular Linux and Unix operating systems, namely the following:

- GNU/Linux Debian, Ubuntu, and derivatives
- Fedora
- FreeBSD

It is not possible to provide detailed instructions for every operating system out there, but the concepts presented in the following sections should prove insightful regardless.

Installing PostgreSQL 12 on GNU/Linux Debian, Ubuntu, and derivatives

The **PostgreSQL Global Developers Group** (**PGDG**) provides binary packages for Debian and its derivatives, including the Ubuntu operating system family. In order to use the PGDG repositories, it is required for you to first install the source and signature of the repository:

1. To import the repository on an Ubuntu 19.10 disc, you need to run the following commands:

```
$ sudo /bin/sh -c '/bin/echo "deb
http://apt.postgresql.org/pub/repos/apt/ $(lsb_release -cs)-pgdg
main" > /etc/apt/sources.list.d/pgdg.list'
$ wget --quiet -O -
https://www.postgresql.org/media/keys/ACCC4CF8.asc | sudo apt-key
add -

$ sudo apt-get update
...
```

This will ensure the repository sources for your operating system are up to date so that you can install the PostgreSQL 12 packages. In the Debian/Ubuntu repositories, the packages are named after the component and the version, and the `postgresql-12` package includes the server and the `contrib` module.

2. Install the modules needed:

```
$ sudo apt install postgresql-12 postgresql-client-12 postgresql-
  contrib-12
```

Debian and Ubuntu provide their own command to control the cluster, `pg_ctlcluster(1)`. The rationale for that is that on a Debian/Ubuntu operating system, every PostgreSQL version is installed in its own directory with separate configuration files, so there is a way to run different versions concurrently and manage them via the operating system. For example, configuration files are under the `/etc/postgresql/12/main` directory, while the `data` directory is set by default to `/var/lib/postgresql/12/main`.

3. Enable PostgreSQL 12 at boot time by executing the following command:

```
$ sudo update-rc.d postgresql enable
```

4. Start the cluster immediately using the `service(1)` command:

```
$ sudo service postgresql start
```

You have thus installed PostgreSQL on GNU/Linux Debian, Ubuntu, and derivatives.

Installing PostgreSQL 12 on Linux Fedora

Fedora PostgreSQL packages are provided by the PostgreSQL community. In order to allow `dnf(8)` to find PostgreSQL packages, you need to install the PGDG repository, and then proceed with the installation as a distribution package:

1. Add the repository using the following command:

```
$ sudo dnf install
https://download.postgresql.org/pub/repos/yum/reporpms/F-30-x86_64/
pgdg-fedora-repo-latest.noarch.rpm
```

The list of available repositories can be obtained by the PostgreSQL official website at the download page (see the *References* section).

2. Install the PostgreSQL packages using the following command. Please note that the `postgresql12` package installs only the client part of the product, not the server:

```
$ sudo dnf -y install \
            postgresql12-server  \
```

```
                    postgresql12-contrib \
                    postgresql12-docs      \
                    postgresql12
...
Installed:
  postgresql12-contrib-12.1-2PGDG.f30.x86_64          postgresql12-
docs-12.1-2PGDG.f30.x86_64            postgresql12-
server-12.1-2PGDG.f30.x86_64
  postgresql12-12.1-2PGDG.f30.x86_64

Complete!
```

3. Configure the system specifying the PGDATA directory and enabling the option to start the service at boot time. In order to specify the PGDATA directory, you need to use systemd(1) to edit an overriding configuration file for the postgresql-12 service:

```
$ sudo systemctl edit postgresql-12
```

The preceding command will open your default text editor with an empty file; you can, therefore, set the PGDATA variable as follows and then save and exit the editor to apply changes:

```
[Service]
Environment=PGDATA=/postgres/12
```

4. Initialize the database directory; this can be done with a specific Fedora installation command named postgresql-12-setup, as follows:

```
$ sudo /usr/pgsql-12/bin/postgresql-12-setup initdb
Initializing database ... OK
```

5. Enable PostgreSQL 12 to start at boot time and launch the server immediately:

```
$ sudo systemctl enable postgresql-12

Created symlink /etc/systemd/system/multi-
user.target.wants/postgresql-12.service
                → /usr/lib/systemd/system/postgresql-12.service.

$ sudo systemctl start postgresql-12.service
```

If your Fedora installation contains the `service(8)` command, you can also start the service with the following:

```
$ sudo service postgresql-12 start

Redirecting to /bin/systemctl start postgresql-12.service
```

You have now successfully installed PostgreSQL 12 on Linux Fedora.

Installing PostgreSQL 12 on FreeBSD

PostgreSQL 12 is available on FreeBSD by means of ports and packages. Thanks to the `pkg(1)` command, it is very easy to install PostgreSQL 12 here, as shown in the following steps:

1. Search for available packages (execute an `update` command in order to scan for new packages):

```
$ pkg update
...
$ pkg search postgresql12
pgtcl-postgresql12-2.1.1_2        TCL extension for accessing a
PostgreSQL server (PGTCL-NG)
postgresql12-client-12.1          PostgreSQL database (client)
postgresql12-contrib-12.1         The contrib utilities from the
PostgreSQL distribution
postgresql12-docs-12.1            The PostgreSQL documentation set
postgresql12-plperl-12.1          Write SQL functions for PostgreSQL
using Perl5
postgresql12-plpython-12.1        Module for using Python to write SQL
functions
postgresql12-pltcl-12.1           Module for using Tcl to write SQL
functions
postgresql12-server-12.1          PostgreSQL is the most advanced
open-source database available anywhere
```

2. Install packages by executing `pkg(1)` and specify the set of packages you need. Of course, the installation must be executed as a user with administrative privileges, as follows:

```
$ sudo pkg install  postgresql12-server-12.1  \
                    postgresql12-client-12.1  \
                    postgresql12-contrib-12.1 \
                    postgresql12-docs-12.1
...
```

3. Initialize the directory to serve the database and to enable the server startup at the machine boot. The minimal parameters to set are `postgresql_enable` and `postgresql_data`. For example, to edit (as an administrative user) the `/etc/rc.conf` file, add the options as follows:

```
# to enable PostgreSQL at boot time
postgresql_enable="YES"

# PGDATA to use
postgresql_data="/postgres/12"
```

4. Then, run the following command to create and initialize the directory where PostgreSQL 12 will store the data:

```
$ sudo /usr/local/etc/rc.d/postgresql initdb

The files belonging to this database system will be owned by user
"postgres".
This user must also own the server process.

The database cluster will be initialized with locale "C".
The default text search configuration will be set to "english".

Data page checksums are disabled.

creating directory /postgres/12 ... ok
creating subdirectories ... ok
selecting dynamic shared memory implementation ... posix
selecting default max_connections ... 100
selecting default shared_buffers ... 128MB
selecting default time zone ... Europe/Rome
creating configuration files ... ok
running bootstrap script ... ok
performing post-bootstrap initialization ... ok
syncing data to disk ... ok

initdb: warning: enabling "trust" authentication for local
connections
You can change this by editing pg_hba.conf or using the option -A,
or
--auth-local and --auth-host, the next time you run initdb.

Success. You can now start the database server using:

    /usr/local/bin/pg_ctl -D /postgres/12 -l logfile start
```

5. Start the PostgreSQL 12 instance with the following command:

```
$ sudo service postgresql start

2019-12-09 14:20:50.344 CET [67267] LOG:  starting PostgreSQL 12.1
on amd64-portbld-freebsd12.0, compiled by FreeBSD clang version
6.0.1 (tags/RELEASE_601/final 335540) (based on LLVM 6.0.1), 64-bit
2019-12-09 14:20:50.344 CET [67267] LOG:  listening on IPv6 address
"::1", port 5432
2019-12-09 14:20:50.344 CET [67267] LOG:  listening on IPv4 address
"127.0.0.1", port 5432
2019-12-09 14:20:50.345 CET [67267] LOG:  listening on Unix socket
"/tmp/.s.PGSQL.5432"
2019-12-09 14:20:50.352 CET [67267] LOG:  ending log output to
stderr
2019-12-09 14:20:50.352 CET [67267] HINT:  Future log output will
go to log destination "syslog".
```

If the server cannot be started, for any reason, the command output will print out an error message that should help you to understand what went wrong.

Installing PostgreSQL from sources

Installing PostgreSQL from sources requires downloading a tarball, which is a compressed package with all the source code files, and starting the compilation. Usually, this takes several minutes, depending on the power of the machine and the I/O bandwidth. In order to compile PostgreSQL from source, you will need tar(1), GNU make(1) (at least at version 3.80), and a C compiler compliant to the C99 standard (or higher). Usually, you already have these tools on a Linux or Unix system; otherwise, please refer to your operating system documentation on how to install these tools.

Once you have all the dependencies installed, follow the steps given here to compile and install PostgreSQL:

1. The very first step is to download the PostgreSQL tarball related to the version you want to install, verifying that it is correct. For instance, to download version 12.1, you can do the following:

```
$ wget
https://ftp.postgresql.org/pub/source/v12.1/postgresql-12.1.tar.bz2
...
$ wget
https://ftp.postgresql.org/pub/source/v12.1/postgresql-12.1.tar.bz2
.md5
...
```

If you want to install the available second beta version of PostgreSQL 13, you can repeat the preceding steps with a different tarball URL:

```
$ wget
https://ftp.postgresql.org/pub/source/v13beta2/postgresql-13beta2.t
ar.bz2
...
$ wget
https://ftp.postgresql.org/pub/source/v13beta2/postgresql-13beta2.t
ar.bz2.md5
...
```

2. Before starting the compilation, check that the downloaded tarball is intact:

```
$ md5sum --check postgresql-12.1.tar.bz2.md5
postgresql-12.1.tar.bz2: OK
```

3. Once you are sure that the downloaded tarball is not corrupt, you can extract its content and start the compilation (please consider that the extracted archive will take around 200 MB of disk space, and the compilation will add some more extra space):

```
$ tar xjvf postgresql-12.1.tar.bz2
$ cd postgresql-12.1
$ ./configure --prefix=/usr/local
...
$ make && sudo make install
...
PostgreSQL installation complete.
```

If you want or need the systemd(1) service file, add the --with-systemd option to the configure line.

4. Once the database has been installed, you need to create a user to run the database with, usually named postgres, and initialize the database directory:

```
$ sudo useradd postgres
$ sudo mkdir /postgres/12
$ sudo chown postgres:postgres /postgres/12
$ /usr/local/bin/initdb -D /postgres/12
...
```

Installing PostgreSQL via pgenv

pgenv is a nice and small tool that allows you to download and manage several instances of different versions of PostgreSQL on the same machine. The idea behind pgenv is to let you explore different PostgreSQL versions—for instance, to test your application against different major versions. pgenv does not aim to be an enterprise-class tool to manage in-production instances; rather, it is a tool to let developers and DBAs experiment with different versions of PostgreSQL and keep them under control easily.

Of course, being an external tool, pgenv must be installed before it can be used. The installation, however, is very simple, since the application is made by a single Bash script:

1. The fastest way to get pgenv installed is to clone the GitHub repository and set the PATH environment variable to point to the executable directory, as follows:

```
$ git clone https://github.com/theory/pgenv

Cloning into 'pgenv'...
remote: Enumerating objects: 79, done.
remote: Counting objects: 100% (79/79), done.
remote: Compressing objects: 100% (34/34), done.
remote: Total 642 (delta 34), reused 72 (delta 29), pack-reused 563
Receiving objects: 100% (642/642), 173.78 KiB | 801.00 KiB/s, done.
Resolving deltas: 100% (300/300), done.

$ export PATH=$PATH:./pgenv/bin
```

2. Now, the pgenv command is at your fingertips, and you can run the command to get a help prompt and see the available commands:

```
$ pgenv
Using PGENV_ROOT /home/luca/git/pgenv
Usage: pgenv <command> [<args>]

The pgenv commands are:
    use        Set and start the current PostgreSQL version
    clear      Stop and unset the current PostgreSQL version
    start      Start the current PostgreSQL server
    stop       Stop the current PostgreSQL server
    restart    Restart the current PostgreSQL server
    build      Build a specific version of PostgreSQL
    rebuild    Re-build a specific version of PostgreSQL
    remove     Remove a specific version of PostgreSQL
    version    Show the current PostgreSQL version
    current    Same as 'version'
    versions   List all PostgreSQL versions available to pgenv
```

```
        help       Show this usage statement and command summary
        available  Show which versions can be downloaded
        check      Check all program dependencies
        config     View, edit, delete the program configuration

    For full documentation, see: https://github.com/theory/pgenv#readme

    This is 'pgenv' version [72faf1a]
```

The idea behind pgenv is pretty simple: it is a tool to automate the "boring" stuff—that is, downloading, compiling, installing, and start/stopping a cluster. In order to let pgenv manage a specific instance, you have to "use" it. When you use an instance, pgenv detects whether the instance has been initialized or not, and in the latter case, it does the initialization for you.

3. In order to install versions 12.0 and 12.1 of PostgreSQL, you simply have to run the following commands:

```
$ pgenv build 12.0
...
PostgreSQL 12.0 built


$ pgenv build 12.1
...

PostgreSQL 12.1 built
```

The preceding commands will download and compile the two versions of PostgreSQL, and the time required for the operations to complete depends on the power and speed of the machine you are running on.

4. After that, you can decide which instance to start with the use command:

```
$ pgenv use 12.0
...
server started
PostgreSQL 12.0 started
Logging to /home/luca/git/pgenv/pgsql/data/server.log
```

pgenv is smart enough to see whether the instance you are starting has been already initialized, or it will initialize (only the first time) for you.

5. Once you have started the instance, you can connect to it with any client tool you like, most notably `psql` (you will learn more about `psql` in the next chapters):

```
$ psql -U postgres -h localhost template1
psql (12.1 (Ubuntu 12.1-1.pgdg18.04+1), server 12.0)
Type "help" for help.

template1=#
```

6. If you need to stop and change the PostgreSQL version to use, you can issue a `stop` command followed by a `use` command with the targeted version. For instance, to stop running the 12.0 instance and start a 12.1 instance, you can use the following:

```
$ pgenv stop
...
PostgreSQL 12.0 stopped

$ pgenv use 12.1
...
PostgreSQL 12.1 started
Logging to /home/luca/git/pgenv/pgsql/data/server.log
```

7. `pgenv` allows you to see which instances are currently installed and which one is currently active—that is, "in use"—and this does not mean it is running:

```
$ pgenv versions
Using PGENV_ROOT /home/luca/git/pgenv
        11.5        pgsql-11.5
        11beta4     pgsql-11beta4
        12.0        pgsql-12.0
  *     12.1        pgsql-12.1
```

If you are searching for a quick way to test and run different PostgreSQL versions on the same machine, `pgenv` is a good tool.

Installing PostgreSQL 13 beta 2 using `pgenv` is really simple—just repeat the preceding process, changing the version number of the cluster you want to build:

```
$ pgenv build 13beta2
...
PostgreSQL 13beta2 built

$ pgenv use 13beta1
...
server started
PostgreSQL 13beta1 started
```

```
$ psql -U postgres -c "SELECT version();" template1
                                              version
------------------------------------------------------------------------
-----------------------------------
 PostgreSQL 13beta2 on x86_64-unknown-freebsd12.1, compiled by gcc (FreeBSD
Ports Collection) 9.2.0, 64-bit
(1 row)
```

You now know how to use your preferred method to install the version of PostgreSQL that you need.

Summary

This chapter has introduced you to PostgreSQL, its history, and its main features. You have learned about PostgreSQL terminology, as well as how to install a cluster on Unix-like operating systems, such as GNU/Linux Debian, Fedora, and FreeBSD, as well as installing the tool from various sources.

In the following chapters, you will start using this great database engine and learn details about every main single feature it provides.

References

- PostgreSQL 12 release note: https://www.postgresql.org/docs/12/release-12.html
- Upgrading documentation: https://www.postgresql.org/docs/current/upgrading.html
- PostgreSQL version policy: https://www.postgresql.org/support/versioning/
- PostgreSQL initdb official documentation: https://www.postgresql.org/docs/12/app-initdb.html
- PostgreSQL pg_ctl official documentation: https://www.postgresql.org/docs/12/app-pg-ctl.html
- pgenv GitHub repository and documentation: https://github.com/theory/pgenv

Getting to Know Your Cluster

2

In order to be a proficient user and administrator of a PostgreSQL cluster, you first have to know and understand how PostgreSQL works. A database system is a very complex beast, and PostgreSQL, being an enterprise-level **Database Management System** (**DBMS**), is in no way a simple software system. However, thanks to a very good design and implementation, once you understand the basic concepts and terminology of PostgreSQL, things will quickly become comprehensive and clear.

This chapter will introduce you to the main PostgreSQL terminology and concepts, as well as teach you how to connect to the cluster. This chapter will also introduce you to the `psql` client, which ships with PostgreSQL and is the recommended way to connect to your database. You can, of course, use any client that supports PostgreSQL to connect to the database, and the rules explained here will also be valid for other clients supporting PostgreSQL. The main free graphical client available for PostgreSQL is `pgAdmin4`, but you can really choose the one you like the most.

This chapter covers the following topics:

- Managing your cluster
- Connecting to the cluster
- Exploring the disk layout of PGDATA
- Exploring configuration files and parameters

Technical requirements

What you need to know for this chapter is as follows:

- How to install binary packages on your Unix machine
- Basic Unix command-line usage
- Basic SQL statements

You can find the code for this chapter in the following GitHub repository: `https://github.com/PacktPublishing/Learn-PostgreSQL`.

Managing your cluster

From an operating system point of view, PostgreSQL is a service that can be started, stopped, and, of course, monitored. As you saw in the previous chapter, usually when you install PostgreSQL, you also get a set of operating system-specific tools and scripts to integrate PostgreSQL with your operating system service management (for example, `systemd` service files).

In particular, PostgreSQL ships with a tool called `pg_ctl` that helps in managing the cluster and the related running processes. This section introduces you to the basic usage of `pg_ctl` and to the processes that you can encounter in a running cluster.

pg_ctl

The `pg_ctl` command-line utility is a tool that allows you to perform different actions on a cluster, mainly initialize it, start it, restart and stop it, and so on. `pg_ctl` accepts the command to execute as the first argument, followed by other specific arguments—the main commands are as follows:

- `start`, `stop`, and `restart` execute the corresponding actions on the cluster.
- `status` reports the current status (running or not) of the cluster.
- `initdb` (or `init` for short) executes the initialization of the cluster, possibly removing any previously existing data.
- `reload` causes the PostgreSQL server to reload the configuration, which is useful when you want to apply configuration changes.
- `promote` is used when the cluster is running as a subordinate server (named `standby`) in a replication setup and, from now on, must be detached from the original master and become independent (replication will be explained in later chapters).

Let's see a possible usage of each of these commands now.

The `status` command just queries the cluster to get information, so it is pretty safe as a starting point to understand what is happening:

```
$ pg_ctl status
pg_ctl: no server running
```

As we can see, it's not currently running, which in hindsight makes sense given that we haven't started it up explicitly. We can then start the cluster with the `start` command:

```
$ pg_ctl start
waiting for server to start....
2019-12-17 19:31:48.421 CET [96724] LOG:   starting PostgreSQL 12.1 on
amd64-portbld-freebsd12.0, compiled by FreeBSD clang version 6.0.1
(tags/RELEASE_601/final 335540) (based on LLVM 6.0.1), 64-bit
2019-12-17 19:31:48.421 CET [96724] LOG:   listening on IPv6 address "::1",
port 5432
2019-12-17 19:31:48.422 CET [96724] LOG:   listening on IPv4 address
"127.0.0.1", port 5432
2019-12-17 19:31:48.423 CET [96724] LOG:   listening on Unix socket
"/tmp/.s.PGSQL.5432"
2019-12-17 19:31:48.429 CET [96724] LOG:   ending log output to stderr
2019-12-17 19:31:48.429 CET [96724] HINT:   Future log output will go to log
destination "syslog".
 done
server started
```

The `pg_ctl` command launches the `postmaster` process, which prints out a few log lines before redirecting the logs to the appropriate log file, and the `server started` message at the end confirms that the server is started. Now, if you run `pg_ctl` again to check the server, you will see that it has been started:

```
$ pg_ctl status
pg_ctl: server is running (PID: 96724)
/usr/local/bin/postgres
```

As you can see, the server is now running and `pg_ctl` shows the **Process Identifier (PID)** of the running process, as well as the command line that launched the process—in this case, `/usr/local/bin/postgres`. This process is the `postmaster`, which is the "root" of all PostgreSQL processes. But wait a minute: why is it called `postmaster` if the launched process is a `postgres` executable? The name `postmaster` is just that: a name used to identify a process among the others. Both backend processes and the postmaster are run starting from the `postgres` executable, and the postmaster is just **the root of all PostgreSQL processes**, with the main aim of keeping all the other processes under control.

Now that the cluster is running, let's stop it. As you can imagine, `stop` is the command used to instruct `pg_ctl` about which action to perform:

```
$ pg_ctl stop
waiting for server to shut down.... done
server stopped
```

However, stopping a cluster can be much more problematic than starting it, and for that reason, it is possible to pass extra arguments to the `stop` command in order to let `pg_ctl` act accordingly. In particular, there are three ways of stopping a cluster:

- The `smart` mode means that the PostgreSQL cluster will gently wait for all the connected clients to disconnect and only then it will shut the cluster down.
- The `fast` mode will immediately disconnect every client and will shut down the server without having to wait.
- The `immediate` mode will abort every PostgreSQL process, including client connections, and shut down the cluster in a dirty way, meaning that data integrity is not guaranteed and the server needs a crash recovery at start up time.

Once you issue a `stop` command through `pg_ctl`, the server will not accept any new incoming connections from clients, and depending on the stop mode you have selected, existing connections will be terminated. The default stop mode, if none is specified, is `fast`, which forces an immediate disconnection of the clients but ensures data integrity.

If you want to change the stop mode, you can use the `-m` flag, specifying the mode name, as follows:

```
$ pg_ctl stop -m smart
waiting for server to shut down...................... done
server stopped
```

In the preceding example, the `pg_ctl` command will wait, printing a dot every second until all the clients disconnect from the server. In the meantime, if you try to connect to the same cluster from another client, you will receive an error, because the server has entered the stopping procedure:

```
$ psql template1
psql: error: could not connect to server: FATAL:  the database system is
shutting down
```

It is possible to specify just the first letter of the stop mode instead of the whole word; so, for instance, s for smart, i for immediate, and f for fast. Interacting with a cluster status, for example, to stop it is an action that not every user must be able to perform; usually, only an operating system administrator must be able to interact with services including PostgreSQL.

pg_ctl must be run by the same unprivileged operating system user that is going to run the cluster. PostgreSQL does not allow a cluster to be run by privileged users, such as *root*, in order to mitigate the side effects of privilege escalation. Therefore, PostgreSQL is run by a "normal" user, usually named postgres on all the operating systems. This unprivileged user will own the PGDATA directory and run the postmaster process, and therefore also all the processes launched by the postmaster itself.

pg_ctl, having to interact with PostgreSQL processes, must be run by the very same unprivileged user, and in fact, if you try to run pg_ctl as a privileged user, you get a warning message:

```
$ sudo pg_ctl stop
pg_ctl: cannot be run as root
Please log in (using, e.g., "su") as the (unprivileged) user that will
own the server process.
```

A better approach is to either log in as the postgres user or ask sudo to use this user—for instance, specifying the user to run the command via the -u flag and keeping the environment with -E:

```
$ sudo -E -u postgres pg_ctl stop
waiting for server to shut down.... done
server stopped
```

On the other hand, if you are going to manage the PostgreSQL cluster without pg_ctl and with operating system tools (such as service scripts), you will need to run the commands as a privileged user:

```
$ sudo service postgresql start
2019-12-17 19:24:43.928 CET [28738] LOG:  starting PostgreSQL 12.1 on
amd64-portbld-freebsd12.0, compiled by FreeBSD clang version 6.0.1
(tags/RELEASE_601/final 335540) (based on LLVM 6.0.1), 64-bit
2019-12-17 19:24:43.929 CET [28738] LOG:  listening on IPv6 address "::1",
port 5432
2019-12-17 19:24:43.929 CET [28738] LOG:  listening on IPv4 address
"127.0.0.1", port 5432
2019-12-17 19:24:43.930 CET [28738] LOG:  listening on Unix socket
"/tmp/.s.PGSQL.5432"
2019-12-17 19:24:43.935 CET [28738] LOG:  ending log output to stderr
```

```
2019-12-17 19:24:43.935 CET [28738] HINT:  Future log output will go to log
destination "syslog".
```

Therefore, it is important to keep in mind that while pg_ctl provides you with all of the possible interactions with your cluster, you need to use the same unprivileged user that the cluster is running. This is particularly important when dealing with your own automation scripts and programs that can control the cluster.

Every tool that interacts with the cluster must know something about the latter—in particular, it must know where the data and the configuration is stored on the disk. So, how can pg_ctl know where the PGDATA directory is? The trick is that almost **every** PostgreSQL-related command searches for the value of PGDATA as an environmental variable or as a -D command-line option.

If a command cannot find the PGDATA directory, it will display it clearly:

```
$ pg_ctl status
pg_ctl: no database directory specified and environment variable PGDATA
unset
Try "pg_ctl --help" for more information.

$ export PGDATA=/postgres/12
$ pg_ctl status
pg_ctl: server is running (PID: 91393)
/usr/local/bin/postgres "-D" "/postgres/12"
```

As you can see from the preceding example, once you *erase* the PGDATA environment variable, the command is no longer able to operate and asks for a PGDATA variable either on the command line or in the environment. The command-line argument, specified with -D, always has precedence against any environment variable, so if you don't set or misconfigure the PGDATA variable, but instead pass the right value on the command line, everything works fine:

```
$ export PGDATA=/postgres/11  # wrong PGDATA!
$ pg_ctl status -D /postgres/12
pg_ctl: server is running (PID: 91393)
/usr/local/bin/postgres "-D" "/postgres/12"
```

The same concepts of PGDATA and the -D optional argument is true for pretty much any "low-level" commands that act against a cluster and makes clear that with the same set of executables, you can run multiple instances of PostgreSQL on the same machine, as long as you keep the PGDATA directory of each one separate.

Do not use the same `PGDATA` directory for multiple versions of PostgreSQL. While it could be tempting to have, on your own test machine, a single `PGDATA` directory that can be used in turn by a PostgreSQL 12 and a PostgreSQL 13 instance, this will not work and you risk losing all your data. Luckily, PostgreSQL is smart enough to see that `PGDATA` has been created and used by a different version and refuses to operate, but please be careful in not sharing the same `PGDATA` directory with different instances.

It is worth reiterating one more time: the `PGDATA` directory can be named whatever you like and it is a common habit to have it named after the PostgreSQL major version it is used by. However, this is not mandatory, and the choice of name and location is up to you.

PostgreSQL processes

You have already learned how the postmaster is the root of all PostgreSQL processes, but as explained in `Chapter 1`, *Introduction to PostgreSQL*, PostgreSQL will launch multiple different processes at startup. These processes are in charge of keeping the cluster in good health, as well as observing and instructing the cluster. This section provides a glance at the main processes you can find in a running cluster, allowing you to recognize each of them and their respective purposes.

If you inspect a running cluster from the operating system point of view, you will see a bunch of processes tied to PostgreSQL:

```
$ pstree
-+= 00001 root /sbin/init --
...
 |-+= 91393 postgres /usr/local/bin/postgres -D /postgres/12
 | |--= 91839 postgres postgres: checkpointer     (postgres)
 | |--= 92351 postgres postgres: background writer    (postgres)
 | |--= 92752 postgres postgres: walwriter    (postgres)
 | |--= 92978 postgres postgres: autovacuum launcher    (postgres)
 | |--= 93359 postgres postgres: stats collector    (postgres)
 | \--= 93739 postgres postgres: logical replication launcher    (postgres)
```

As you can see, the `postmaster` process with PID `91393` is one that owns all other subprocesses. The maintenance processes are as follows:

- `checkpointer` is a process responsible for executing the checkpoints, which are points in time where the database ensures that all the data is actually stored persistently on the disk.

- `background writer` is responsible for helping to push the data out of the memory to permanent storage.
- `walwriter` is responsible for writing out the **Write-Ahead Logs** (**WALs**), the logs that are needed to ensure data reliability even in the case of a database crash.
- `stats collector` is a process that monitors the amount of data PostgreSQL is handling, storing it for further elaboration, such as deciding on which indexes to use to satisfy a query.
- `logical replication launcher` is a process responsible for handling logical replication.

Depending on the exact configuration of the cluster, there could be other processes active:

- **Background workers**: These are processes that can be customized by the user to perform background tasks.
- **WAL receiver or WAL sender**: These are processes involved in receiving from or sending data to another cluster in replication scenarios.

Many of the concepts and aims of the preceding process list will become clearer as you progress through the book's chapters, but for now, it is sufficient that you know that PostgreSQL has a few other processes that are always active without any regard to incoming client connections.

When a client connects to your cluster, a new process is spawned: this process, named the *backend process*, is responsible for serving the client requests (meaning executing the queries and returning the results). You can see and count connections by inspecting the process list:

```
$ pstree
-+= 00001 root /sbin/init --
.../
 |-+= 91393 postgres /usr/local/bin/postgres -D /postgres/12
 | |--= 14530 postgres postgres: postgres template1 [local]  (postgres)
 | |--= 91839 postgres postgres: checkpointer    (postgres)
 | |--= 92351 postgres postgres: background writer    (postgres)
 | |--= 92752 postgres postgres: walwriter    (postgres)
 | |--= 92978 postgres postgres: autovacuum launcher    (postgres)
```

```
| |--= 93359 postgres postgres: stats collector     (postgres)
| \--= 93739 postgres postgres: logical replication launcher   (postgres)
```

If you compare the preceding list with the previous one, you will see that there is another process with PID `14530`: this process is a backend process. In particular, this process represents a client connection to the database named `template1`.

 PostgreSQL uses a process approach to concurrency instead of a multi-thread approach. There are different reasons, most notably the isolation and portability that a multi-process approach offers. Moreover, on modern hardware and software, forking a process is no longer so much of an invasive operation.

Therefore, once PostgreSQL is running, there is a tree of processes that root at `postmaster`. The aim of the latter is to spawn new processes when there is the need to handle new database connections, as well as to monitor all maintenance processes to ensure that the cluster is running fine.

Connecting to the cluster

Once PostgreSQL is running, it awaits incoming database connections to serve; as soon as a connection comes in, PostgreSQL serves it by connecting the client to the right database. This means that in order to interact with the cluster, you need to connect to it. However, you don't connect to the whole cluster; rather, you ask PostgreSQL to interact with one of the databases the cluster is serving. Therefore, when you connect to the cluster, you need to connect to a specific database. This also means that the cluster must have at least one database from the very beginning of its life. That is the role of the so-called *template databases*, which, among other duties, serve as a common database to which you can connect on a freshly installed cluster.

When you initialize the cluster with the `initdb` command, PostgreSQL builds the filesystem layout of the `PGDATA` directory and builds two template databases, named `template0` and `template1`. The aim of these databases is to provide an initialization point for later operations—for instance, to allow users to connect to one of them in order to interact with the cluster.

In order to connect to one of the databases, either a template or a user-defined one, you need a *client* to connect with. PostgreSQL ships with `psql`, a command-line client that allows you to interact with, connect, and administer databases and the cluster itself. However, other clients do exist, but they will not be discussed in this chapter. You could also connect your own applications to a database, which is an important task in a day-to-day database activity: to this end, you also need a set of parameters that can be "composed" into a connection string (something similar to a URL, for what it matters) that your application can use to gain access to PostgreSQL.

This section will explain all of the preceding concepts, starting from the template databases and then showing the basic usage of `psql` and the connection string.

The template databases

The `template1` database is the first database created when the system is initialized, and then it is cloned into `template0`. This means that the two databases are, at least initially, identical, and the aim of `template0` is to act as a safe copy for rebuilding in case it is accidentally damaged or removed.

You can inspect available databases using the `psql -l` command:

```
$ psql -l                              List of databases
   Name    |  Owner   | Encoding | Collate | Ctype |   Access privileges
-----------+----------+----------+---------+-------+-----------------------
 postgres  | postgres | UTF8     | C       | C     |
 template0 | postgres | UTF8     | C       | C     | =c/postgres          +
           |          |          |         |       | postgres=CTc/postgres
 template1 | postgres | UTF8     | C       | C     | =c/postgres          +
```

It is interesting to note that there's a third database that is created during the installation process: the `postgres` database. That database belongs to the `postgres` user, which is, by default, the only database administrator created during the initialization process. This database is a *common space* to be used for connections instead of the template databases.

The name **template** indicates the real aim of these two databases: when you create a new database, PostgreSQL clones a template database as a **common base**. This is somewhat similar to creating a user home directory on Unix systems: the system clones a **skeleton** directory and assigns the new copy to the user. PostgreSQL does the same—it clones `template1` and assigns the newly created database to the user that requested it.

What this also means is that whatever object you put into `template1`, you will find the very same object in freshly created databases. This can be really useful for providing a common **base database** and having all other databases brought to life with the same set of attributes and objects.

Nevertheless, you are not forced to use `template1` as the base template and, in fact, you can create your own databases and use them as templates for other databases. However, please keep in mind that by default (and most notably on a newly initialized system), the `template1` database is the one that is cloned for the first databases you will create.

Another difference between `template1` and `template0`, apart from the former being the default for new databases, is that you cannot connect to the latter. This is in order to prevent accidental damage to `template0` (the safety copy).

It is important to note that the cluster (and all user-defined databases) can work even without the template databases—the `template1` and `template0` databases are not fundamental for the other databases to run. However, if you lose the templates, you will be required to use another database as a template every time you perform an action that requires it, such as creating a new database.

The psql command-line client

The `psql(1)` command is the command-line interface that ships with every installation of PostgreSQL. While you can certainly use a graphical user interface to connect and interact with the databases, a basic knowledge of `psql` is mandatory in order to administer the cluster. In fact, as `psql(1)` is shipped with PostgreSQL, it is the most updated client, especially when a new major version is released, and therefore provides a consistent way to access your cluster. Moreover, the client is lightweight and useful even in emergency situations when a GUI is not available. `psql` accepts several options to connect to a database, mainly the following:

- `-d`: The database name
- `-U`: The username
- `-h`: The host (either an IPv4 or IPv6 address or a hostname)

If no option is specified, `psql` assumes your operating system user is trying to connect to a database with the same name, and a database user with a name that matches the operating system on a local connection. Take the following connection:

```
$ id
uid=770(postgres) gid=770(postgres) groups=770(postgres)
```

```
$ psql
psql (12.1)
Type "help" for help.

postgres=#
```

This means that the current operating system user (`postgres`) has required `psql` to connect to a database named `postgres` via the PostgreSQL user named `postgres` on the local machine. Explicitly, the connection could have been requested as follows:

```
$ psql -U postgres -d postgres
psql (12.1)
Type "help" for help.

postgres=#
```

The first thing to note is that once a connection has been established, the command prompt changes: `psql` reports the database to which the user has been connected (`postgres`) and a sign to indicate they are a superuser (#). In the case that the user is not a database administrator, a > sign is placed at the end of the prompt.

If you need to connect to a database that is named differently by your operating system username, you need to specify it:

```
$ psql -d template1
psql (12.1)
Type "help" for help.

template1=#
```

Similarly, if you need to connect to a database that does not correspond to your operating username with a PostgreSQL user that is different from your operating system username, you have to explicitly pass both parameters to `psql`:

```
$ id
uid=770(postgres) gid=770(postgres) groups=770(postgres)

$ psql -d template1 -U luca
psql (12.1)
Type "help" for help.

template1=>
```

As you can see from the preceding example, the operating system user `postgres` has connected to the `template1` database with the PostgreSQL user `luca`. Since the latter is not a system administrator, the command prompt ends with the > sign.

In order to quit from `psql` and close the connection to the database, you have to type `\q` or `quit` and press *Enter* (you can also press *CTRL + D* to exit on any Unix and Linux machines):

```
$ psql -d template1 -U luca
psql (12.1)
Type "help" for help.

template1=> \q
$
```

Entering SQL statements via psql

Once you are connected to a database via `psql`, you can issue any statement you like. Statements must be terminated by a semicolon, indicating that the next *Enter* key will execute the statement. The following is an example where the *Enter* key has been emphasized:

```
$ psql -d template1 -U luca
psql (12.1)
Type "help" for help.

template1=> SELECT current_date; <ENTER>
 current_date
--------------
 2019-12-23
(1 row)
```

Another way to execute the statement is to issue a `\g` command, again followed by <ENTER>. This is useful when connecting via a terminal emulator that has keys remapped:

```
template1=> SELECT current_date \g <ENTER>
 current_date
--------------
 2019-12-23
(1 row)
```

Until you end a statement with a semicolon or `\g`, `psql` will keep the content you are typing in the *query buffer*, so you can also edit multiple lines of text as follows:

```
template1=> SELECT
template1-> current_date
template1-> ;
 current_date
--------------
```

```
2019-12-23
(1 row)
```

Note how the `psql` command prompt has changed on the lines following the first one: the difference is there to remind you that you are editing a multi-line statement and `psql` has not (yet) found a statement terminator.

One useful feature of the `psql` query buffer is the capability to edit the content of the query buffer in an external editor. If you issue the \e command, your favorite editor will pop up with the content of the last-edited query. You can then edit and refine your SQL statement as much as you want, and once you exit the editor, `psql` will read what you have produced and execute it. The editor to use is chosen with the EDITOR operating system environment variable.

It is also possible to execute all the statements included in a file or edit a file before executing it. As an example, assume the `test.sql` file has the following content:

```
$ cat test.sql

SELECT current_date;
SELECT current_time;
SELECT current_role;
```

The file has three very simple SQL statements. In order to execute all of the file at once, you can use the \i special command followed by the name of the file:

```
template1=> \i test.sql
 current_date
---------------
 2019-12-23
(1 row)

    current_time
--------------------
 17:56:05.015434+01
(1 row)

 current_role
---------------
 luca
(1 row)
```

As you can see, the client has executed, one after the other, every statement within the file. If you need to edit the file without leaving `psql`, you can issue \e `test.sql` to open your favorite editor, make changes, and come back to the `psql` connection.

SQL is case-insensitive and space-insensitive: you can write it in all uppercase or all lowercase, with however many horizontal and vertical spaces you want. In this book, SQL keywords will be written in uppercase and the statements will be formatted to read cleanly.

A glance at the psql commands

Every command specific to `psql` starts with a backslash character (\). It is possible to get some help about SQL statements and PostgreSQL commands via the special \h command, after which you can specify the specific statement you want help for:

```
template1=> \h SELECT
Command:     SELECT
Description: retrieve rows from a table or view
Syntax:
[ WITH [ RECURSIVE ] with_query [, ...] ]
SELECT [ ALL | DISTINCT [ ON ( expression [, ...] ) ] ]
    [ * | expression [ [ AS ] output_name ] [, ...] ]
...
URL: https://www.postgresql.org/docs/12/sql-select.html
```

 The displayed help is, for space reasons, concise. You can find much more verbose description and usage examples in the online documentation. For this reason, at the end of the help screen, there is a link reference to the online documentation.

If you need help with the `psql` commands, you can issue a \? command:

```
template1=> \?
General
  \copyright              show PostgreSQL usage and distribution terms
  \crosstabview [COLUMNS] execute query and display results in crosstab
  \errverbose             show most recent error message at maximum
verbosity
  \g [FILE] or ;          execute query (and send results to file or |pipe)
  \gdesc                  describe result of query, without executing it
...
```

There are also a lot of *introspection* commands, such as, for example, \d to list all user-defined tables. These special commands are, under the hood, a way to execute queries against the PostgreSQL system catalogs, which are in turn registries about all objects that live in a database. The introspection commands will be shown later in the book, and are useful as shortcuts to get an idea of which objects are defined in the current database.

Many `psql` features will be detailed as you move on through the book, but it is worth spending some time trying to get used to this very efficient and rich command-line client.

Introducing the connection string

In the previous section, you learned how to specify basic connection options, such as `-d` and `-U` for a database and user, respectively. `psql` also accepts a LibPQ connection string.

LibPQ is the underlying library that every application can use to connect to a PostgreSQL cluster and is, for example, used in C and C++ clients, as well as non-native connectors.

A connection string in LibPQ is a URI made up of several parts:

```
postgresql://username@host:port/database
```

Here, we have the following:

- `postgresql` is a fixed string that specifies the protocol the URI refers to.
- `username` is the PostgreSQL username to use when connecting to the database.
- `host` is the hostname (or IP address) to connect to.
- `port` is the TCP/IP port the server is listening on (by default, `5432`).
- `database` is the name of the database to which you want to connect.

The username, port, and database parts can be omitted if they are set to their default (the username is the same as the operating system username).

The following connections are all equivalent:

```
$ psql -d template1 -U luca -h localhost

$ psql postgresql://luca@localhost/template1

$ psql postgresql://luca@localhost:5432/template1
```

Solving common connection problems

There are a few common problems when dealing with database connections, and this section explains them in order to ease your task of getting connected to your cluster.

Please note that the solutions provided here are just for testing purposes and not for production usage. All the security settings will be explained in later chapters, so the aim of the following subsection is just to help you get your test environment usable.

Database "foo" does not exist

This means either you misspelled the name of the database in the connection parameter or you are trying to connect without specifying the database name.

For instance, the following connection fails because, by default, *X* is assuming that user `luca` is trying to connect to a database with the same name (meaning, `luca`) since none has been explicitly set:

```
$ psql
psql: error: could not connect to server: FATAL:  database "luca" does not
exist
```

The solution is to provide an existing database name via the `~-d~` option or to create a database with the same name as the user.

Connection refused

This usually means there is a network connection problem, so either the host you are trying to connect to is not reachable or the cluster is not listening on the network.

As an example, imagine PostgreSQL is running on a machine named `miguel` and we are trying to connect from another host on the same network:

```
$ psql -h miguel -U luca template1
psql: error: could not connect to server: could not connect to server:
Connection refused
        Is the server running on host "miguel" (192.168.222.123) and
accepting
        TCP/IP connections on port 5432?
```

In this case, the database cluster is running on the remote host but is not accepting connections from the outside. Usually, you have to fix the server configuration or connect (via SSH, for instance) to the remote machine and open a local connection from there.

In order to quickly solve the problem, you have to edit the `postgresql.conf` file and ensure the `listen_address` option has an asterisk (or the name of your external network card) so that the server will listen on any available network address:

```
listen_addresses = '*'
```

After a restart of the service, the client will be able to connect. Please note that enabling the server to listen on any available network address could not be the optimal solution and can expose the server to risks in a production environment. However, you will learn later in the book how to specifically configure the connection properties for your server.

No pg_hba.conf entry

This error means the server is up and running and able to accept your request, but the **Host-Based Access** (**HBA**) control does not permit you to enter.

As an example, the following connection is refused:

```
$  psql -h localhost -U luca template1
psql: error: could not connect to server: FATAL:  no pg_hba.conf entry for
host "127.0.0.1", user "luca", database "template1", SSL off
```

The reason for this is that, inspecting the pg_hba.conf file, there is no rule to let the user luca in on the localhost interface. So, for instance, adding a single line such as the following to the pg_hba.conf file can fix the problem:

```
host all luca 127.0.0.1/32 trust
```

You need to reload the configuration in order to apply changes. The format of every line in the pg_hba.conf file will be discussed later, but for now, please assume that the preceding line instruments the cluster to accept any connection incoming from localhost by means of user luca.

Exploring the disk layout of PGDATA

In the previous sections, you have seen how to install PostgreSQL and connect to it, but we have not looked at the storage part of a cluster. Since the aim of PostgreSQL, as well as the aim of any relational database, is to permanently store data, the cluster needs some sort of permanent storage. In particular, PostgreSQL exploits the underlying filesystem to store its own data. All of the PostgreSQL-related stuff is contained in a directory known as PGDATA.

The PGDATA directory acts as the disk container that stores all the data of the cluster, including the users' data and cluster configuration.

The following is an example of the content of PGDATA for a running PostgreSQL 12 cluster (it looks the same for a PostgreSQL 13 instance):

```
$ sudo ls -1 /postgres/12
PG_VERSION
base
global
pg_commit_ts
pg_dynshmem
pg_hba.conf
pg_ident.conf
pg_logical
pg_multixact
pg_notify
pg_replslot
pg_serial
pg_snapshots
pg_stat
pg_stat_tmp
pg_subtrans
pg_tblspc
pg_twophase
pg_wal
pg_xact
postgresql.auto.conf
postgresql.conf
postmaster.opts
postmaster.pid
```

The PGDATA directory is structured in several files and subdirectories. The main files are as follows:

- postgresql.conf is the main configuration file, used as default when the service is started.
- postgresql.auto.conf is the automatically included configuration file used to store dynamically changed settings via SQL instructions.
- pg_hba.conf is the HBA file that provides the configuration regarding available database connections.
- PG_VERSION is a text file that contains the major version number (useful when inspecting the directory to understand which version of the cluster has managed the PGDATA directory).
- postmaster.pid is the PID of the running cluster.

The main directories available in PGDATA are as follows:

- base is a directory that contains all the users' data, including databases, tables, and other objects.
- global is a directory containing cluster-wide objects.
- pg_wal is the directory containing the WAL files.
- pg_stat and pg_stat_tmp are, respectively, the storage of the permanent and temporary statistical information about the status and health of the cluster.

Of course, all files and directories in PGDATA are important for the cluster to work properly, but so far, the preceding is the "core" list of objects that are fundamental in PGDATA itself. Other files and directories will be discussed in later chapters.

Objects in the PGDATA directory

PostgreSQL does not name objects on disk, such as tables, in a mnemonic or human-readable way; instead, every file is named after a numeric identifier. You can see this by having a look, for instance, at the base subdirectory:

```
$ sudo ls -1 /postgres/12/base
1
13777
13778
```

As you can see from the preceding, the base directory contains three objects, named 1, 13777, and 13778, respectively. In particular, each of the preceding is a directory that contains other files, as shown here:

```
$ sudo ls -1 /postgres/12/base/13777 | head
112
113
1247
1247_fsm
1247_vm
1249
1249_fsm
1249_vm
1255
1255_fsm
```

As you can see, each file is named with a numeric identifier. Internally, PostgreSQL holds a specific catalog that allows the database to match a mnemonic name to a numeric identifier and vice versa. The integer identifier is named OID (**Object Identifier**); this name is a historical term that today corresponds to the so-called *filenode*. The two terms will be used interchangeably in this section.

There is a specific utility that allows you to inspect a PGDATA directory and extract mnemonic names: oid2name. For example, if you executed the oid2name utility, you'd get a list of all available databases:

```
$ oid2name
All databases:
    Oid  Database Name  Tablespace
  ----------------------------------
  13778         postgres  pg_default
  13777        template0  pg_default
      1        template1  pg_default
```

As you can see, the Oid numbers in the oid2name output reflect the same directory names listed in the base directory; every subdirectory has a name corresponding to the database. You can even go further and inspect a single file going into the database directory, specifying the database where you are going to search for an object name with the -d flag:

```
$ cd /postgres/12/base/1
$ oid2name -d template1 -f 3395
From database "template1":
  Filenode                 Table Name
  ----------------------------------
      3395  pg_init_privs_o_c_o_index
```

As you can see from the preceding, the 3395 file in the /postgres/12/base/1 directory corresponds to the table named pg_init_privs_o_c_o_index. Therefore, when PostgreSQL needs to interact with a table like this, it will seek the disk to the /postgres/12/base/1/3395 file.

From the preceding, it should be clear that every SQL table is stored as a file with a numeric name. However, PostgreSQL does not allow a single file to be greater than 1 GB in size, so what happens if a table grows beyond that limit? PostgreSQL "attaches" another file with a numeric extension that indicates the next chunk of 1 GB of data. In other words, if your table is stored in the 123 file, the second gigabyte will be stored in the 123.1 file, and if another gigabyte of storage is needed, another file, 123.2, will be created. Therefore, the filenode refers to the very first file related to a specific table, but more than one file can be stored on disk.

Tablespaces

PostgreSQL pretends to find all its data within the PGDATA directory, but that does not mean that your cluster is "jailed" to this directory. In fact, PostgreSQL allows "escaping" the PGDATA directory by means of *tablespaces*. A tablespace is a storage space that can be outside the PGDATA directory. Tablespaces are dragged into the PGDATA directory by means of symbolic links stored in the pg_tblspc subdirectory. In this way, the PostgreSQL processes do not have to seek outside PGDATA, still being able to access "external" storage. A tablespace can be used to achieve different aims, such as enlarging the storage data or providing different storage performances for specific objects. For instance, you can create a tablespace on a slow disk to contain infrequently accessed objects and tables, keeping fast storage within another tablespace for frequently accessed objects.

You don't have to make links by yourself: PostgreSQL provides the TABLESPACE feature to manage this and the cluster will create and manage the appropriate links under the pg_tblspc subdirectory.

For instance, the following is a PGDATA directory that has three different tablespaces:

```
$ sudo ls -l /postgres/12/pg_tblspc/
total 0
lrwx------  1 postgres  postgres  22 Dec 23 19:47 16384 ->
/data/tablespaces/ts_a
lrwx------  1 postgres  postgres  22 Dec 23 19:47 16385 ->
/data/tablespaces/ts_b
lrwx------  1 postgres  postgres  22 Dec 23 19:47 16386 ->
/data/tablespaces/ts_c
lrwx------  1 postgres  postgres  22 Dec 23 19:47 16387 ->
/data/tablespaces/ts_d
```

As you can see from the preceding example, there are four tablespaces that are attached to the /data storage. You can inspect them with oid2name and the -s flag:

```
$ oid2name -s
All tablespaces:
    Oid  Tablespace Name
-----------------------
    1663        pg_default
    1664        pg_global
   16384              ts_a
   16385              ts_b
   16386              ts_c
   16387              ts_d
```

As you can see, the numeric identifiers of the symbolic links are mapped to mnemonic names of the tablespaces. From the preceding example, you can observe that there are also two particular tablespaces:

- `pg_default` is the default tablespace corresponding to "none," the default storage to be used for every object when nothing is explicitly specified. In other words, every object stored directly under the `PGDATA` directory is attached to the `pg_default` tablespace.
- `pg_global` is the tablespace used for system-wide objects.

By default, both of the preceding tablespaces refer directly to the `PGDATA` directory, meaning any cluster without a custom tablespace is totally contained within the `PGDATA` directory.

Exploring configuration files and parameters

The main configuration file for PostgreSQL is `postgresql.conf`, a text-based file that drives the cluster when it starts.
Usually, when changing the configuration of the cluster, you have to edit the `postgresql.conf` file to write the new settings and, depending on the context of the settings you have edited, to issue a cluster `SIGHUP` signal (that is, *reload* the configuration) or restart it.

Every configuration parameter is associated with a *context*, and depending on the context, you can apply changes with or without a cluster restart. In particular, available contexts are the following:

- `internal`: A group of parameters that are set at compile-time and therefore cannot be changed at runtime.
- `postmaster`: All the parameters that require the cluster to be restarted (that is, to kill the `postmaster` process and start it again) to activate them.
- `sighup`: All the configuration parameters that can be applied with a `SIGHUP` signal sent to the `postmaster` process, which is equivalent to issuing a `reload` signal in the operating system service manager.
- `backend` and `superuser-backend`: All the parameters that can be set at run time but will be applied to the next normal or administrative connection.
- `user` and `superuser`: A group of settings that can be changed at run time and are immediately active for normal and administrative connection.

The configuration parameters will be explained later in the book, but the following is an example of a minimal configuration file with some different settings:

```
$ cat postgresql.conf
shared_buffers = 512MB
maintenance_work_mem = 128MB
checkpoint_completion_target = 0.7
wal_buffers = 16MB
work_mem = 1310kB
min_wal_size = 1GB
max_wal_size = 2GB
```

The `postgrsql.auto.conf` file has the very same syntax of the main `postgresql.conf` file but is automatically overwritten by PostgreSQL when the configuration is changed at run time directly within the system, by means of specific administrative statements such as `ALTER SYSTEM`.

You are not tied to having a single configuration file, and, in fact, there are specific directives that can be used to include other configuration files. The configuration of the cluster will be detailed in a later chapter.

The PostgreSQL HBA file (`pg_hba.conf`) is another text file that contains the connection allowance: it lists the databases, the users, and the networks that are allowed to connect to your cluster. As an example, the following is an excerpt from a `pg_hba.conf` file:

```
hosts    all luca 192.168.222.1/32 md5
hostssl all enrico 192.168.222.1/32 md5
```

In short, the preceding lines mean that user `luca` can connect to any database in the cluster by the machine with IPv4 address `192.168.222.1`, while user `enrico` can connect to any database from the same machine but only on an SSL-encrypted connection. All the available `pg_hba.conf` rules will be detailed in a later chapter, but for now, it is sufficient to know that this file acts as a "firewall" for incoming connections.

Summary

PostgreSQL can handle several databases within a single cluster, served out of disk storage contained in a single directory named PGDATA. The cluster runs many different processes; one, in particular, is named postmaster and is in charge of spawning other processes, one per client connection, and keeping track of the status of maintenance processes.

The configuration of the cluster is managed via text-based configuration files, the main one being postgresql.conf. It is possible to drive the cluster, by means of postmaster, to recognize allowed user connections by means of rules placed in the pg_hba.conf text file.

You can interact with the cluster status by means of the pg_ctl tool or, depending on your operating system, by other provided programs, such as service.

This chapter has presented you with all of the preceding information so that you are able not only to install PostgreSQL but also to start and stop it regularly, integrate it with your operating system, connect to the cluster.

In the following chapter, you will learn how to manage users and connections.

References

- PostgreSQL PGDATA disk layout: https://www.postgresql.org/docs/12/storage-file-layout.html
- PostgreSQL initdb official documentation: https://www.postgresql.org/docs/12/app-initdb.html
- PostgreSQL pg_ctl official documentation: https://www.postgresql.org/docs/12/app-pg-ctl.html
- The pgAdmin4 graphical client for PostgreSQL: https://www.pgadmin.org/

3
Managing Users and Connections

PostgreSQL is a complex system that includes users, databases, and data. In order to be able to interact with a database in the cluster, you need to have at least one user. By default, when installing a new cluster, a single administrator user (named `postgres`) is created. While it is possible to handle all the connections, applications, and databases with that single administrative user, it is much better to create different users with different properties and privileges, as well as login credentials, for every specific task.

PostgreSQL provides a very rich user-management structure, and single users can be grouped into a variety of different groups at the same time. Moreover, groups can be nested within other groups, so that you can have a very accurate representation of your account model. Thanks to this accurate representation, and thanks to the fact that every user and group can be assigned different properties and privileges, it is possible to apply fine-grained permissions to each user in the database, depending on the specific task and activity involved.

This chapter introduces you to the concepts behind users and groups and their relationships. The chapter will focus mainly on the login properties of roles (either users or groups) and how PostgreSQL can prevent specific users from connecting to specific databases.

This chapter covers the following main topics:

- Introduction to users and groups
- Managing roles
- Managing incoming connections at the role level

Introduction to users and groups

In order to connect interactively or via an application to a PostgreSQL database, you need to have login credentials. In particular, a database user, a user who is allowed to connect to that specific database, must exist.

Database users are somewhat similar to operating system users: they have a username and a password (usually encrypted) and are known to the PostgreSQL cluster. Similarly to operating system users, database users can be grouped into **user groups** in order to ease the massive administration of users.

In SQL, and therefore even in PostgreSQL, both concepts of a single user account and a group of accounts are encompassed by the concept of a **role**.

A role can be a single account, a group of accounts, or even both depending on how you configure it; however, in order to ease management, a role should express one and only one concept at a time: that is, it should be either a single user or a single group, but not both.

 While a role can be used simultaneously as a group or a single user, we strongly encourage you to keep the two concepts of user and group separated—it will simplify the management of your infrastructure.

Every role must have a unique name or identifier, usually called the username.

A role represents a collection of database permissions and connection properties. The two elements are orthogonal. You can set up a role simply as a container for other roles, configuring the contained roles to hold the assigned permissions, or you can have a role that holds all the permissions for contained roles, or mix and match these two approaches.

It is important to understand that a role is defined at the cluster level. This means that the same role can have different privileges and properties depending on the database it is using (for instance, being allowed to connect to one database and not to another).

 Since a role is defined at a cluster level, it must have a unique name within the entire cluster, not just the group!

Managing roles

Roles can be managed by means of three main SQL statements: CREATE ROLE to create a role from scratch, ALTER ROLE to change some role properties (for example, the login password), and DROP ROLE to remove an existing role.

In order to use the SQL statements to create new roles and then manage them, it is necessary to connect to a database in the cluster. The superuser role postgres can be used to that aim, at least initially; such a role is created when the database cluster is initialized. Using the postgres role and a template database is the most common way to create your initial roles.

 PostgreSQL ships with a set of shell scripts that can be used to create, modify, and delete roles without connecting directly to the cluster. Under the hood, those scripts connect to the template database and perform the same SQL commands found in this section.

A role is identified by a string that represents the role name, or better, the account name of that role. Such a name must be unique across the system, meaning that you cannot have two different roles with identical names, and must consist of letters, digits, and some symbols, such as underscore.

Creating new roles

In order to create a new role, either a single user account or a group container, you need to use the CREATE ROLE statement. The statement has the following short synopsis and requires a mandatory parameter that is the role username:

```
CREATE ROLE name [ [ WITH ] option [ ... ] ]
```

The options that you can specify in the statement range from the account password, the ability to log in interactively, and the superuser privileges. Please remember that, unlike other systems, in PostgreSQL, you can have as many superusers as you want, and everyone has the same live-or-die rights on the cluster.

Almost every option of the CREATE ROLE statement has a positive form that adds the ability to the role, and a negative form (with a NO prefix) that excludes the ability from the role. As an example, the SUPERUSER option adds the ability to act as a cluster superuser, while the NOSUPERUSER option removes it from the role.

In this chapter, we will focus on the login abilities, which is a restricted set of options that allows a role to log in to the cluster. Other options will be discussed in Chapter 10, *Users, Roles, and Database Security*, since they are more related to the security features of the role.

 What if you forgot an option at the CREATE ROLE time? And what if you changed your mind and want to remove an option from an existing role? There is an ALTER ROLE statement that allows you (as a cluster superuser) to modify an existing role without having to drop and recreate it. The statement will be shown in Chapter 10, *Users, Roles, and Database Security*, along with some other interesting options for roles.

Role passwords, connections, and availability

Connecting to a database in the cluster means that the role must authenticate itself, and therefore there must be an authentication mechanism, the username and password being the most classical ones.

When a user attempts to connect to a database, PostgreSQL checks the login credentials and a few other properties of the user to ensure that it is allowed to log in and has valid credentials.

The main options that allow you to manipulate and manage the login attempts are as follows:

- PASSWORD or ENCRYPTED PASSWORD are equivalent options and allow you to set the login password for the role. Both options exist for backward compatibility with older PostgreSQL versions, but nowadays, the cluster always stores role passwords in an encrypted form, so the use of ENCRYPTED PASSWORD does not add any value to the PASSWORD option.
- PASSWORD NULL explicitly forces a null (not empty) password, preventing the user from logging in with any password. This option can be used to deny password-based authentication.
- CONNECTION LIMIT <n> allows the user to open no more than <n> simultaneous connections to the cluster, without any regard to a specific database. This is often useful to prevent a user from wasting resources on the cluster.
- VALID UNTIL allows you to specify an instant (in the future) when the role will expire.

Setting the password for a specific role does not mean that that role will be able to connect to the cluster: in order to be allowed to interactively log in, the role must also have the LOGIN option. In other words, the following statement will not allow the user to log in:

```
template1=# CREATE ROLE luca
           WITH PASSWORD 'xxx';
```

The default option is NOLOGIN (which prevents interactive login). Therefore, in order to define interactive users, remember to add the LOGIN option when creating the role:

```
template1=# CREATE ROLE luca
           WITH LOGIN PASSWORD 'xxx';
```

Multiple options can be written in any order, so the preceding code represents the same statement, but in a form that is less human readable:

```
template1=# CREATE ROLE luca
           WITH PASSWORD 'xxx' LOGIN;
```

The VALID UNTIL option allows you to define a date or even a timestamp (that is, an instant) in the future when the role password will expire, and will no longer be allowed to log in to the cluster.

Of course, this option only makes sense for interactive roles, meaning those who have the LOGIN capability. As an example, the following role will be prevented from logging in after Christmas 2020:

```
template1=# CREATE ROLE luca
           WITH LOGIN PASSWORD 'xxx'
           VALID UNTIL '2020-12-25 23:59:59';
```

Using a role as a group

A group is a role that contains other roles. It's that simple!

Usually, when you want to create a group, all you need to do is create a role without the LOGIN option and then add all the members one after the other to the **containing role**. Adding a role to a containing role makes the latter a group.

In order to create a role as a member of a specific group, the IN ROLE option can be used. This option accepts the name of the group (which, in turn, is another role) to which the newly created role will become a member. As an example, in the following code block, you can see the creation of the book_authors group and the addition of the role members luca and enrico:

```
template1=# CREATE ROLE book_authors
           WITH NOLOGIN;
CREATE ROLE
template1=# CREATE ROLE luca
 WITH LOGIN PASSWORD 'xxx'
 IN ROLE book_authors;
CREATE ROLE
template1=# CREATE ROLE enrico
           WITH LOGIN PASSWORD 'xxx'
           IN ROLE book_authors;
CREATE ROLE
```

 The IN GROUP clause of CREATE ROLE is an obsolete synonym for the IN ROLE clause.

It is also possible to add members to a group using the special GRANT statement. The GRANT statement is the general SQL statement that allows for fine privilege tuning (more on this in Chapter 10, *Users, Roles, and Database Security*); PostgreSQL extends the SQL syntax allowing the *granting of a role to another role*. When you grant a role to another, the former becomes a member of the latter. In other words, assuming that all roles already exist without any particular association, the following adds the role enrico to the book_authors group:

```
template1=# GRANT ROLE book_authors
           TO enrico;
```

Every group can have one or more **admin** members, which are allowed to add new members to the group. The ADMIN option allows a user to specify the member that will be associated as an administrator of the newly created group. For instance, in the following code block, you can see the creation of the new group called book_reviewers with luca as administrator; this means that the user luca, even if they are not a cluster superuser, will be able to add new members to the book_reviewers group:

```
template1=# CREATE ROLE book_reviewers
           WITH NOLOGIN
           ADMIN luca;
CREATE ROLE
```

As you can see, the ADMIN option can be used in CREATE ROLE only if the administrator role already exists; in the example, the luca role must have been created before the group, as he is going to be the administrator.

The GRANT statement can solve the problem—the WITH ADMIN OPTION clause allows the membership of a role with administrative privileges.

As an example, the following piece of code shows how to make the user enrico also an administrator of the book_reviewers group. Please note that the full WITH ADMIN OPTION has to be spelled out:

```
template1=# GRANT book_reviewers
            TO enrico
            WITH ADMIN OPTION;
GRANT ROLE
```

What happens if a group role has the LOGIN option? The group will still be a role container, but it can act also as a single user account with the ability to log in. While this is possible, it is a more common practice to deny group roles access to login to avoid confusion.

Removing an existing role

In order to remove an existing role, you need to use the DROP ROLE statement. The statement has a very simple synopsis:

```
DROP ROLE [ IF EXISTS ] name [, ...]
```

You need to specify only the role name you want to delete, or, if you need to delete multiple roles, you can specify them as a comma-separated list.

In order to be deleted, the role must exist; therefore, if you try to remove a nonexistent role, you will receive an error:

```
template1=# DROP ROLE this_role_does_not_exists;
ERROR:  role "this_role_does_not_exists" does not exist
```

As you can see, PostgreSQL warns you that it cannot delete a role if the role does not exist.

> You cannot break PostgreSQL! PostgreSQL will protect itself from your mistakes, and does a very good job of keeping your data safe! The preceding example about the deletion of a nonexistent role is an example of how PostgreSQL protects itself from your own mistakes in order to ensure an always-stable service.

The DROP ROLE statement supports the IF EXISTS clause, which stops PostgreSQL from complaining about the deletion of a role that is missing:

```
template1=# DROP ROLE IF EXISTS this_role_does_not_exists;
NOTICE:  role "this_role_does_not_exists" does not exist, skipping
DROP ROLE
```

As you can see, this time PostgreSQL does not raise an error; instead, it displays a notice about the fact that the role does not exist. However, it executes the statement, doing nothing, but reporting success instead of failure. Why could this be useful? Imagine that you have an automated task that is in charge of deleting several roles: if the DROP ROLE reports a failure, your task could be interrupted, while with IF EXISTS, you will rest assured that PostgreSQL will not cause an abort due to a missing role.

> There are several statements that support the IF EXISTS clause, as you will see in later chapters. The idea is to avoid reporting an error when you are not interested in catching it, and you should use, whenever possible, this clause in automating programs.

What happens if you drop a group? Member roles will stay in place, but of course, the association with the group will be lost (since the group has been deleted). In other words, deleting a group does not cascade to its members.

Inspecting existing roles

Now that you know how to manage roles, how can you inspect existing roles, including yours? There are different ways to get information about existing roles, and all rely on the PostgreSQL catalogs, the only source of introspection into the cluster.

In order to get information about what role you are running, use the special keyword CURRENT_ROLE: you can query it via a SELECT statement (such statements will be presented in later chapters, so for now, just blindly use it as shown here):

```
template1=# SELECT current_role;
 current_role
--------------
 postgres
(1 row)
```

If you connect to the database with another user, you will see different results:

```
$ psql -U luca template1
psql (12.1)
Type "help" for help.
```

```
template1=> SELECT current_role;
 current_role
 --------------
 luca
(1 row)
```

Knowing your own role is important, but getting information about existing roles and their properties can be even more illuminating. `psql` provides the special `\du` (describe users) command to list all the available roles within the system:

```
$ psql -U postgres template1
psql (12.1)
Type "help" for help.
template1=# \du
                                 List of roles
  Role name    |                       Attributes
 |    Member of
 ---------------+--------------------------------------------------------------
 +-----------------
 book_authors | Cannot login
 | {}
 enrico        |
 | {book_authors}
 luca          | 1 connection
 | {book_authors}
 postgres      | Superuser, Create role, Create DB, Replication, Bypass RLS
 | {}
```

The `Attributes` column shows the options and properties of the role, many of which will be discussed in `Chapter 10`, *Users, Roles, and Database Security*. With regard to the login properties, if a role is prevented from connecting interactively to the cluster, a `Cannot login` information will be displayed in the `book_authors` line in the preceding example.

You can get information about a specific role by directly querying the `pg_roles` catalog, a catalog that contains information about all PostgreSQL roles. For example, to get the basic connection information for the `luca` role, you can execute the following query:

```
template1=# SELECT rolname, rolcanlogin, rolconnlimit, rolpassword
            FROM pg_roles
            WHERE rolname = 'luca';
-[ RECORD 1 ]--+----------
rolname        | luca
rolcanlogin    | t
rolconnlimit   | 1
rolpassword    | ******
```

As you can see, the password is not displayed for security reasons, even if the cluster superuser is asking for it. It is not possible to get the password in plain text; as we have already explained, the passwords are always stored encrypted. The special catalog `pg_authid` represents the backbone for the `pg_roles` information, and can be queried with the very same statement, but reports the user password (as encrypted text). The following code shows the result of querying `pg_authid` for the very same user as in the fourth listing; note how the `rolpassword` field contains some more useful information this time:

```
template1=# SELECT rolname, rolcanlogin, rolconnlimit, rolpassword
           FROM pg_authid WHERE rolname = 'luca';
-[ RECORD 1 ]--+-------------------------------------
rolname        | luca
rolcanlogin    | t
rolconnlimit   | 1
rolpassword    | md5bd18b4163ec8a322833d8d7a6633c8ec
```

The password is represented as a hash and the initial part specifies the encryption algorithm used—MD5. You will learn more about password encryption in Chapter 10, *Users, Roles, and Database Security*.

Managing incoming connections at the role level

When a new connection is established to a cluster, PostgreSQL validates the incoming request at the role level. The fact that the role has the LOGIN property is not enough for it to open a new connection to any database within the cluster. This is because PostgreSQL checks the incoming connection request against a kind of firewall table, formerly know as **host-based access**, that is defined within the `pg_hba.conf` file.

If the table states that the role can open the connection to the specified database, the connection is granted (assuming it has the LOGIN property); otherwise, it is rejected.

Every time you modify the `pg_hba.conf` file, you need to instruct the cluster to reload the new rules via a HUP signal or by means of a `reload` command in `pg_ctl`. Therefore, the usual workflow when dealing with `pg_hba.conf` is similar to the following:

```
$ $EDITOR $PGDATA/pg_hba.conf
... modify the file as you wish ...

$ sudo -u postgres pg_ctl reload -D $PGDATA
```

The syntax of pg_hba.conf

The `pg_hba.conf` file contains the firewall for incoming connections. Every line within the file has the following structure:

```
<connection-type> <database> <role> <remote-machine> <auth-method>
```

Here, we see the following values:

- `connection-type` is the type of connection supported by PostgreSQL, and is either `local` (meaning via operating system sockets), `host` (TCP/IP connection), or `hostssl` (TCP/IP encrypted connection).
- `database` is a name of a specific database that the line refers to or the special keyword `all`, which means every available database.
- `role` is the specific role username that the line refers to or the special keyword `all`, which means all available roles (and groups).
- `remote-machine` is the hostname, IP address, or subnet from which the connection is expected. The special keyword `all` matches with any remote machine that the connection is established from, while the special keywords `samehost` and `samenet` match the localhost or the whole network that the PostgreSQL cluster is running within, respectively.
- `auth-method` dictates how the connection must be handled; more generally, it deals with how the login credentials have to be checked. The main methods are `scram-sha-256` (the most robust method, available since PostgreSQL 10), `md5` (the method used in older versions), `reject` to always refuse the connection, and `trust` to always accept the connection without any regard to supplied credentials.

In order to better understand how the system works, the following is an excerpt of a possible `pg_hba.conf` file:

```
host     all       luca     carmensita      scram-sha-256
hostssl  all       test     192.168.222.1/32 scram-sha-256
host     digikamdb pgwatch2 192.168.222.4/32 trust
host     digikamdb enrico   carmensita      reject
```

The first line indicates that the user `luca` can connect to every database within the cluster (`all` clause) via a TCP/IP connection (`host` clause) coming from a host named `carmensita`, but he must provide a valid username/password to verify the SCRAM authentication method.

The second line states that the user `test` can connect to every database in the system over an SSL-encrypted connection (see the `hostssl` clause), but only from a machine that has the IPv4 address of `192.168.222.1`; again, the credentials must pass the SCRAM authentication method.

The third line states that access to the `digikamdb` database is granted only to the `pgwatch2` user over a nonencrypted connection from the host `192.168.222.4`; this time, the access is granted (`trust`) without any credential being required.

Finally, the last line rejects any incoming connection from the host named `carmensita`, opened by the user `enrico` against the `digikamdb`; in other words, `enrico` is not able to connect to `digikamdb` from the `carmensita` host.

 The authentication method `trust` should never be used; it allows any role to connect to the database if the HBA has a rule that matches the incoming connection. This is the method that is used when the cluster is initialized in order to enable the freshly created superuser to connect to the cluster. You can always use this trick as a last resort if you get yourself locked out of your own cluster.

Order of rules in pg_hba.conf

The order by which the rules are listed in the `pg_hba.conf` file matters. The first rule that satisfies the logic is applied, and the others are skipped. In order to better understand this, imagine that we want to allow `luca` to connect to any database in the cluster except `forumdb`. The following does not make this happen:

```
host all     luca all scram-sha-256
 host forumdb luca all reject
```

Why does the preceding code not work?

Imagine that the user `luca` tries to open a connection to the `forumdb` database: the machine from which the connection is attempted is matched against the `all` keyword with the line containing `luca`, and then the database name is matched against the `all` keyword for the database field.

Since both the remote machine and the database name are subsets of `all`, the connection is passed through the SCRAM-256 authentication method; if the user succeeds in the authentication, the connection is opened. The `reject` line is therefore skipped because the first line matches. On the other hand, exchanging the order of the rules as shown in the following code does work:

```
host forumdb luca all reject
host all      luca all scram-sha-256
```

In this way, when `luca` tries to connect to a database, he gets rejected if the database is `forumdb`; otherwise, he can connect (if he passes the required authentication method).

Merging multiple rules into a single one

One line declares at least one rule, but it is possible to merge multiple lines into a single one. In fact, the role, database, and remote-machine fields allow for the definition of multiple matches, each one separated by a , (comma).

As an example, suppose we want to give access to both `luca` and `enrico` roles (from the same network that the cluster is running into) to the `forumdb` and `digikamdb` databases, so that `pg_hba.conf` looks like the following:

```
host forumdb   luca   samenet scram-sha-256
host forumdb   enrico samenet scram-sha-256
host digikamdb luca   samenet scram-sha-256
host digikamdb enrico samenet scram-sha-256
```

Since the database and the role fields can list more than one item, the preceding code can be compressed into the following one:

```
host forumdb,digikamdb   luca   samenet scram-sha-256
host forumdb,digikamdb   enrico samenet scram-sha-256
```

We can shrink the rules one step further since the machine from which the database connection can be established is literally the same for both the rules, and therefore the final code is as follows:

```
host forumdb,digikamdb   luca, enrico   samenet scram-sha-256
```

It should now be clear to you that if more rules have the same authentication method and connection protocol, then it is possible to collapse them into an aggregation. This can help you manage the host-based access configuration.

Using groups instead of single roles

The role field in every `pg_hba.conf` rule can be substituted by the name of a group (remember that a group is itself a role); however, in order to make the rule valid for every member of the group, you have to prefix the group name with a + (plus) sign.

To better understand this, consider the example of the `forum_stats` group that includes the `luca` member. The following rule will not allow the `luca` role to access the `forumdb` database:

```
host forumdb forum_stats all scram-sha-256
```

Even if the user is a member the `forum_stats` role, it will be denied the ability to log in to the database; the cluster host-based access policy requires the `forum_stats` role to be exactly matched by a rule, and in the following, the `luca` role does not match any rule:

```
$ psql -U luca forumdb
psql: error: could not connect to server:
FATAL:  no pg_hba.conf entry for host "192.168.222.1", user "luca",
database "forumdb", SSL off
```

On the other hand, if we clearly state that we want to use the `forum_stat` role as a group name, and therefore allow all of its members, the connection can be established by any role that is a member of the group, including `luca`. Therefore, we change the rule to the following:

```
host forumdb +forum_stats all scram-sha-256
```

This, in turn, (bearing in mind the plus sign) makes the connection possible, as shown here:

```
$ psql   -U luca forumdb
psql (12.1)
Type "help" for help.

forumdb=>
```

The `pg_hba.conf` rules, when applied to a group name (that is, with the + preceding the role name) include all the direct and indirect members.

What if we want to allow every group member except one to access the database? Remembering that the rule engine stops at the first match, it is possible to place a reject rule before the group acceptance rule. For example, to allow every member of the `forum_stats` group to access the database while preventing the single `luca` role from connecting, you can use the following:

```
host forumdb luca          all reject
host forumdb +forum_stats all scram-sha-256
```

The first line will prevent the `luca` role from connecting, even if the following one allows every member of the `forum_stats` (including `luca`) to connect: the first match wins and so `luca` is locked out the database.

Using files instead of single roles

The role field of a rule can also be specified as a text file, both line or comma-separated. This is handy when you deal with long usernames or group names, or with lists produced automatically from batch processes.

If you specify the role field with an at sign prefix (@), the name is interpreted as a line-separated text file (as a relative name to the `PGDATA` directory). For instance, in order to reject connections to all the users and groups listed in the file `rejected_users.txt`, while allowing connection to all the usernames and groups specified in the `allowed_users.txt` file, the `pg_hba.conf` file has to look like the following snippet:

```
host forumdb @rejected_users.txt   all reject
host forumdb @allowed_users.txt    all scram-sha-256
```

The following is the content of the `rejected_users.txt` file, followed by the `allowed_users.txt` file:

```
$ sudo cat $PGDATA/rejected_users.txt
luca
enrico

$ sudo cat $PGDATA/allowed_users.txt
+forum_stats, postgres
```

As you can see, it is possible to specify the file contents as either a line-separated list of usernames or a comma-separated list. It is also possible to specify which roles to use as a group by placing a + sign in front of the role name.

Summary

Roles are a powerful tool to represent both single users and a group of users. When a database connection attempt is made, PostgreSQL processes the connection credential information through the host-based access control so that it can immediately establish or reject the connection depending on firewall-like rules. Moreover, single users and groups can have other limitations to the number of connections they can open against a database or against the whole cluster.

In this chapter, you have seen how to create and manage roles, as well as how to allow single roles to connect to the cluster and to specific databases. In `Chapter 10`, *Users, Roles, and Database Security,* you will see how to deal with the security properties of users and groups, but before you proceed further, you need to know how PostgreSQL objects can be created and managed.

In the following chapter, you will learn how to interact with the PostgreSQL database using SQL statements.

References

- `CREATE ROLE` statement official documentation: https://www.postgresql.org/docs/12/sql-createrole.html
- `DROP ROLE` statement official documentation: https://www.postgresql.org/docs/12/sql-droprole.html
- PostgreSQL `pg_roles` catalog details: https://www.postgresql.org/docs/12/view-pg-roles.html
- PostgreSQL `pg_authid` catalog details: https://www.postgresql.org/docs/12/catalog-pg-authid.html
- PostgreSQL host-based access rule details: https://www.postgresql.org/docs/12/auth-pg-hba-conf.html

Section 2: Interacting with the Database

2

In this section, you will learn how to interact with a live cluster, creating database objects and using tools to connect to it. You will also learn about various basic and advanced commands that can be used for managing databases.

This section contains the following chapters:

- *Chapter 4, Basic Statements*
- *Chapter 5, Advanced Statements*
- *Chapter 6, Window Functions*
- *Chapter 7, Server-Side Programming*
- *Chapter 8, Triggers and Rules*
- *Chapter 9, Partitioning*

4

Basic Statements

In this chapter, we will discuss basic SQL commands for PostgreSQL; these are **Data Definition Language** (**DDL**) commands and **Data Manipulation Language** (**DML**) commands. In basic terms, DDL commands are used to manage databases and tables, and DML commands are used to insert, delete, update, and select data inside databases. In this chapter, we will also discuss the `psql` environment, which refers to the interactive terminal for working with PostgreSQL. `psql` can be described as PostgreSQL's shell environment; it is the gate we have to go through in order to start writing commands natively in PostgreSQL. We have to remember that `psql` is always present in any PostgreSQL installation we work with. `psql` is a powerful environment in which to manage our data and our databases.

Basic statements and `psql` are therefore the foundations on which we will build our knowledge of PostgreSQL. Therefore, reading and understanding this chapter is essential if you are going to understand some of the more complex topics we are going to talk about later.

Let's show a list of what we're going to learn in this chapter:

- Setting up our developing environment
- Creating and managing databases
- Managing tables
- Understanding basic table manipulation statements

Technical requirements

You can find the code for this chapter in the following GitHub repository: `https://github.com/PacktPublishing/Learn-PostgreSQL`.

Setting up our developing environment

At this point in the book, we have learned how to install PostgreSQL and how to configure users. Let's now see how to connect to our database. In the next four steps, we will see how easy it is to do this:

1. Start by connecting to your `psql` environment:

   ```
   postgres@pgdev:~$ psql
   psql (12.1 (Debian 12.1-1.pgdg100+1))
   Type "help" for help.
   postgres=#
   ```

2. Next, switch on the expanded mode using the \x command:

   ```
   postgres=# \x
   Expanded display is on.
   ```

3. Then list all the databases that are present in the cluster:

   ```
   postgres=# \l
   List of databases
   -[ RECORD 1 ]-----+----------------------
   Name              | forumdb
   Owner             | postgres
   Encoding          | UTF8
   Collate           | en_US.UTF-8
   Ctype             | en_US.UTF-8
   Access privileges |
   ```

4. Finally, connect to the `forumdb` database:

   ```
   postgres=# \c forumdb
   You are now connected to database "forumdb" as user "postgres".
   forumdb=#
   ```

Now, that we have finished setting up our developing environments, we can move on to creating databases in them.

Creating and managing databases

In this section, we will start by creating our first database, then we will learn how to delete a database and, finally, how to create a new database from an existing one. We will also analyze the point of view of the DBA. We will see what happens behind the scenes when we create a new database and learn some basic functions useful to the DBA to get an idea of the real size of the databases.

Let's see how to create a database from scratch and what happens behind the scenes when a database is created.

Creating a database

To create the `forumdb` database from scratch, you will need to execute this simple statement:

```
CREATE DATABASE databasename
```

SQL is a case insensitive language, so we can write all the commands with uppercase or lowercase letters.

Now, let's see what happens behind the scenes when we create a new database. PostgreSQL performs the following steps:

1. Makes a physical copy of the template database, `template1`.
2. Assigns the database name to the database just copied.

The `template1` database is a database that is created by the `initdb` process during the initialization of the PostgreSQL cluster.

Managing databases

In the previous section, we created a new database called `forumdb`. In this section, we will see how to manage databases, how to list all the databases present on a cluster, how to create a database starting from an existing database, how to drop a database, and what happens internally, behind the scenes, when we create and drop the database.

Listing all databases

To list all the tables present in the database forumdb, we have to use the psql command \d. The \d command makes a list of all the tables present in the forumdb database:

```
forumdb=# \d
 List of relations
 Schema |       Name        |   Type   |  Owner
--------+-------------------+----------+----------
 public | categories        | table    | postgres
 public | categories_pk_seq | sequence | postgres
 public | j_posts_tags      | table    | postgres
 public | posts             | table    | postgres
 public | posts_pk_seq      | sequence | postgres
 public | tags              | table    | postgres
 public | tags_pk_seq       | sequence | postgres
 public | users             | table    | postgres
 public | users_pk_seq      | sequence | postgres
(9 rows)
```

Making a new database from a modified template

Now that we've learned how to list all tables in a database, let's see that any changes made to the template1 database will be seen by all the databases that will be created later. Now we will perform these steps:

1. Connect to the template1 database.
2. Create a table called dummytable inside the template1 database.
3. Create a new database called dummydb.

So let's start making the database using the following steps:

1. Connect to the template1 database:

   ```
   forumdb=# \c template1
   You are now connected to database "template1" as user "postgres".
   ```

2. Create a table called dummytable. For now, we don't need to worry about the exact syntax for creating tables; this will be explained in more detail later on:

   ```
   template1=# create table dummytable (dummyfield integer not null
   primary key);
   CREATE TABLE
   ```

3. Use the `\d` command to show a list of tables that are present in the `template1` database:

```
template1=# \d
 List of relations
 Schema | Name        | Type  | Owner
--------+-------------+-------+----------
 public | dummytable  | table | postgres
(1 row)
```

4. So, we have successfully added a new table to the `template1` database. Now let's try to create a new database called `dummydb` and make a list of all the tables in the `dummydb` database:

```
template1=# create database dummydb;
CREATE DATABASE
template1=# \c dummydb
You are now connected to database "dummydb" as user "postgres".
```

The `dummydb` database contains the following tables:

```
dummydb=# \d
 Schema | Name        | Type  | Owner
--------+-------------+-------+----------
 public | dummytable  | table | postgres
(1 row)
```

As expected, in the `dummydb` database, we can see the table created previously in the `template1` database.

It is important to remember that any changes made to the `template1` database will be present in all databases created after this change.

Now we will delete the `dummydb` database and the dummy table in the `template1` database.

Dropping tables and databases

In the next section, you will learn how to delete tables and databases. The commands we are going to learn are the following:

- DROP TABLE: This is used to drop a table in the database.
- DROP DATABASE: This is used to drop a database in the cluster.

Dropping tables

In PostgreSQL, the command needed to drop a table is simply DROP TABLE tablename. To do this, we have to connect to the database to which the table belongs, and then run the command DROP TABLE tablename.

For example, if we want to drop the dummytable table from database template1, we have to take the following steps.

We connect to database template1 using the following command:

```
dummydb=# \c template1
You are now connected to database "template1" as user "postgres".
```

And we can drop the table using the following command:

```
template1=# drop table dummytable;
DROP TABLE
```

Dropping databases

In PostgreSQL the command needed to drop a table is simply DROP DATABASE databasename; for example, if we want to drop the dummydb database, we have to execute the following command:

```
template1=# drop database dummydb ;
DROP DATABASE
```

With this, everything has now been returned to how it was at the beginning of the chapter.

Making a database copy

The following steps show you how to make a new database out of a template database:

1. Make a copy of the `forumdb` database on the same PostgreSQL cluster by performing the following command:

    ```
    template1=# create database forumdb2 template forumdb;
    CREATE DATABASE
    ```

    By using this command, you are simply telling PostgreSQL to create a new database called `forumdb2` using the `forumdb` database as a template.

2. Connect to the `forumdb2` database:

    ```
    template1=# \c forumdb2
    You are now connected to database "forumdb2" as user "postgres".
    ```

3. List all the tables in the `forumdb2` database:

    ```
    forumdb2=# \d
     List of relations
     Schema | Name             | Type     | Owner
    --------+------------------+----------+----------
     public | categories       | table    | postgres
     public | categories_pk_seq | sequence | postgres
     public | j_posts_tags     | table    | postgres
     public | posts            | table    | postgres
     public | posts_pk_seq     | sequence | postgres
     public | tags             | table    | postgres
     public | tags_pk_seq      | sequence | postgres
     public | users            | table    | postgres
     public | users_pk_seq     | sequence | postgres
    (9 rows)
    ```

You can see that the same tables that are present in the `forumdb` database are now present in this database.

Confirming the database size

We are now going to address the question of how one can determine the real size of a database. There are two methods you can use to do this: `psql` and SQL. Let's compare the two in the following sections.

The psql method

We can check the database size using the `psql` method, using the following steps:

1. First, we return to expanded mode:

   ```
   forumdb=# \x
   Expanded display is on.
   ```

2. Then, execute the following command:

   ```
   forumdb=# \l+ forumdb
   List of databases
   -[ RECORD 1 ]-----+------------
   Name              | forumdb
   Owner             | postgres
   Encoding          | UTF8
   Collate           | en_US.UTF-8
   Ctype             | en_US.UTF-8
   Access privileges |
   Size              | 8369 kB
   Tablespace        | pg_default
   Description       |
   ```

As you can see, in the `Size` field, you can now see the real size of the database at that moment.

The SQL method

When trying to use the method outlined above, you may find that you cannot connect to your database through the `psql` command. This happens when we only have web access to the database; for example, if we only have `pgadmin4` server-side installation access. If this happens, the SQL method is an alternative approach that will allow you to find the same information. To use this method, complete the following steps:

1. Execute the following command:

   ```
   forumdb=# select pg_database_size('forumdb');
   -[ RECORD 1 ]----+--------
   pg_database_size | 8569711
   ```

   The `pg_database_size(name)` function returns the disk space used by the database called `forumdb`. This means that the result is the number of bytes used by the database.

2. If you wanted a more readable result in "human" terms, you could use the `pg_size_pretty` function and write the following:

```
forumdb=# select pg_size_pretty(pg_database_size('forumdb'));
-[ RECORD 1 ]--+--------
pg_size_pretty | 8369 kB
```

As you can see, both methods give the same result.

Creating a database

We have just learned what commands are used to create a new database, but what happens behind the scenes when a database is created? In this section, we will see the relationships that exist between what we perform at the SQL level and what happens physically in the filesystem.

To understand this, we need to introduce the `pg_database` system table:

1. Go back to the expanded mode and execute the following:

```
forumdb=# select * from pg_database where datname='forumdb';
-[ RECORD 1 ]-+------------
oid           | 16630
datname       | forumdb
datdba        | 10
encoding      | 6
datcollate    | en_US.UTF-8
datctype      | en_US.UTF-8
datistemplate | f
datallowconn  | t
datconnlimit  | -1
datlastsysoid | 14049
datfrozenxid  | 479
datminmxid    | 1
dattablespace | 1663
datacl
```

This query gives us all the information about the `forumdb` database. The first field is an **object identifier** (**OID**), which is a number that uniquely identifies the database called `forumdb`.

2. Exit the `psql` environment and go to the `$PGDATA` directory (as shown in previous chapters). In a Linux Debian environment, we have to execute the following:

   ```
   cd /var/lib/postgresql/12/main/
   ```

   For the PostgreSQL 13 version, the path is as follows:

   ```
   cd /var/lib/postgresql/13/main/
   ```

3. Use the `ls` command to see what is inside the main directory:

   ```
   postgres@pgdev:~/12/main$ ls -l
   total 84
   drwx------ 6 postgres postgres 4096 Dec 8 20:28 base
   drwx------ 2 postgres postgres 4096 Dec 10 11:05 global
   drwx------ 2 postgres postgres 4096 Dec 6 18:47 pg_commit_ts
   drwx------ 2 postgres postgres 4096 Dec 6 18:47 pg_dynshmem
   drwx------ 4 postgres postgres 4096 Dec 10 11:09 pg_logical
   drwx------ 4 postgres postgres 4096 Dec 6 18:47 pg_multixact
   drwx------ 2 postgres postgres 4096 Dec 10 11:04 pg_notify
   drwx------ 2 postgres postgres 4096 Dec 6 18:47 pg_replslot
   drwx------ 2 postgres postgres 4096 Dec 6 18:47 pg_serial
   drwx------ 2 postgres postgres 4096 Dec 6 18:47 pg_snapshots
   drwx------ 2 postgres postgres 4096 Dec 10 11:04 pg_stat
   drwx------ 2 postgres postgres 4096 Dec 6 18:47 pg_stat_tmp
   drwx------ 2 postgres postgres 4096 Dec 6 18:47 pg_subtrans
   drwx------ 2 postgres postgres 4096 Dec 6 18:47 pg_tblspc
   drwx------ 2 postgres postgres 4096 Dec 6 18:47 pg_twophase
   -rw------- 1 postgres postgres 3 Dec 6 18:47 PG_VERSION
   drwx------ 3 postgres postgres 4096 Dec 6 18:47 pg_wal
   drwx------ 2 postgres postgres 4096 Dec 6 18:47 pg_xact
   -rw------- 1 postgres postgres 88 Dec 6 18:47 postgresql.auto.conf
   -rw------- 1 postgres postgres 130 Dec 10 11:04 postmaster.opts
   -rw------- 1 postgres postgres 107 Dec 10 11:04 postmaster.pid
   ```

   As you can see, the first directory is called `base`. It contains all the databases that are in the cluster.

4. Go inside the `base` directory in order to see the contents:

   ```
   postgres@pgdev:~/12/main$ cd base
   postgres@pgdev:~/12/main/base$
   ```

5. List all files that are present in the directory:

   ```
   postgres@pgdev:~/12/main/base$ ls -l
   total 40
   ```

```
drwx------ 2 postgres postgres 12288 Dec 10 11:04 1
drwx------ 2 postgres postgres 4096 Dec 6 18:47 14049
drwx------ 2 postgres postgres 12288 Dec 10 11:05 14050
drwx------ 2 postgres postgres 12288 Dec 10 11:05 16630
```

As you can see, there is a directory called `16630`; its name is exactly the same as the OID in the `pg_database` catalog.

 When PostgreSQL creates a new database, it copies the directory relative to the database `template1` and then gives it a new name. In PostgreSQL, databases are directories.

In this section, we have learned how to manage databases. In the next section, we will learn how to manage tables.

Managing tables

In this section, we will learn how to manage tables in the database.

PostgreSQL has three types of tables:

- **Temporary tables**: Very fast tables, visible only to the user who created them
- **Unlogged tables**: Very fast tables to be used as support tables common to all users
- **Logged tables**: Regular tables

We will now use the following steps to create a user table from scratch:

1. Create a new database using the following command:

   ```
   forumdb=# create database forumdb2;
   CREATE DATABASE
   ```

2. Execute the following command:

   ```
   forumdb=# \c forumdb2
   You are now connected to database "forumdb2" as user "postgres".

   forumdb2=# CREATE TABLE users (
    pk int GENERATED ALWAYS AS IDENTITY
    , username text NOT NULL
    , gecos text
    , email text NOT NULL
   ```

```
, PRIMARY KEY( pk )
, UNIQUE ( username )
) ;
CREATE TABLE
```

The CREATE TABLE command creates a new table. The command GENERATED AS IDENTITY, automatically assigns a unique value to a column.

3. Observe what was created on the database using the /d command:

```
forumdb2=# \d users
                        Table "public.users"
  Column  | Type | Collation | Nullable | Default
----------+---------+-----------+----------+----------------------
-------
 pk       | integer |         | not null | generated always as
identity
 username | text    |         | not null |
 gecos    | text    |         |          |
 email    | text    |         | not null |
Indexes:
    "users_pkey" PRIMARY KEY, btree (pk)
    "users_username_key" UNIQUE CONSTRAINT, btree (username)
```

Something to note is that PostgreSQL has created a unique index. Later in this book, we will analyze indexes in more detail and address what they are, what kind of indexes exist, and how to use them. For now, we will simply say that a unique index is an index that does not allow the insertion of duplicate values for the field where the index was created.

In PostgreSQL, primary keys are implemented using unique indexes.

4. Use the following command to drop a table:

```
forumdb=# drop table users ;
```

The preceding command simply drops the table users. The CREATE TABLE command, as we've seen before, has some useful options:

*   IF NOT EXISTS
*   TEMP
*   UNLOGGED

We'll cover each of these in the following subsections.

The EXISTS option

The EXISTS option can be used in conjunction with entity create or drop commands to check whether the object already exists or the object doesn't exist. An example of its use may be combined with the CREATE TABLE or CREATE DATABASE command. We can use also this option when we create or drop sequences, indices, roles, and schemas. The use case is very simple – the create or drop command is executed if the EXISTS clause is true; for example, if we want to create a table named users, if the table exists, we have to execute this SQL statement:

```
forumdb=# create table if not exists users (
    pk int GENERATED ALWAYS AS IDENTITY
   ,username text NOT NULL
   ,gecos text
   ,email text NOT NULL
   ,PRIMARY KEY( pk )
   ,UNIQUE ( username )
);
NOTICE: relation "users" already exists, skipping
CREATE TABLE
```

The command described above will only create the users table if the users table does not exist already, otherwise, the command will be skipped. The DROP command works similarly; the DROP table command is used to drop tables; the if exists option also exists for the DROP table command; for example, if we want to drop the users table, if it exists, we have to execute the following:

```
forumdb=# drop table if exists users;
DROP TABLE
```

This command will delete the users table if the users table exists in the database. Now if we run it for the second time, we will have the following:

```
forumdb=# drop table if exists users;
NOTICE: table "users" does not exist, skipping
DROP TABLE
```

You can see that the command is skipped because the table does not exist. This option can be useful because, if the table does not exist, PostgreSQL does not block any other subsequent instructions.

Managing temporary tables

Later in this book, we will explore sessions, transactions, and concurrency in more depth. For now, you simply need to know that a session is a set of transactions, each session is isolated, and that a transaction is isolated from everything else. In other words, anything that happens inside the transaction cannot be seen from outside the transaction until the transaction ends. Due to this, we might need to create a data structure that is visible only within the transaction that is running. In order to do this, we have to use the `temp` option.

We will now explore two possibilities. The first possibility is that we could have a table visible only in the session where it was created. The second is that we might have a table visible in the same transaction where it was created.

The following is an example of the first possibility where there is a table visible within the session:

```
forumdb=# create temp table if not exists temp_users  (
    pk int GENERATED ALWAYS AS IDENTITY
  ,username text NOT NULL
  ,gecos text
  ,email text NOT NULL
  ,PRIMARY KEY( pk )
  ,UNIQUE ( username )
);
CREATE TABLE
```

The preceding command will create the `temp_users` table, which will only be visible within the session where the table was created.

If instead, we wanted to have a table visible only within our transaction, then we would have to add the `on commit drop` options. To do this, we would have to do the following:

1. Start a new transaction.
2. Create the table `temp_users`.
3. Commit or rollback the transaction started at point one.

Let's start from the first point:

1. Start the transaction with the following code:

   ```
   forumdb=# begin work;
   BEGIN
   ```

2. Create a table visible only inside the transaction:

```
forumdb# create temp table if not exists temp_users (
 pk int GENERATED ALWAYS AS IDENTITY
 ,username text NOT NULL
 ,gecos text
 ,email text NOT NULL
 ,PRIMARY KEY( pk )
 ,UNIQUE ( username )
) on commit drop;
```

Now check that the table is present inside the transaction and not outside the transaction:

```
forumdb=# \d temp_users;
                     Table "pg_temp_4.temp_users"
  Column  | Type    | Collation | Nullable | Default
----------+---------+-----------+----------+----------------------
-------
 pk       | integer |           | not null | generated always as
identity
 username | text    |           | not null |
 gecos    | text    |           |          |
 email    | text    |           | not null |
Indexes:
    "temp_users_pkey" PRIMARY KEY, btree (pk)
    "temp_users_username_key" UNIQUE CONSTRAINT, btree (username)
```

3. You can see the structure of the `temp_users` table, so now commit the transaction:

```
forumdb=# commit work;
COMMIT
```

If you re-execute the DESCRIBE command \d temp_users, PostgreSQL responds in this way :

```
forumdb=# \d temp_users;
Did not find any relation named "temp_users".
```

This happens because the `on commit drop` option drops the table once the transaction is completed.

Managing unlogged tables

We will now address the topic of unlogged tables. For now, we will simply note that unlogged tables are much faster than classic tables (also known as logged tables) but are not crash-safe. This means that the consistency of the data is not guaranteed in the event of a crash.

The following snippet shows how to create an unlogged table:

```
forumdb=# create unlogged table if not exists unlogged_users (
    pk int GENERATED ALWAYS AS IDENTITY
   ,username text NOT NULL
   ,gecos text
   ,email text NOT NULL
   ,PRIMARY KEY( pk )
   ,UNIQUE ( username )
);
CREATE TABLE
```

Unlogged tables are a fast alternative to permanent and temporary tables. This performance increase comes at the expense of losing data in the event of a server crash, however. This is something you may be able to afford under certain circumstances.

Creating a table

We will now explore what happens behind the scenes when a new table is created. Also, for tables, PostgreSQL assigns an object identifier called OID. An OID is simply a number that internally identifies an object inside a PostgreSQL cluster. Let's now see the relationship between the tables created at the SQL level and what happens behind the scenes in the filesystem:

1. To do this, we will use the OIDs and a system table called `pg_class`, which collects information about all the tables that are present in the database. So let's run this query:

```
forumdb=# select oid,relname from pg_class where relname='users';
  oid  | relname
-------+---------
 16630 | users
(1 row)
```

Here, the `oid` field is the object identifier field, and `relname` represents the relation name of the object. As seen here, the `forumdb` database is stored in the `16630` directory.

2. Now, let's see where the `users` table is stored. To do this, go to the `16630` directory using the following code:

```
postgres@pgdev:~/12/main/base/16630$ cd
/var/lib/postgresql/12/main/base/16630
```

3. Once here, execute the following command:

```
postgres@pgdev:~/12/main/base/16630$ ls -l | grep 16633
-rw------- 1 postgres postgres 0 Dec 6 23:33 16633
```

As you can see, in the directory `16630`, there is a file called `16633`. In PostgreSQL, each table is stored in one or more files. If the table size is less than 1 GB, then the table will be stored in a single file. If the table has a size greater than 1 GB, then the table will be stored in two files and the second file will be called `16633.1`. If the `users` table has a size greater than 2 GB, then the table will be stored in three files, called `16633`, `16633.1`, `16633.2`, and so on; the same thing happens for the index `users_username_key`.

In PostgreSQL, each table or index is stored in one or more files. When a table or index exceeds 1 GB, it is divided into gigabyte-sized *segments*.

In this section, we've learned how to manage tables, and we've seen what happens internally. In the next section, we will learn how to manipulate data inside tables.

Understanding basic table manipulation statements

Now that you have learned how to create tables, you need to understand how to insert, view, modify, and delete data in the tables. This will help you update any incorrect entries, or update existing entries, as needed. There are a variety of commands that can be used for this, which we will look at now.

Inserting and selecting data

In this section, we will learn how to insert data into tables. To insert data into tables, you need to use the INSERT command. The INSERT command inserts new rows into a table. It is possible to insert one or more rows specified by value expressions, or zero or more rows resulting from a query. We will now go through some use cases as follows:

1. To insert a new user in the users table, execute the following command:

```
forumdb=# insert into users (username,gecos,email) values
('myusername','mygecos','myemail');
INSERT 0 1
```

   This result shows that PostgreSQL has inserted one record into the users table.

2. Now, if we want to see the record that we have just entered into the users table, we have to perform the SELECT command:

```
forumdb=# select * from users;
 pk | username   | gecos    | email
----+------------+----------+---------
  1 | myusername | mygecos  | myemail
(1 row)
```

   The select command is executed in order to retrieve rows from a table. With this SQL statement, PostgreSQL returns all the data present in all the fields of the table. The value * specifies all the fields present. The same thing can be expressed in this way:

```
forumdb=# select pk,username,gecos,email from users;
 pk | username   | gecos    | email
----+------------+----------+---------
 1  | myusername | mygecos  | myemail
(1 row)
```

3. Let's now insert another user into the users table; for example, insert the user 'scotty' with all their own fields:

```
forumdb=# insert into users (username,gecos,email) values
('scotty','scotty_gecos','scotty_email');
INSERT 0 1
```

4. If we want to perform the same search as before, ordering data by the `username` field, we have to execute the following:

```
forumdb=# select pk,username,gecos,email from users order by
username;
 pk | username   | gecos        | email
----+------------+--------------+--------------
 1  | myusername | mygecos      | myemail
 2  | scotty     | scotty_gecos | scotty_email
(2 rows)
```

The SQL language, without the `ORDER BY` option, does not return the data in an orderly manner.

In PostgreSQL, this could also be written as follows:

```
forumdb=# select pk,username,gecos,email from users order by 2;
 pk | username   | gecos        | email
----+------------+--------------+--------------
 1  | myusername | mygecos      | myemail
 2  | scotty     | scotty_gecos | scotty_email
(2 rows)
```

PostgreSQL also accepts field positions on a query as sorting options.

5. Let's now see how to insert multiple records using a single-row statement. For example, the following statement will insert three records in the `categories` table:

```
forumdb=# insert into categories (title,description) values
('apple', 'fruits'), ('orange','fruits'),('lettuce','vegetable');
INSERT 0 3
```

This is a slight variation of the `INSERT` command. Our `categories` table will now contain the following values:

```
forumdb=# select * from categories;
 pk | title   | description
----+---------+-------------
 10 | apple   | fruits
 11 | orange  | fruits
 12 | lettuce | vegetable
(3 rows)
```

6. Now if we want to select only the tuples where the description is equal to `vegetable`, use the `WHERE` condition:

```
forumdb=# select * from categories where description ='vegetable';
 pk | title   | description
----+---------+-------------
 12 | lettuce | vegetable
(1 row)
```

7. The `where` condition filters on one or more fields of the table. For example, if we wanted to search for all those topics with `title` as `orange` and `description` as `fruits`, we would have to write the following:

```
forumdb=# select * from categories where description ='fruits' and
title='orange';
 pk | title   | description
----+---------+-------------
 11 | orange  | fruits
(1 row)
```

8. Now, if for example, we want to select all the tuples that have both a `description` field equal to `fruits` and are sorted by title in reverse order, execute the following:

```
forumdb=# select * from categories where description ='fruits'
order by title desc;
 pk | title   | description
----+---------+-------------
 11 | orange  | fruits
 10 | apple   | fruits
(2 rows)
```

Or we could also write this:

```
forumdb=# select * from categories where description ='fruits'
order by 2 desc;
```

```
 pk | title  | description
----+--------+-------------
 11 | orange | fruits
 10 | apple  | fruits
(2 rows)
```

 The ASC or DESC options sort the query in ascending or descending order; if nothing is specified, ASC is the default.

NULL values

In this section, we will talk about NULL values. In the SQL language, the value NULL is defined as follows:

Null (or NULL) is a special marker used in Structured Query Language to indicate that a data value does not exist in the database. Introduced by the creator of the relational database model, E. F. Codd, SQL Null serves to fulfill the requirement that all true relational database management systems (RDBMS) support a representation of missing information.

Now let's check out how it is used in PostgreSQL:

1. Let's start by inserting a tuple in this way:

   ```
   forumdb=# insert into categories (title) values ('lemon');
   INSERT 0 1
   ```

2. Let's see now which tuples are present in the categories table:

   ```
   forumdb=# select * from categories;
    pk | title   | description
   ----+---------+-------------
    10 | apple   | fruits
    11 | orange  | fruits
    12 | lettuce | vegetable
    13 | lemon   |
   (4 rows)
   ```

3. So now, if we want to select all the tuples in which the description is not present, we use the following:

   ```
   forumdb=# select * from categories where description ='';
    pk | title | description
   ```

```
----+-------+-------------
(0 rows)
```

As you can see, PostgreSQL does not return any tuples. This happens because the last insert has entered a NULL value in the `description` field.

4. In order to see the `NULL` values present in the tables, let's execute the following command:

```
forumdb=# \pset null NULL
Null display is "NULL".
```

5. This tells `psql` to show `NULL` values that are present in the table as NULL, as shown here:

```
forumdb=# select * from categories;
 pk | title  | description
----+--------+-------------
 10 | apple  | fruits
 11 | orange | fruits
 12 | lettuce| vegetable
 13 | lemon  | NULL
(4 rows)
```

As you can see, the `description` value associated with the `title` `lemon` is not an empty string; it is a NULL value.

6. Now, if we want to see all records that have NULL values in the `description` field, we have to use the `IS NULL` operator:

```
forumdb=# select title,description from categories where
description is null;
 title | description
-------+-------------
 lemon | NULL
(1 row)
```

The preceding query looks for all tuples for which there is no value in the `description` field.

7. Now, we will search for all tuples for which there *is* a value in the `description` field using the following query:

```
forumdb=# select title,description from categories where
description is not null;
 title  | description
--------+-------------
```

```
apple  | fruits
orange | fruits
lettuce| vegetable
(3 rows)
```

 To perform searches on NULL fields, we have to use the operators IS NULL / IS NOT NULL. The empty string is different from a NULL value.

Sorting with NULL values

Now let's see what happens when ordering a table where there are NULL values present:

1. Before we do this, let's insert another tuple into the table:

   ```
   insert into categories (title,description) values
   ('apricot','fruits');
   ```

2. Now let's repeat the sorting query that you performed previously:

   ```
   forumdb=# select * from categories order by description NULLS last;
    pk | title   | description
   ----+---------+-------------
    10 | apple   | fruits
    11 | orange  | fruits
    14 | apricot | fruits
    12 | lettuce | vegetable
    13 | lemon   | NULL
   (5 rows)
   ```

   As you can see, all description values are sorted and NULL values are positioned at the end of the result set. The same thing can be achieved by running the following:

   ```
   forumdb=# select * from categories order by description;
    pk | title   | description
   ----+---------+-------------
    10 | apple   | fruits
    11 | orange  | fruits
    14 | apricot | fruits
    12 | lettuce | vegetable
    13 | lemon   | NULL
   (5 rows)
   ```

3. If we want to place NULL values at the beginning, we have to perform the following:

```
forumdb=# select * from categories order by description NULLS
first;
 pk | title   | description
----+---------+-------------
 13 | lemon   | NULL
 10 | apple   | fruits
 11 | orange  | fruits
 14 | apricot | fruits
 12 | lettuce | vegetable
(5 rows)
```

 If not specified, the default action for ORDER BY type queries are: ORDER BY NULLS LAST is the default for ASC (which also is the default), and NULLS FIRST for DESC.

Creating a table starting from another table

We will now examine how to create a new table using data from another table.

To do this, you need to create a temporary table with the data present in the categories table as follows:

```
forumdb=# create temp table temp_categories as select * from categories;
SELECT 5
```

This command creates a table called temp_data with the same data structure and data as the table called categories:

```
forumdb=# select * from temp_categories ;
 pk | title   | description
----+---------+-------------
 10 | apple   | fruits
 11 | orange  | fruits
 12 | lettuce | vegetable
 13 | lemon   | NULL
 14 | apricot | fruits
(5 rows)
```

Updating data

Now let's try updating some data:

1. If you wanted to change the `apricot` value to the `peach` value, you would need to run the following statement:

    ```
    forumdb=# update temp_categories set title='peach' where pk = 14;
    UPDATE 1
    ```

    This statement will modify the value `apricot` to the value `peach` in the `title` field for all rows of the `temp_categories` table that have `pk=14`, as seen here:

    ```
    forumdb=# select * from temp_categories where pk=14;
     pk | title | description
    ----+-------+------------
     14 | peach | fruits
    (1 row)
    ```

2. If you wanted to change the `title` value of all the lines for which the `description` value is `vegetable`, you would need to run the following statement:

    ```
    forumdb=# update temp_categories set title = 'no title' where
    description = 'vegetable';
    UPDATE 1
    ```

    `UPDATE 1` means that only 1 row has been modified, as shown here:

    ```
    forumdb=# select * from temp_categories order by description;
     pk | title    | description
    ----+----------+------------
     10 | apple    | fruits
     11 | orange   | fruits
     14 | peach    | fruits
     12 | no title | vegetable
     13 | lemon    | NULL
    (5 rows)
    ```

You must be careful when using the `UPDATE` command. If you work in auto-commit mode, there is no chance of turning back after the update is complete.

Deleting data

In this section, we will see how to delete data from a table. The command needed to delete data is `delete`. Let's get started:

1. If we want to delete all records in the `temp_categories` table that have `pk=10`, we have to perform the following command:

```
forumdb=# delete from temp_categories where pk=10;
DELETE 1
```

The preceding statement deletes all the records that have `pk=10`. `DELETE 1` means that one record has been deleted. As you can see here, the row with the value of `pk=10` is no longer present in `temp_categories`:

```
forumdb=# select * from temp_categories order by description;
 pk | title    | description
----+----------+-------------
 11 | orange   | fruits
 14 | peach    | fruits
 12 | no title | vegetable
 13 | lemon    | NULL
(4 rows)
```

2. Now if we want to delete all rows that have a description value equal to `NULL`, we have to execute this statement:

```
forumdb=# delete from temp_categories where description is null;
DELETE 1
```

The preceding statement used a `DELETE` command combined with the `IS NULL` operator.

3. If you want to delete all records from a table, you have to execute the following:

```
forumdb=# delete from temp_categories ;
DELETE 3
```

 Be very careful when you use this command – all records present in the table will be deleted!

Now the `temp_categories` table is empty, as shown here:

```
forumdb=# select * from temp_categories order by description;
 pk | title | description
----+-------+-------------
(0 rows)
```

4. If we want to reload all the data from the `categories` table to the
    `temp_categories` table, we have to execute this statement:

```
forumdb=# insert into temp_categories select * from categories;
INSERT 0 5
```

The preceding statement takes all values from the `categories` table and puts
them in the `temp_categories` table, as you can see here:

```
forumdb=# select * from temp_categories order by description;
 pk | title   | description
----+---------+-------------
 10 | apple   | fruits
 11 | orange  | fruits
 14 | apricot | fruits
 12 | lettuce | vegetable
 13 | lemon   | NULL
(5 rows)
```

5. Another way to delete data is by using the TRUNCATE command. When we want
    to delete all the data from a table without providing a `where` condition, we can
    use the TRUNCATE command:

```
forumdb=# truncate table temp_categories ;
TRUNCATE TABLE
```

The TRUNCATE command deletes all data in a table. As you can see here, the
`temp_categories` table is now empty:

```
forumdb=# select * from temp_categories order by description;
 pk | title | description
----+-------+-------------
(0 rows)
```

Here is some key information about the TRUNCATE command:

*   TRUNCATE deletes all the records in a table similar to the DELETE command.
*   In the TRUNCATE command, it is not possible to use WHERE conditions.
*   The TRUNCATE command deletes records much faster than the DELETE command.

Summary

This chapter introduced you to the basic SQL/PostgreSQL statements and some basic SQL commands. You learned how to create and delete databases, how to create and delete tables, what types of tables exist, which basic statements to use to insert, modify, and delete data, and the first basic queries to query the database.

In the next chapter, you will learn how to write more complex queries that relate to multiple tables in different ways.

References

- The CREATE DATABASE official documentation: https://www.PostgreSQL.org/docs/12/sql-createdatabase.html
- The CREATE TABLE official documentation: https://www.PostgreSQL.org/docs/12/sql-createtable.html
- The SELECT official documentation: https://www.PostgreSQL.org/docs/12/sql-select.html
- The INSERT official documentation: https://www.PostgreSQL.org/docs/12/sql-insert.html
- The DELETE official documentation: https://www.PostgreSQL.org/docs/12/sql-delete.html
- The UPDATE official documentation: https://www.PostgreSQL.org/docs/12/sql-update.html
- The TRUNCATE official documentation: https://www.PostgreSQL.org/docs/12/sql-truncate.html

5
Advanced Statements

In the previous chapter, we started taking our first steps with PostgreSQL. In this chapter, we will analyze the SQL language more deeply and write more complex queries. We will talk about SELECT/INSERT/UPDATE again, but this time, we will use the more advanced options surrounding them. We will then cover JOIN and **common table expressions** (**CTEs**) in depth.

The topics we will talk about will be the following:

- Exploring the SELECT statement
- Using UPSERT
- Exploring CTEs

Exploring the SELECT statement

As we saw in the previous chapter, we can use the SELECT statement to filter our datasets using the equality condition. In the same way, we can filter records using > or < conditions, such as in the following example:

```
forumdb=# select * from categories where pk > 12 order by title;
 pk | title   | description
----+---------+--------------
 14 | apricot | fruits
 13 | lemon   |
(2 rows)
```

The preceding query returns all records that have pk> 12.

Another condition that we can use with the SELECT statement is the like condition. Let's take a look at this next.

Using the like clause

Suppose we wanted to find all records that have a `title` field value starting with the letter `'a'`.

To do this, we would have to use the `like` condition:

```
forumdb=# select * from categories where title like 'a%';

 pk  | title  | description
-----+--------+------------
 10  | apple  | fruits
 14  | apricot | fruits
(2 rows)
```

As shown, the preceding query returns all records that have a title beginning with the letter a. In a similar vein, if we wanted to find all records with titles ending with the letter e, we would have to write the following:

```
forumdb=# select * from categories where title like '%e';
 pk | title  | description
-----+--------+------------
 10 | apple  | fruits
 11 | orange | fruits
(2 rows)
```

The two kinds of searches can also be combined. For example, if we wanted to search all records that contain the letters `'ap'`, we would write the following:

```
forumdb=# select * from categories where title like '%ap%';
 pk | title  | description
-----+---------+------------
 10 | apple  | fruits
 14 | apricot | fruits
(2 rows)
```

The query given here will return all records whose titles contain the string ap.

Now let's try to run the following query and see what happens:

```
forumdb=# select * from categories where title like 'A%';
 pk | title | description
-----+-------+------------
(0 rows)
```

As we can see, the search does not return any results. This happens because `like` searches are case-sensitive.

Now let's introduce the `upper (text)` function. The `upper` function, given an input string, returns the same string with all characters in uppercase, as here:

```
forumdb=# select upper('orange');
 upper
--------
 ORANGE
(1 row)
```

 In PostgreSQL, it is possible to call functions without writing `FROM`. PostgreSQL does not need dummy tables to perform the `SELECT` function. If we were in Oracle, the same query would have to be written this way: `select upper('orange') from DUAL;`.

Returning to our preceding example, if we wanted to perform a `like` case-insensitive search, we would have to write this statement:

```
forumdb=# select * from categories where upper(title) like 'A%';
 pk | title   | description
----+---------+-------------
 10 | apple   | fruits
 14 | apricot | fruits
(2 rows)
```

We have now covered all of the functions that can be performed using the `like` operator.

Using ilike

In PostgreSQL, it is possible to perform a case-insensitive `like` query by using the `ilike` operator. In this situation, our query would become the following:

```
forumdb=# select * from categories where title ilike 'A%';
 pk | title   | description
----+---------+-------------
 10 | apple   | fruits
 14 | apricot | fruits
(2 rows)
```

This is the PostgreSQL way of solving the case-insensitive `like` query issue that we encountered previously.

Using distinct

We will now discuss another kind of query: the `distinct` query. Firstly, however, we need to introduce another very useful function for the DBA called the `coalesce` function. The `coalesce` function, given two or more parameters, returns the first value that is not `NULL`.

For example, let's use the `coalesce` function for the `test` value:

```
forumdb=# select coalesce(NULL,'test');
 coalesce
----------
 test
(1 row)
```

In the preceding query, the `coalesce` function returns `test` because the first argument is `NULL` and the second argument is not `NULL`.

In the following query, we can see that the `coalesce` function returns `orange` because the first argument is not `NULL`:

```
forumdb=# select coalesce('orange','test');
 coalesce
----------
 orange
(1 row)
```

Now let's perform the following query:

```
forumdb=# \pset null (NULL)
Null display is "(NULL)".

forumdb=# select description,coalesce(description,'No description') from categories order by 1;
 description | coalesce
-------------+----------------
 fruits      | fruits
 fruits      | fruits
 fruits      | fruits
 vegetable   | vegetable
 (NULL)      | No description
(5 rows)
```

In the preceding code, the `coalesce` function transforms any `NULL` value into the string `No description`. Another thing that isn't very user-friendly about the `coalesce` function is that the name of the field that is given when a function is called is not the name we would want for our query. In this case, the second field of the result set is called `coalesce`, which is not the name we would prefer.

In PostgreSQL, an alias can be assigned to any field in a query. For example, we can assign an alias to the `coalesce` field as follows:

```
forumdb=# select coalesce(description,'No description') as description from
categories order by 1;
  description
----------------
 fruits
 fruits
 fruits
 No description
 vegetable
(5 rows)
```

Now the result set has the `description` field instead of the `coalesce` field.

If we want to use an alias with spaces or capital letters, we have to quote the alias using " ", as in the following example:

```
forumdb=# select coalesce(description,'No description') as Description from
categories order by 1;
  description
----------------
 fruits
 fruits
 fruits
 No description
 vegetable
(5 rows)
```

The resultset doesn't have an alias of `Description` but does have an alias of `description`, which doesn't seem right. The correct way to perform this is as follows:

```
forumdb=# select coalesce(description,'No description') as "Description"
from categories order by 1;
  Description
----------------
 fruits
 fruits
 fruits
 No description
```

```
  vegetable
(5 rows)
```

Now let's perform the following query:

```
forumdb=# select distinct coalesce(description,'No description') as
description from categories order by 1;
  description
----------------
 fruits
 No description
 vegetable
(3 rows)
```

In the preceding query, we have used the `select distinct` statement. The `select distinct` statement is used to return only distinct (different) values. Internally, the `distinct` statement involves a data sort for large tables, which means that if a query uses the `DISTINCT` statement, the query may become slower as the number of records increases.

Using limit and offset

The `limit` clause is the PostgreSQL way to limit the number of rows returned by a query, whereas the `offset` clause is used to skip a specific number of rows returned by the query.

`limit` and `offset` are used to return a portion of data from a resultset generated by a query; the `limit` clause is used to limit the number of records in output and the `offset` clause is used to provide PostgreSQL with the position on the resultset from which to start returning data.

They can be used independently or together.

Now let's test `limit` and `offset` using the following queries:

```
forumdb=# select * from categories order by pk limit 1;
 pk | title | description
----+-------+-------------
 10 | apple | fruits
(1 row)
```

The preceding query returns only the first record that we have inserted; this is because the `pk` field is an integer type with a default value generated always as the identity.

If we want the two first records that were inserted, we have to perform the following query:

```
forumdb=# select * from categories order by pk limit 2;
 pk | title  | description
----+--------+-------------
 10 | apple  | fruits
 11 | orange | fruits
(2 rows)
```

If we only want the second record that was inserted, we have to perform the following query:

```
forumdb=# select * from categories order by pk offset 1 limit 1;
 pk | title  | description
----+--------+-------------
 11 | orange | fruits
(1 row)
```

offset and limit are very useful when we want to return data in a paged way.

Another valuable function of limit is that it can create a new table from an existing table. For example, if we want to create a table called new_categories starting from the categories table, we have to perform the following statement:

```
forumdb=# create table new_categories as select * from categories limit 0;
SELECT 0
```

This statement will copy into the new_categories table only the data structure of the table categories.

The SELECT 0 clause means that no data has been copied into the new_categories table; only the data structure has been replicated, as we can see here:

```
forumdb=# \d new_categories
     Table "public.new_categories"
   Column    | Type    | Collation | Nullable | Default
-------------+---------+-----------+----------+---------
 pk          | integer |           |          |
 title       | text    |           |          |
 description | text    |           |          |
```

Using subqueries

Subqueries can be described as nested queries – they are where we can nest a query inside another query using parentheses. Subqueries can return a single value or a recordset, just like regular queries. We will start by introducing subqueries using the IN/NOT IN operator.

Using the IN/NOT IN condition

Let's start with the IN operator; we can use the IN operator inside a where clause instead of using multiple OR conditions. For example, if you wanted to search for all categories that have the value pk=10 or the value pk=11, we would have to perform the following statement:

```
forumdb=# select * from categories where pk=10 or pk=11;
 pk | title  | description
----+--------+-------------
 10 | apple  | fruits
 11 | orange | fruits
(2 rows)
```

Another way to reach the same outcome is the following:

```
forumdb=# select * from categories where pk in (10,11);
 pk | title  | description
----+--------+-------------
 10 | apple  | fruits
 11 | orange | fruits
(2 rows
```

If we wanted to return the records that don't have pk=10 or pk=11, we would have to perform the following:

```
forumdb=# select * from categories where not (pk=10 or pk=11);
 pk | title   | description
----+---------+-------------
 12 | tomato  | vegetable
 13 | lemon   |
 14 | apricot | fruits
(3 rows)
```

An operator similar to the IN operator but with reverse functionality is the NOT IN operator. For example, if we wanted to search for all categories that do not have pk=10 or pk=11, we would have to execute the following:

```
forumdb=# select * from categories where pk not in (10,11);
 pk | title   | description
----+---------+-------------
 12 | tomato  | vegetable
 13 | lemon   |
 14 | apricot | fruits
(3 rows)
```

Now, we can insert some data into the posts table:

```
forumdb=# insert into posts(title,content,author,category) values('my
orange','my orange is the best orange in the world',1,11);
forumdb=# insert into posts(title,content,author,category) values('my
apple','my apple is the best orange in the world',1,10);
forumdb=# insert into posts(title,content,author,category,reply_to)
values('Re:my orange','No! It''s my orange the best orange in the
world',2,11,2);
forumdb=# insert into posts(title,content,author,category) values('my
tomato','my tomato is the best orange in the world',2,12);
```

The records present in the posts table are now as follows:

```
forumdb=# select pk,title,content,author,category from posts;
 pk | title        | content                                            |
author | category
----+--------------+----------------------------------------------------+-----
---+----------
  2 | my orange    | my orange is the best orange in the world          | 1
| 11
  3 | my apple     | my apple is the best orange in the world           | 1
| 10
  4 | Re:my orange | No! It's my orange the best orange in the world | 2
| 11
  5 | my tomato    | my tomato is the best orange in the world          | 2
| 12
(4 rows)
```

Suppose we now want to search for all posts that belong to the orange category. To do this, we can use several methods.

The following method uses subqueries:

```
forumdb=# select pk,title,content,author,category from posts where category
in (select pk from categories where title ='orange');
 pk | title       | content                                               |
author | category
----+-------------+-------------------------------------------------------+-----
---+----------
  2 | my orange   | my orange is the best orange in the world             | 1
| 11
  4 | Re:my orange | No! It's my orange the best orange in the world | 2
| 11
(2 rows)
```

The subquery is represented by the following:

```
forumdb=# select pk from categories where title ='orange'
```

This statement extracts the values pk=2 and pk=4 from the category table and the external query searches the records in the posts table that have pk=2 or pk=4. Similarly, if you wanted to search for all post values that do not belong to the orange category, you would have to perform the following statement:

```
forumdb=# select pk,title,content,author,category from posts where category
not in (select pk from categories where title ='orange');
 pk | title     | content                                          | author |
category
----+-----------+--------------------------------------------------+--------+-----
-----
  3 | my apple  | my apple is the best orange in the world   | 1      | 10
  5 | my tomato | my tomato is the best orange in the world  | 2      | 12
(2 rows)
```

Using the EXISTS/NOT EXISTS condition

The EXISTS statement is used when we want to check whether a subquery returns (TRUE), and the NOT EXISTS statement is used when we want to check whether a subquery does not return (FALSE). For example, if we wanted to write the same conditions written previously using the EXISTS/NOT EXISTS conditions, we'd have to perform the following:

```
forumdb=# select pk,title,content,author,category from posts where exists
(select 1 from categories where title ='orange' and posts.category=pk);
 pk | title       | content                                               |
author | category
----+-------------+-------------------------------------------------------+-----
---+----------
```

```
    2 | my orange   | my orange is the best orange in the world       | 1
 | 11
    4 | Re:my orange | No! It's my orange the best orange in the world | 2
 | 11
(2 rows)
```

The preceding query returns the same results as the query written with the IN condition.

Similarly, if we wanted to search for all post values that do not belong to the orange category using the NOT EXISTS condition, we'd have to write the following:

```
forumdb=# select pk,title,content,author,category from posts where not
exists (select 1 from categories where title ='orange' and
posts.category=pk);
 pk | title     | content                                            | author |
category
----+-----------+----------------------------------------------------+--------+-----
-----
    3 | my apple | my apple is the best orange in the world       | 1      | 10
    5 | my tomato | my tomato is the best orange in the world | 2      | 12
(2 rows)
```

Both queries written with the IN condition and with the EXISTS condition are called **semi-join queries**, and we will be looking at joins in the next section.

Learning joins

We will now explore joins in more detail. We will address what a join is, how many types of joins exist, and what they are used for. We can think of a join as a combination of rows from two or more tables.

For example, the following query returns all the combinations from the rows of the category table and the rows of the posts table:

```
forumdb=# select c.pk,c.title,p.pk,p.category,p.title from categories
c,posts p;
 pk | title   | pk | category | title
----+---------+----+----------+--------------
 10 | apple   | 2  | 11       | my orange
 10 | apple   | 3  | 10       | my apple
 10 | apple   | 4  | 11       | Re:my orange
 10 | apple   | 5  | 12       | my tomato
 11 | orange  | 2  | 11       | my orange
 11 | orange  | 3  | 10       | my apple
 11 | orange  | 4  | 11       | Re:my orange
```

```
11 | orange  | 5 | 12 | my tomato
12 | tomato  | 2 | 11 | my orange
12 | tomato  | 3 | 10 | my apple
12 | tomato  | 4 | 11 | Re:my orange
12 | tomato  | 5 | 12 | my tomato
13 | lemon   | 2 | 11 | my orange
13 | lemon   | 3 | 10 | my apple
13 | lemon   | 4 | 11 | Re:my orange
13 | lemon   | 5 | 12 | my tomato
14 | apricot | 2 | 11 | my orange
14 | apricot | 3 | 10 | my apple
14 | apricot | 4 | 11 | Re:my orange
14 | apricot | 5 | 12 | my tomato
(20 rows)
```

This query makes a Cartesian product between the `category` table and the `posts` table. It can also be called a **cross join**:

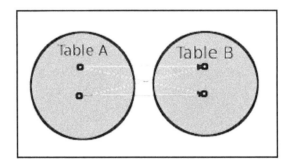

The same query can also be written in the following way:

```
forumdb=# select c.pk,c.title,p.pk,p.category,p.title from categories c
CROSS JOIN posts p;
 pk | title   | pk | category | title
----+---------+----+----------+--------------
 10 | apple   | 2  | 11 | my orange
 10 | apple   | 3  | 10 | my apple
 10 | apple   | 4  | 11 | Re:my orange
 10 | apple   | 5  | 12 | my tomato
 11 | orange  | 2  | 11 | my orange
 11 | orange  | 3  | 10 | my apple
 11 | orange  | 4  | 11 | Re:my orange
 11 | orange  | 5  | 12 | my tomato
 12 | tomato  | 2  | 11 | my orange
 12 | tomato  | 3  | 10 | my apple
 12 | tomato  | 4  | 11 | Re:my orange
 12 | tomato  | 5  | 12 | my tomato
```

```
13 | lemon   | 2 | 11      | my orange
13 | lemon   | 3 | 10      | my apple
13 | lemon   | 4 | 11      | Re:my orange
13 | lemon   | 5 | 12      | my tomato
14 | apricot | 2 | 11      | my orange
14 | apricot | 3 | 10      | my apple
14 | apricot | 4 | 11      | Re:my orange
14 | apricot | 5 | 12      | my tomato
(20 rows)
```

Using INNER JOIN

Now suppose that starting with all the possible combinations that exist between the rows of the `category` table and the rows of the `posts` table, we want to filter all the rows that have the same value as the `category` field (`category.pk = posts.category`). We want to have a result like the one described in the following diagram:

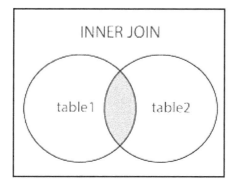

 The `INNER JOIN` keyword selects records that have matching values in both tables.

To achieve this, we need to run the following code:

```
forumdb=# select c.pk,c.title,p.pk,p.category,p.title from categories
c,posts p where c.pk=p.category;
 pk | title  | pk | category | title
----+--------+----+----------+--------------
 11 | orange | 2  | 11       | my orange
 10 | apple  | 3  | 10       | my apple
 11 | orange | 4  | 11       | Re:my orange
 12 | tomato | 5  | 12       | my tomato
(4 rows)
```

We can also write the same query using the explicit JOIN operation:

```
forumdb=# select c.pk,c.title,p.pk,p.category,p.title from categories c
inner join posts p on c.pk=p.category;
 pk | title  | pk | category | title
----+--------+----+----------+--------------
 11 | orange | 2  | 11       | my orange
 10 | apple  | 3  | 10       | my apple
 11 | orange | 4  | 11       | Re:my orange
 12 | tomato | 5  | 12       | my tomato
(4 rows)
```

INNER JOIN versus EXISTS/IN

If we wanted to search for all posts that belong to the orange category using the INNER JOIN condition, we would have to rewrite the query in this way:

```
forumdb=# select distinct p.pk,p.title,p.content,p.author,p.category from
categories c inner join posts p on c.pk=p.category where c.title='orange';
 pk | title       | content                                             |
author | category
----+-------------+-----------------------------------------------------+-----
---+----------
  2 | my orange     | my orange is the best orange in the world         | 1
| 11
  4 | Re:my orange | No! It's my orange the best orange in the world | 2
| 11
(2 rows)
```

 Using the INNER JOIN condition, we can rewrite all queries that can be written using the IN or EXISTS condition.

It is preferable to use JOIN conditions whenever possible instead of IN or EXISTS conditions, because they perform better in terms of the execution speed, as we will see in the following chapters.

Using LEFT JOINS

We will now explore what a left join is. As an example, we can perform the following query:

```
forumdb=# select c.*,p.category,p.title from categories c left join posts p
on c.pk=p.category;
 pk | title  | description | category | title
```

```
----+---------+-------------+----------+--------------
 11 | orange  | fruits      | 11       | my orange
 10 | apple   | fruits      | 10       | my apple
 11 | orange  | fruits      | 11       | Re:my orange
 12 | tomato  | vegetable   | 12       | my tomato
 13 | lemon   |             |          |
 14 | apricot | fruits      |          |
(6 rows)
```

This query returns all records of the `categories` table and returns the matched records from the `posts` table. As we can see, if the second table (the `posts` table, in this example) has no matches, the result is NULL.

 The LEFT JOIN keyword returns all records from the left table (**table1**), and all the records from the right table (**table2**). The result is NULL from the right side if there is no match.

This diagram gives us an idea of how a left join works:

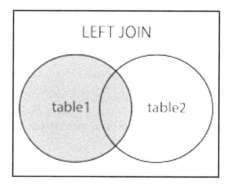

Suppose now that we want to search for all categories that do not have posts – we could write the following:

```
forumdb=# select * from categories c where c.pk not in (select category
from posts);
 pk | title   | description
----+---------+--------------
 13 | lemon   |
 14 | apricot | fruits
(2 rows)
```

This query, written using the NOT IN condition, looks for all records in the categories table for which the pk value does not match in the category field of the posts table. As we have already seen, another way to write the same query would be to use the NOT EXISTS condition:

```
forumdb=# select * from categories c where not exists (select 1 from posts
where category=c.pk);
 pk | title   | description
----+---------+-------------
 13 | lemon   |
 14 | apricot | fruits
(2 rows)
```

If we now wanted to use a left join in order to achieve the same purpose, we would start by writing the following left join query:

```
forumdb=# select c.*,p.category from categories c left join posts p on
p.category=c.pk;
 pk | title   | description | category
----+---------+-------------+----------
 11 | orange  | fruits      | 11
 10 | apple   | fruits      | 10
 11 | orange  | fruits      | 11
 12 | tomato  | vegetable   | 12
 13 | lemon   |             |
 14 | apricot | fruits      |
(6 rows)
```

From the result, it is immediately clear that all the values we are looking for are those for which the value of p.category is NULL.

So, we rewrite the query in the following way:

```
forumdb=# select c.* from categories c left join posts p on p.category=c.pk
where p.category is null;
 pk | title   | description
----+---------+-------------
 13 | lemon   |
 14 | apricot | fruits
(2 rows)
```

As shown here, we get the same result we had using the NOT EXISTS or NOT IN condition.

 Using the LEFT JOIN condition, we can rewrite some queries that can be written using the IN or EXISTS conditions.

It is preferable to use `JOIN` conditions whenever possible instead of `IN` or `EXISTS` conditions, because they perform better in terms of execution speed, as we will see in the following chapters.

Using RIGHT JOIN

The right join is the twin of the left join; it takes data from the right table, reverses the order of the tables, and uses a right join instead of a left join. For example, we can obtain the same results as the preceding query using a right join instead of a left join:

```
forumdb=# select c.*,p.category,p.title from posts p right join categories
c on c.pk=p.category;
 pk | title | description | category | title
----+---------+-------------+----------+--------------
 11 | orange  | fruits      | 11       | my orange
 10 | apple   | fruits      | 10       | my apple
 11 | orange  | fruits      | 11       | Re:my orange
 12 | tomato  | vegetable   | 12       | my tomato
 13 | lemon   |             |          |
 14 | apricot | fruits      |          |
(6 rows)
```

 The `RIGHT JOIN` keyword returns all records from the right table (**table2**) and all the records from the left table (**table1**) that match the right table (**table2**). The result is `NULL` from the left side when there is no match.

This diagram illustrates how `RIGHT JOIN` works:

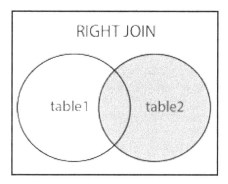

Using FULL OUTER JOIN

In SQL, FULL OUTER JOIN is the combination of what we would have if we put together the right join and the left join. We will check it out using the following steps:

1. Let's insert some data:

```
forumdb=# insert into tags (tag,parent) values ('fruits',NULL);
INSERT 0 1
forumdb=# insert into tags (tag,parent) values ('vegetables',NULL);
INSERT 0 1
forumdb=# insert into j_posts_tags values (1,2),(1,3);
INSERT 0 2
```

2. Having inserted some data into the tags table and some data into the j_posts_tags table, the j_tags_posts table relates the tags table to the posts table. So, the current situation is as follows:

```
forumdb=# select * from tags;
 pk | tag | parent
----+------------+--------
  1 | fruits |
  2 | vegetables |
(2 rows)

forumdb=# select * from j_posts_tags ;
 tag_pk | post_pk
--------+---------
      1 | 2
      1 | 3
(2 rows)
```

3. Now let's try to write this JOIN query:

```
forumdb=# select jpt.*,t.*,p.title from j_posts_tags jpt
inner join tags t on jpt.tag_pk=t.pk
inner join posts p on jpt.post_pk = p.pk;

 tag_pk | post_pk | pk | tag | parent | title
--------+---------+----+--------+--------+-----------
      1 | 2       | 1  | fruits |        | my orange
      1 | 3       | 1  | fruits |        | my apple
(2 rows)
```

This query returns all the records that have posts and tags. It's a JOIN query between three tables: tags, j_posts_tags, and posts.

4. If we wanted to have the left and right joins between the `tags`, `j_posts_tags`, and `posts` tables, we'd have to use the full outer join and write the following:

```
forumdb=# select jpt.*,t.*,p.title from j_posts_tags jpt full outer
join tags t on jpt.tag_pk=t.pk full outer join posts p on
jpt.post_pk = p.pk;
 tag_pk | post_pk | pk | tag        | parent | title
--------+---------+----+------------+--------+-------------
      1 | 2       | 1  | fruits     |        | my orange
      1 | 3       | 1  | fruits     |        | my apple
        |         | 2  | vegetables |        |
        |         |    |            |        | my tomato
        |         |    |            |        | Re:my orange
(5 rows)
```

This diagram illustrates how the full outer join works:

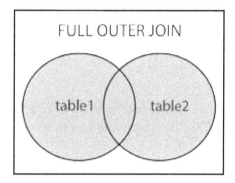

One question we need to consider is, *What is the difference between a full join and a cross join, which we saw at the beginning of this section on joins?*

Well, a full outer join is different from a cross join because a cross join makes a Cartesian product from all the records present in the tables.

For example, in a cross join with the same data as the preceding full join, we would get the following result:

```
forumdb=# select jpt.*,t.*,p.title from j_posts_tags jpt
cross join tags t
cross join posts p ;
 tag_pk | post_pk | pk | tag    | parent | title
--------+---------+----+--------+--------+-------------
 1      | 2       | 1  | fruits |        | my orange
 1      | 3       | 1  | fruits |        | my orange
 1      | 2       | 1  | fruits |        | my apple
 1      | 3       | 1  | fruits |        | my apple
```

```
1        | 2      | 1 | fruits     |    | Re:my orange
1        | 3      | 1 | fruits     |    | Re:my orange
1        | 2      | 1 | fruits     |    | my tomato
1        | 3      | 1 | fruits     |    | my tomato
1        | 2      | 2 | vegetables |    | my orange
1        | 3      | 2 | vegetables |    | my orange
1        | 2      | 2 | vegetables |    | my apple
1        | 3      | 2 | vegetables |    | my apple
1        | 2      | 2 | vegetables |    | Re:my orange
1        | 3      | 2 | vegetables |    | Re:my orange
1        | 2      | 2 | vegetables |    | my tomato
1        | 3      | 2 | vegetables |    | my tomato
(16 rows)
```

Using SELF JOIN

A self join is a regular join, but the table is joined with itself. Let's start by inserting some rows into the posts table:

```
forumdb=# insert into posts (title,content,author,category) values ('my new
orange','this my post
on my new orange',1,11);
```

Suppose we wanted to find all posts that belong to author 2 that have the same category as those entered by author 1. Our first step would be to search for all the records that belong to author 1:

```
forumdb=# select distinct p1.title,p1.author,p1.category from posts p1
where p1.author=1;
     title      | author | category
----------------+--------+----------
 my apple       | 1      | 10
 my new orange  | 1      | 11
 my orange      | 1      | 11
(3 rows)
```

The second step would be to search for all the records that belong to author 2:

```
forumdb=# select distinct p2.title,p2.author,p2.category from posts p2
where p2.author=2;
     title      | author | category
----------------+--------+----------
 my tomato      | 2      | 12
 Re:my orange   | 2      | 11
(2 rows)
```

The result that we want would be as follows:

title	author	category
Re:my orange	2	11

The following snippet is the query that realizes what we want:

```
forumdb=# select distinct p2.title,p2.author,p2.category from posts
p1,posts p2 where p1.category=p2.category and p1.author<>p2.author and
p1.author=1 and p2.author=2;
    title     | author | category
--------------+--------+----------
 Re:my orange | 2      | 11
(1 row)
```

We can also write the same query this way:

```
forumdb=# select distinct p2.title,p2.author,p2.category from posts p1
inner join posts p2 on ( p1.category=p2.category and p1.author<>p2.author)
where p1.author=1 and p2.author=2;
    title     | author | category
--------------+--------+----------
 Re:my orange | 2      | 11
(1 row)
```

 Aliases must be used for table names when a self join is performed, otherwise, PostgreSQL will not know which table the column names belong to.

Aggregate functions

Aggregate functions perform a calculation on a set of rows and return a single row. PostgreSQL provides all the standard SQL aggregate functions:

- AVG(): This function returns the average value.
- COUNT(): This function returns the number of values.
- MAX(): This function returns the maximum value.
- MIN(): This function returns the minimum value.
- SUM(): This function returns the sum of values.

Aggregate functions are used in conjunction with the GROUP BY clause. A GROUP BY clause splits a resultset into groups of rows and aggregate functions perform calculations on them. For example, if we wanted to count how many records there are for each category, PostgreSQL first groups the data and then counts it. The following diagram illustrates the process:

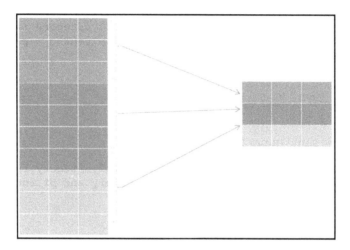

This diagram illustrates that PostgreSQL, before grouping the data, sorts it internally. Therefore, we must remember that a grouping operation always implies an ordering operation; this will become more clear when we discuss performance later on.

Now that we have understood the theory, let's address how to actually calculate how many records there are for each category:

```
forumdb=# select category,count(*) from posts group by category;
 category | count
----------+-------
 11       | 3
 10       | 1
 12       | 1
(3 rows)
```

The preceding query counts how many records there are for each category in the posts table.

Another way to write the same query is as follows:

```
forumdb=# select category,count(*) from posts group by 1;
 category | count
----------+-------
 11       | 3
 10       | 1
 12       | 1
(3 rows)
```

In PostgreSQL, we can write the GROUP BY condition using the name of the fields or their position in the query.

Another condition that we can use is the `having` condition. Suppose that we want to count how many records there are for each category that have a count greater than 2. To do this, we would have to add the `having` condition after the `group by` condition, thus writing the following:

```
forumdb=# select category,count(*) from posts group by category having
count(*) > 2;
 category | count
----------+-------
       11 | 3
(1 row)
```

Similarly, we could do this:

```
forumdb=# select category,count(*) from posts group by 1 having count(*) >
2;
 category | count
----------+-------
 11 | 3
(1 row)
```

Now let's see how the aggregation functions work if we add aliases. Let's resume the first query and write the following:

```
forumdb=# select category,count(*) as category_count from posts group by
category;
 category | category_count
----------+----------------
       11 | 3
       10 | 1
       12 | 1
(3 rows)
```

As seen here, we can use an alias on aggregate functions.

However, what do we do if we want to use an alias inside a query that has a `having` condition too? To answer this question, let's try the following statement:

```
forumdb=# select category,count(*) as category_count from posts group by
category having category_count > 2;
ERROR: column "category_count" does not exist
```

As we can see, we can't use an alias on a `having` condition. The correct way to write the preceding query is as follows:

```
forumdb=# select category,count(*) as category_count from posts group by
category having count(*) > 2;
 category | category_count
----------+----------------
       11 | 3
(1 row)
```

In the next chapter, we will discuss aggregates in more detail.

UNION/UNION ALL

The `UNION` operator is used to combine the resultset of two or more `SELECT` statements. We can use the `UNION` statement only if the following rules are respected:

- Each `SELECT` statement within `UNION` must have the same number of columns.
- The columns must have similar data types.
- The columns in each `SELECT` statement must be in the same order.

Let's explore an example.

First, we need to insert some data:

```
forumdb=# insert into tags (tag,parent) values ('apple',1);
INSERT 0 1

forumdb=# select * from tags;
 pk | tag        | parent
----+------------+--------
  1 | fruits     |
  2 | vegetables |
  3 | apple      | 1
(3 rows)
```

```
forumdb=# select * from categories;
 pk | title   | description
----+---------+-------------
 10 | apple   | fruits
 11 | orange  | fruits
 12 | tomato  | vegetable
 13 | lemon   |
 14 | apricot | fruits
(5 rows)
```

Suppose now that we want to have a result set that is a union of `tags` and `categories`; in other words, we want to reach this result:

title
apple
apricot
fruits
lemon
orange
tomato
vegetables

To achieve this, we have to use the `UNION` operator:

```
forumdb=# select title from categories union select tag from tags order by
title;
   title
------------
 apple
 apricot
 fruits
 lemon
 orange
 tomato
 vegetables
(7 rows)
```

The `union` operator combines the values of the two tables and removes duplicates. If we don't want duplicates to be removed and instead have them remain in the resultset, we have to use the `UNION ALL` operator:

```
forumdb=# select title from categories union all select tag from tags order
by title;
 title
------------
 apple
```

```
apple
apricot
fruits
lemon
orange
tomato
vegetables
(8 rows)
```

The UNION operator always implies DISTINCT before returning the data, and, as we have seen previously, for large tables, DISTINCT always implies sorting. Therefore, UNION ALL is a much faster operation than UNION. It is recommended to use UNION instead of UNION ALL only when you do not want duplicates in the resultset and only when you are sure that duplicates exist.

EXCEPT/INTERSECT

The EXCEPT operator returns rows by comparing the resultsets of two or more queries. The EXCEPT operator returns distinct rows from the first (left) query that is not in the output of the second (right) query. Similar to the UNION operator, the EXCEPT operator can also compare queries that have the same number and the same datatype of fields.

For example, say we have the following:

```
forumdb=# select * from tags;
 pk | tag        | parent
----+------------+--------
  1 | fruits     |
  2 | vegetables |
  3 | apple      | 1
(3 rows)

forumdb=# select * from categories;
 pk | title   | description
----+---------+-------------
 10 | apple   | fruits
 11 | orange  | fruits
 12 | tomato  | vegetable
 13 | lemon   |
 14 | apricot | fruits
(5 rows)
```

Say we want to reach this result:

title
apricot
lemon
orange
tomato

We would need to order all records that are present in the `categories` table but that are not present in the `tags` table by the `title` field. To do this, we would use the following query:

```
forumdb=# select title from categories except select tag from tags order by
1;
 title
---------
 apricot
 lemon
 orange
 tomato
(4 rows)
```

The `INTERSECT` operator performs the reverse operation. It searches for all the records present in the first table that are also present in the second table:

```
forumdb=# select title from categories intersect select tag from tags order
by 1;
 title
-------
 apple
(1 row)
```

In this section, we have taken a detailed look at the instructions needed to search data in tables using various statements and joins. In the next section, we will see how to modify the data in the tables in more advanced ways.

Using UPSERT

In this section, we will look at some interesting features of the `insert` statement:

- How to make an `upsert` statement starting from an `insert` statement
- How to make a SQL query that inserts the same data and returns the record inserted

UPSERT – the PostgreSQL way

In PostgreSQL, the `upsert` statement does not exist as in other DBMSes. An `upsert` statement is used when we want to insert a new record on top of the existing record or update an existing record. To do this in PostgreSQL, we can use the `ON CONFLICT` keyword:

```
INSERT INTO table_name(column_list) VALUES(value_list)
ON CONFLICT target action;
```

Here, `ON CONFLICT` means that the target action is executed when the record already exists (meaning when a record with the same primary key exists). The target action could be this:

```
DO NOTHING
```

Alternatively, it could be the following:

```
DO UPDATE SET { column_name = { expression | DEFAULT } |
    ( column_name [, ...] ) = [ ROW ] ( { expression | DEFAULT } [, ...] )
|
    ( column_name [, ...] ) = ( sub-SELECT )
    } [, ...]
[ WHERE condition ]
```

Now, let's look at an example to better understand how `upsert` works:

1. For example, start with the `j_posts_tags` table:

   ```
   forumdb=# \d j_posts_tags ;
                Table "public.j_posts_tags"
    Column  | Type    | Collation | Nullable | Default
   ---------+---------+-----------+----------+---------
    tag_pk  | integer |           | not null |
    post_pk | integer |           | not null |
   Foreign-key constraints:
       "j_posts_tags_post_pk_fkey" FOREIGN KEY (post_pk) REFERENCES
   posts(pk)
       "j_posts_tags_tag_pk_fkey" FOREIGN KEY (tag_pk) REFERENCES
   tags(pk)
   ```

2. First, let's add a primary key to the `j_posts_add` table:

   ```
   forumdb=# alter table j_posts_tags add constraint j_posts_tags_pkey
   primary key (tag_pk,post_pk);
   ALTER TABLE

   forumdb=# \d j_posts_tags;
   ```

```
    Table "public.j_posts_tags"
    Column  | Type     | Collation | Nullable | Default
    --------+----------+-----------+----------+---------
    tag_pk  | integer  |           | not null |
    post_pk | integer  |           | not null |
    Indexes:
     "j_posts_tags_pkey" PRIMARY KEY, btree (tag_pk, post_pk)
    Foreign-key constraints:
     "j_posts_tags_post_pk_fkey" FOREIGN KEY (post_pk) REFERENCES
    posts(pk)
     "j_posts_tags_tag_pk_fkey" FOREIGN KEY (tag_pk) REFERENCES
    tags(pk)
```

3. Next, let's see what the records of j_posts_tags are:

```
forumdb=# select * from j_posts_tags ;
 tag_pk | post_pk
--------+---------
      1 | 2
      1 | 3
(2 rows)
```

4. Now let's try to insert another record with the same primary key. If we perform a standard insert statement, as follows, we can see that PostgreSQL returns an error because we are trying to insert a record that already exists:

```
forumdb=# insert into j_posts_tags values(1,2);
ERROR: duplicate key value violates unique constraint
"j_posts_tags_pkey"
DETAIL: Key (tag_pk, post_pk)=(1, 2) already exists.
```

5. Let's now try using the ON CONFLICT DO NOTHING option:

```
forumdb=# insert into j_posts_tags values(1,2) ON CONFLICT DO
NOTHING;
INSERT 0 0
forumdb=# select * from j_posts_tags ;
 tag_pk | post_pk
--------+---------
      1 | 2
      1 | 3
(2 rows)
```

In this case, PostgreSQL doesn't return an error; instead, it simply does nothing.

6. Now let's try the DO UPDATE set option. This option realizes the upsert statement, as in the following example:

```
forumdb=# insert into j_posts_tags values(1,2) ON CONFLICT
(tag_pk,post_pk) DO UPDATE set tag_pk=excluded.tag_pk+1;
INSERT 0 1

forumdb=# select * from j_posts_tags ;
 tag_pk | post_pk
--------+---------
      1 | 3
      2 | 2
```

The fields inside the ON CONFLICT condition must have a unique or exclusion constraint. The previous statement simply replaces the following statement:

```
INSERT INTO  j_posts_tags values (1,2)
```

It gets replaced with this statement:

```
UPDATE set tag_pk=tag_pk+1 where tag_pk=1 and post_pk=2
```

Learning the RETURNING clause for INSERT

In PostgreSQL, we can add the RETURNING keyword to the insert statement. The RETURNING keyword in PostgreSQL provides an opportunity to return the values of any columns from an insert or update statement after the insert or update was run. For example, if we want to return all the fields of the record that we have just inserted, we have to perform a query as follows:

```
forumdb=# insert into j_posts_tags values(1,2) returning *;
 tag_pk | post_pk
--------+---------
      1 | 2
(1 row)
```

The * means that we want to return all the fields of the record that we have just inserted; if we want to return only some fields, we have to specify what fields the query has to return:

```
forumdb=# insert into j_posts_tags values (1,6) returning tag_pk;
 tag_pk
--------
 1
(1 row)
```

This feature will show itself to be particularly useful at the end of the chapter when we talk about CTEs.

Returning tuples out of queries

In previous chapters, we have looked at simple update queries, such as the following:

```
forumdb=# update posts set title = 'my new apple' where pk = 3;
UPDATE 1
```

Now we will look at something more complicated. What if we want to update some records in the posts table that are related in some way?

UPDATE related to multiple tables

Let's start with the following scenario:

1. Consider the categories table:

    ```
    forumdb=# select * from categories order by pk;
     pk | title   | description
    ----+---------+-------------
     10 | apple   | fruits
     11 | orange  | fruits
     12 | tomato  | vegetable
     13 | lemon   |
     14 | apricot | fruits
    (5 rows)
    ```

2. Consider the `posts` table (only the `pk`, `title`, and `category` fields):

```
forumdb=# select pk,title,category from posts order by pk;
 pk |    title       | category
----+----------------+----------
  2 | my orange      | 11
  3 | my new apple   | 10
  4 | Re:my orange   | 11
  5 | my tomato      | 12
  6 | my new orange  | 11
(5 rows)
```

Now we want to modify all the records of the `posts` table that belong to the `apple` category. The only record in the table that belongs to the `apple` category is this:

3	my new apple	10

We want this result:

3	my new apple last updated current date	10

Thus, we want to add a string that contains the words `last update + current date`, where the current date is the effective current date. We can reach our goal in three different ways. The first two ways are SQL standard queries, but the third is not. Let's look at the non-standard option in detail.

First of all, let's create a temporary table so as not to modify the data for subsequent tests, as we've seen in the previous chapter. For this, let's perform the following statement:

```
drop table if exists t_posts;
create temp table t_posts as select * from posts;
```

Now, let's start using the first way:

```
update t_posts p
set title=p.title||' last updated '||current_date::text
where p.category in (select pk from categories c where c.title='apple');
```

The preceding query searches all records in the `posts` table that have a value of the `category` field equal to the `pk` values of the `categories` table, which was already filtered by the `where` condition. The part of the preceding query that executes `p.title||' last updated '||current_date::text` appends the `' last update ' + current_date` string, where `current_date` is the effective current date, as seen here:

```
forumdb=# select current_date;
current_date
```

```
--------------
2020-01-09
(1 row)
```

So, the result of the `update` query is as follows:

```
forumdb=# select pk,title,category from t_posts order by pk;
 pk | title                               | category
----+-------------------------------------+----------
  2 | my orange                           | 11
  3 | my new apple last updated 2020-01-09 | 10
  4 | Re:my orange                        | 11
  5 | my tomato                           | 12
  6 | my new orange                       | 11
(5 rows)
```

Another way to make the `update` query is using the `exists` condition:

```
forumdb=# update t_posts p set title=p.title||' last updated
'||current_date::text
where exists (select 1 from categories c where c.pk=p.category and
c.title='apple' limit 1);
```

Now we'll see the third way to make this kind of update. It's a PostgreSQL query, but it isn't a SQL standard query:

```
forumdb=# update t_posts p
  set title=p.title||' last updated '||current_date::text
  from categories c
where c.pk=p.category and c.title='apple';
```

This query is slightly different from those that we've seen before; PostgreSQL allows us to add a `from` condition to the `update` statement. It works very similarly to the inner join mechanism seen previously. For further information, see the official documentation (https://www.postgresql.org/docs/12/sql-update.html).

Exploring UPDATE .... RETURNING

As we've seen in the `INSERT` statement, the `update` statement also has the possibility to add the `RETURNING` keyword. The `update` statement works in the same way as the `INSERT` statement:

```
forumdb=# update t_posts p set title=p.title||' last updated
'||current_date::text
where exists (select 1 from categories c where c.pk=p.category and
c.title='apple' limit 1) returning pk,title,category;
```

```
  pk | title | category
----+------------------------------------+----------
   3 | my new apple last updated 2020-01-09 | 10
(1 row)

UPDATE 1
```

DELETE .... RETURNING

As we've seen, the update statement, like the INSERT statement, has the possibility to add the RETURNING keyword; this feature is also available for the delete statement:

```
forumdb=# delete from t_posts p where exists (select 1 from categories c
where c.pk=p.category and c.title='apple') returning pk,title,category;
  pk | title        | category
----+--------------+----------
   3 | my new apple | 10
(1 row)
DELETE 1

forumdb=# select pk,title,category from t_posts order by 1;
  pk | title         | category
----+---------------+----------
   2 | my orange     | 11
   4 | Re:my orange  | 11
   5 | my tomato     | 12
   6 | my new orange | 11
(4 rows)
```

As we can see, all the records associated with the apple category are not present anymore. In this section, we've seen how to modify the data inside the tables in an advanced way.

In the next section, we'll talk about CTEs, an advanced method to return and modify data.

Exploring CTEs

In this section, we are going to talk about CTEs. This section will be split into three parts. Firstly, we will talk about the concept of CTEs; secondly, we will discuss how CTEs are implemented in PostgreSQL 12; and finally, we will explore some examples of how to use CTEs.

CTE concept

A CTE, or a common table expression, is a temporary result taken from a SQL statement. This statement can contain SELECT, INSERT, UPDATE, or DELETE instructions. The lifetime of a CTE is equal to the lifetime of the query. Here is an example of a CTE definition:

```
WITH cte_name (column_list) AS (
 CTE_query_definition
 )
statement;
```

If, for example, we wanted to create a temporary dataset with all the posts written by the author scotty, we would have to write this:

```
forumdb=# with posts_author_1 as
 (select p.* from posts p
 inner join users u on p.author=u.pk
 where username='scotty')
select pk,title from posts_author_1;
 pk | title
----+--------------
  4 | Re:my orange
  5 | my tomato
(2 rows)
```

We could also write the same thing using an inline view:

```
forumdb=# select pk,title from
(select p.* from posts p inner join users u on p.author=u.pk where
u.username='scotty') posts_author_1;
 pk | title
----+--------------
  4 | Re:my orange
  5 | my tomato
(2 rows)
```

As we can see, the result is the same. The difference is that in the first example, the CTE creates a temporary result set, whereas the second query, the inline view, does not.

CTE in PostgreSQL 12

Starting from PostgreSQL version 12, things have changed, and two new options have been introduced for the execution of a CTE, namely MATERIALIZED and NOT MATERIALIZED. If we want to perform a CTE that materializes a temporary resultset, we have to add the materialized keyword:

```
forumdb=#  with posts_author_1 as materialized
 (select p.* from posts p
 inner join users u on p.author=u.pk
 where username='scotty')
select pk,title from posts_author_1;
 pk | title
----+--------------
 4  | Re:my orange
 5  | my tomato
(2 rows)
```

The query written here materializes a temporary resultset as happened in previous versions of PostgreSQL. If we write the query with the NOT MATERIALIZE option, PostgreSQL will not materialize any temporary resultset:

```
forumdb=# with posts_author_1 as not materialized
 (select p.* from posts p
 inner join users u on p.author=u.pk
 where username='scotty')
select pk,title from posts_author_1;
 pk | title
----+--------------
 4  | Re:my orange
 5  | my tomato
(2 rows)
```

If we don't specify any option, the default is NOT MATERIALIZED, and this could be a problem if we are migrating a database from a minor version to PostgreSQL 12. This is because the behavior of the query planner could change, and the performance could change too.

From version 12, we have to insert the MATERIALIZED option if we want to have our queries display the same behavior that we had with the previous versions.

CTE – some examples

Let's now present some examples of the use of CTEs:

1. Firstly, we will recreate the t_posts table from scratch and then we'll create a
   new table, delete_posts, with the same data structure as the posts table:

```
forumdb=# drop table if exists t_posts;
DROP TABLE
forumdb=# create temp table t_posts as select * from posts;
SELECT 5
forumdb=# create table delete_posts as select * from posts limit 0;
SELECT 0
forumdb=# \d delete_posts
                        Table "public.delete_posts"
        Column       | Type                     | Collation | Nullable |
Default
---------------------+--------------------------+-----------+----------
+---------
 pk                  | integer                  |           |          |
 title               | text                     |           |          |
 content             | text                     |           |          |
 author              | integer                  |           |          |
 category            | integer                  |           |          |
 reply_to            | integer                  |           |          |
 created_on          | timestamp with time zone |           |          |
 last_edited_on      | timestamp with time zone |           |          |
 editable            | boolean                  |           |          |
```

The starting values for the t_posts and delete_posts tables are as follows:

```
forumdb=# select pk,title,category from t_posts ;
 pk | title         | category
----+---------------+----------
  4 | Re:my orange  | 11
  5 | my tomato     | 12
  2 | my orange     | 11
  6 | my new orange | 11
  3 | my new apple  | 10
(5 rows)

forumdb=# select pk,title,category from delete_posts ;
 pk | title | category
----+-------+----------
(0 rows)
```

2. Now suppose that we want to delete some records from the `posts` table, and we want all the records that we have deleted from the `t_posts` table to be inserted into the `delete_posts` table. To reach this goal, we have to use CTEs as follows:

```
forumdb=# with del_posts as (
    delete from t_posts
    where category in (select pk from categories where title
='apple')
returning *)
insert into delete_posts select * from del_posts;
INSERT 0 1
```

The query here deletes all the records from the `t_posts` table that have their category as `'apple'` and, in the same transaction, inserts all the records deleted in the `delete_posts` table, as we can see here:

```
forumdb=# select pk,title,category from t_posts ;
 pk | title | category
----+--------------+----------
  4 | Re:my orange | 11
  5 | my tomato | 12
  2 | my orange | 11
  6 | my new orange | 11
(4 rows)

forumdb=# select pk,title,category from delete_posts ;
 pk | title | category
----+--------------+----------
  3 | my new apple | 10
(1 row)
```

3. Now let's make another example by returning to the starting scenario:

```
forumdb=# drop table if exists t_posts;
DROP TABLE
forumdb=# create temp table t_posts as select * from posts;
SELECT 5
```

4. As we have done before, let's create a new table named `inserted_post` with the same data structure as the `posts` table:

```
forumdb=# create table inserted_posts as select * from posts limit
0;
SELECT 0
```

5. Suppose now that we want to perform a SQL query that moves, in the same transaction, all the records that are present in the `t_posts` table to the `inserted_posts` table. This query will be as follows:

```
forumdb=# with ins_posts as ( insert into inserted_posts select *
from t_posts returning pk) delete from t_posts where pk in (select
pk from ins_posts);
DELETE 5
```

As we can see from the results, the query has achieved our goal:

```
forumdb=# select pk,title,category from t_posts ;
 pk | title | category
----+-------+----------
(0 rows)

forumdb=# select pk,title,category from inserted_posts ;
 pk | title         | category
----+---------------+----------
  4 | Re:my orange  | 11
  5 | my tomato     | 12
  2 | my orange     | 11
  6 | my new orange | 11
  3 | my new apple  | 10
(5 rows)
```

Query recursion

In PostgreSQL, it is possible to create recursive queries. Recursive queries are used in graph databases and in many common use cases, such as querying tables that represent website menus. Recursive CTEs make it possible to have recursive queries in PostgreSQL.

Recursive CTEs

A recursive CTE is a special construct that allows an auxiliary statement to reference itself and, therefore, join itself onto previously computed results. This is particularly useful when we need to join a table an unknown number of times, typically to "explode" a flat tree structure. The traditional solution would involve some kind of iteration, probably by means of a cursor that iterates one tuple at a time over the whole resultset. However, with recursive CTEs, we can use a much cleaner and simpler approach. A recursive CTE is made by an auxiliary statement that is built on top of the following:

- A non-recursive statement, which works as a bootstrap statement and is executed when the auxiliary term is first evaluated.
- A recursive statement, which can either reference the bootstrap statement or itself

These two parts are joined together by means of a UNION predicate. For example, let's see inside the tags table:

```
 pk | tag        | parent
----+------------+--------
  1 | fruits     |
  2 | vegetables |
  3 | apple      | 1
(3 rows)
```

Now we would like to "explode" the flat tree structure and follow the relation between parent and child using the parent field of the tags table. So, we want the result to be something like this:

level	tag
1	fruits
1	vegetable
2	fruits -> apple

To reach this goal, we have to perform the following:

```
forumdb# WITH RECURSIVE tags_tree AS (
 -- non recursive statment
SELECT tag, pk, 1 AS level
FROM tags WHERE parent IS NULL
UNION
-- recursive statement
SELECT tt.tag|| ' -> ' || ct.tag, ct.pk
, tt.level + 1
FROM tags ct
```

```
JOIN tags_tree tt ON tt.pk = ct.parent
)
SELECT level,tag FROM tags_tree
order by level;
level  | tag
-------+-----------------
  1    | fruits
  1    | vegetables
  2    | fruits -> apple
(3 rows)
```

 When we use CTEs, it is important to avoid infinite loops. These can happen if the recursion does not end properly.

Thus, we have learned how to use CTEs to tinker with tables.

Summary

Hopefully, this chapter was full of interesting ideas for the developer and the DBA. In this chapter, we talked about complex queries; we then saw the SELECT statement and the use of the LIKE, ILIKE, DISTINCT, OFFSET, LIMIT, IN, and NOT IN clauses. We then started talking about aggregates through the GROUP BY and HAVING clauses, and we introduced some aggregate functions, such as SUM(), COUNT(), AVG(), MIN(), and MAX().

We then talked in depth about subqueries and joins. Another very interesting set of topics covered in this chapter were the UNION, EXCEPT, and INTERSECT queries. Finally, by looking at the advanced options for the INSERT, DELETE, and UPDATE instructions, and by covering CTEs, we gave you an idea of the power of the SQL language owned by PostgreSQL.

As for the concept of aggregates, in the next chapter, we will see a new way to make aggregates using windows functions. Through the use of windows functions, we will see that we are able to create all the aggregates and aggregation functions described in this chapter, but we will also see that we have the option to create new ones.

References

- PostgreSQL 12 - Subquery expressions official documentation: `https://www.postgresql.org/docs/12/functions-subquery.html`
- PostgreSQL 12 - Joins official documentation: `https://www.postgresql.org/docs/12/tutorial-join.html`
- PostgreSQL 12 - CTEs official documentation: `https://www.postgresql.org/docs/12/queries-with.html`

6
Window Functions

In the previous chapter, we talked about aggregates. In this chapter, we are going to further discuss another way to make aggregates: window functions. The official documentation (`https://www.postgresql.org/docs/12/tutorial-window.html`) describes window functions as follows:

> *A window function performs a calculation across a set of table rows that are somehow related to the current row. This is comparable to the type of calculation that can be done with an aggregate function. However, window functions do not cause rows to become grouped into a single output row as non-window aggregate calls would. Instead, the rows retain their separate identities. Behind the scenes, the window function is able to access more than just the current row of the query result.*

In this chapter, we will talk about window functions, what they are, and how we can use them to improve the performance of our queries.

The following topics will be covered in this chapter:

- Using basic statement window functions
- Using advanced statement window functions

Using basic statement window functions

As we saw in the previous chapter, aggregation functions behave in the following way:

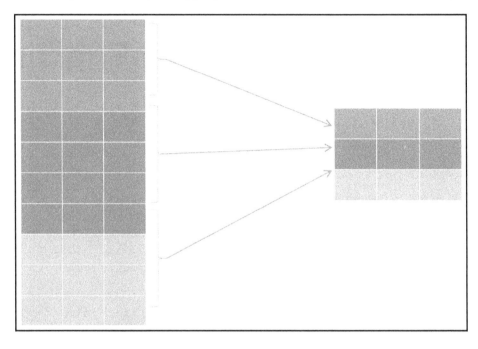

The data is first sorted and then aggregated; the data is then flattened through aggregation. This is what happens when we execute the following statement:

```
forumdb=# select category,count(*) from posts group by category order by
category;
```

Alternatively, we can decide to use window functions by executing the following statement:

```
forumdb=# select category, count(*) over (partition by category) from posts
order by category;
 category | count
----------+-------
       10 | 1
       11 | 3
       11 | 3
       11 | 3
       12 | 1
(5 rows)
```

Window functions create aggregates without flattening the data into a single row. However, they replicate it for all the rows to which the grouping functions refer. The behavior of PostgreSQL is depicted in the following diagram:

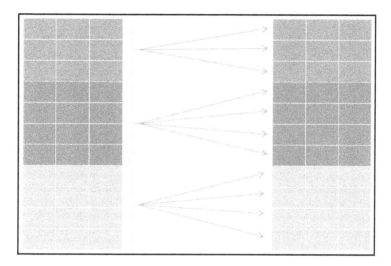

This is the reason that the distinct keyword has to be added to the preceding query if we want to obtain the same result that we get with a classic GROUP BY query.

Using the PARTITION BY function and WINDOW clause

Let's now run some basic queries using the window functions. Suppose that we want to use two over clauses. For example, if, on one column, we want to count the rows relating to the category, and on another column the total count of the columns, then we have to run the following statement:

```
forumdb=# select category, count(*) over (partition by category),count(*)
over () from posts order by category;
 category | count | count
----------+-------+-------
 10       | 1     | 5
 11       | 3     | 5
 11       | 3     | 5
 11       | 3     | 5
 12       | 1     | 5
(5 rows)
```

Or if we want to remove all duplicate rows, we will have to run the following:

```
forumdb=# select distinct category, count(*) over (partition by
category),count(*) over ()
from posts
order by category;
 category | count | count
----------+-------+-------
       10 | 1     | 5
       11 | 3     | 5
       12 | 1     | 5
(3 rows)
```

In the preceding query, the first window function aggregates the data using the category field, while the second one aggregates the data of the whole table.

Using the window functions, it is possible to aggregate the data in different fields in the same query.

As we've seen here, we can define the window frame directly on the query level, but we can also define an alias for the window frame. For example, the preceding query becomes the following:

```
forumdb=# select distinct category, count(*) over w1 ,count(*) over W2
from posts
WINDOW w1 as (partition by category),W2 as ()
order by category;
 category | count | count
----------+-------+-------
 10       | 1     | 5
 11       | 3     | 5
 12       | 1     | 5
(3 rows)
```

The use of aliases is called the WINDOW clause. The WINDOW clause is very useful when we have many aggregates.

Introducing some useful functions

Window functions can use all the aggregation functions that we explored in the previous chapter. In addition to these, window functions introduce new aggregation functions.

Before we examine some of those, let's introduce a unique function – `generate_series`. `generate_series` simply generates a numerical series, for example:

```
forumdb=# select generate_series(1,5);
 generate_series
-----------------
     1
     2
     3
     4
     5
(5 rows)
```

In the following examples, we will use this function for various use cases.

The ROW_NUMBER function

Now let's look at the `ROW_NUMBER()` function. The `ROW_NUMBER()` function assigns a progressive number for each row within the partition:

```
forumdb=# select category, count(*) over w from posts WINDOW w as
(partition by category) order by category;
 category | count | row_number
----------+-------+------------
    10    |   1   |    1
    11    |   3   |    1
    11    |   3   |    2
    11    |   3   |    3
    12    |   1   |    1
(5 rows)
```

In the preceding query, we've used the `PARTITION BY` clause to divide the window into subsets based on the values in the `category` column. As can be seen, we have three category values: `10`, `11`, and `12`. This means that we have three windows and inside each window, the `ROW_NUMBER()` function assigns numbers as we defined before.

The ORDER BY clause

The `ORDER BY` clause sorts the values inside the window. We can also use the `NULLS FIRST` or `NULLS LAST` option to have the null values at the beginning or at the end of the sorting. For example, we can perform a window function query without an `ORDER BY` clause, as we can see in the following snippet, but we have to pay attention to what kind of function we are using, and what our goal is.

If we use aggregation functions that do not depend on the sort order, such as the COUNT function, we can avoid sorting the data, otherwise, it is good practice to sort the data inside the partition in order to avoid the risk of having different results every time the query is launched:

```
forumdb=# select category,row_number() over w,title
from posts WINDOW w as (partition by category) order by category;
 category | row_number | title
----------+------------+---------------
       10 | 1          | my new apple
       11 | 1          | Re:my orange
       11 | 2          | my orange
       11 | 3          | my new orange
       12 | 1          | my tomato
(5 rows)
```

Or, we can use the order by clause and order data inside the partition:

```
forumdb=# select category,row_number() over w,title
from posts WINDOW w as (partition by category order by title) order by
category;
 category | row_number | title
----------+------------+---------------
       10 | 1          | my new apple
       11 | 1          | my new orange
       11 | 2          | my orange
       11 | 3          | Re:my orange
       12 | 1          | my tomato
(5 rows)
```

As we can see in the second example, inside the partition, the data is sorted on the title field.

FIRST_VALUE

The FIRST_VALUE function returns the first value within the partition, for example:

```
forumdb=# select category,row_number() over w,title,first_value(title) over
w
from posts WINDOW w as (partition by category order by category) order by
category;
 category | row_number | title        | first_value
----------+------------+--------------+-------------
       10 | 1          | my new apple | my new apple
       11 | 1          | Re:my orange | Re:my orange
       11 | 2          | my orange    | Re:my orange
```

```
   11 | 3           | my new orange | Re:my orange
   12 | 1           | my tomato     | my tomato
(5 rows)
```

LAST_VALUE

The LAST_VALUE function returns the last value within the partition, for example:

```
forumdb=# select category,row_number() over w,title,last_value(title) over
w
from posts WINDOW w as (partition by category order by category) order by
category;
 category | row_number | title | last_value
----------+------------+---------------+---------------
       10 | 1 | my new apple | my new apple
       11 | 1 | Re:my orange | my new orange
       11 | 2 | my orange | my new orange
       11 | 3 | my new orange | my new orange
       12 | 1 | my tomato | my tomato
(5 rows)
```

It is important to always use the order by clause when we use the first_value() or last_value() functions to avoid incorrect results, as mentioned previously.

RANK

The RANK function ranks the current row within its partition with gaps. If we don't specify a PARTITION BY clause, the function doesn't know how to correlate the current tuple, so the function correlates to itself, as seen here:

```
forumdb=# select pk,title,author,category,rank() over () from posts order
by category;
 pk | title           author | category | rank
----+---------------+--------+----------+------
  3 | my new apple  | 1      | 10       | 1
  4 | Re:my orange  | 2      | 11       | 1
  2 | my orange     | 1      | 11       | 1
  6 | my new orange | 1      | 11       | 1
  5 | my tomato     | 2      | 12       | 1
(5 rows)
```

If we add the `order by` clause, the function ranks in the assigned order, for example, the author with id `1` starts from record 1, and the author with id `2` starts from record 4, as we can see in the following example:

```
forumdb=# select pk,title,author,category,rank() over (order by author)
from posts ;
 pk |    title       | author | category | rank
----+----------------+--------+----------+------
  2 | my orange      | 1      | 11       | 1
  6 | my new orange  | 1      | 11       | 1
  3 | my new apple   | 1      | 10       | 1
  4 | Re:my orange   | 2      | 11       | 4
  5 | my tomato      | 2      | 12       | 4
(5 rows)
```

If we add the `PARTITION BY` clause, the working mechanism is the same, the only difference is that the ranking is calculated within the partition and not on the whole table as in the previous example:

```
forumdb=# select pk,title,author,category,rank() over (partition by author
order by category) from posts order by author;
 pk |    title       | author | category | rank
----+----------------+--------+----------+------
  3 | my new apple   | 1      | 10       | 1
  2 | my orange      | 1      | 11       | 2
  6 | my new orange  | 1      | 11       | 2
  4 | Re:my orange   | 2      | 11       | 1
  5 | my tomato      | 2      | 12       | 2
(5 rows)
```

DENSE_RANK

The `DENSE_RANK` function is similar to the `RANK` function. The difference is that the `DENSE_RANK` function ranks the current row within its partition without gaps:

```
forumdb=# select pk,title,author,category,dense_rank() over (order by
author) from posts order by category;
 pk |    title       | author | category | dense_rank
----+----------------+--------+----------+------------
  3 | my new apple   | 1      | 10       | 1
  2 | my orange      | 1      | 11       | 1
  6 | my new orange  | 1      | 11       | 1
  4 | Re:my orange   | 2      | 11       | 2
  5 | my tomato      | 2      | 12       | 2
(5 rows)
```

The LAG and LEAD functions

In this section, we will show how the LAG and LEAD functions work. First of all, we are going to set up our environment and we are going to generate a sequence of numbers as we did previously:

```
forumdb=# select x from (select generate_series(1,5) as x) V ;
 x
---
 1
 2
 3
 4
 5
(5 rows)
```

This is our starting point for this example. The official documentation (https://www.postgresql.org/docs/12/functions-window.html) defines the LAG function as follows:

The LAG function returns a value evaluated at the row that is offset rows before the current row within the partition; if there is no such row, it instead returns the default (which must be of the same type as the value). Both the offset and the default are evaluated with respect to the current row. If omitted, offset defaults to 1 and default to null.

Now, let's write the following statement:

```
forumdb=# select x,lag(x) over w from (select generate_series(1,5) as x) V
WINDOW w as (order by x) ;
 x | lag
---+-----
 1 |
 2 | 1
 3 | 2
 4 | 3
 5 | 4
(5 rows)
```

As we can see, the lag function returns a result set with an offset value equal to 1. If we introduce an offset parameter, the lag function will return a result set with an offset equal to the number that we have passed as input, as can be seen in the next example:

```
forumdb=# select x,lag(x,2) over w from (select generate_series(1,5) as x)
V WINDOW w as (order by x) ;
 x | lag
---+-----
 1 |
 2 |
```

```
3 | 1
4 | 2
5 | 3
(5 rows)
```

The `lead` function is the opposite of the `lag` function, as described in the official documentation:

> *The* LEAD *function returns the value evaluated at the row that is offset rows after the current row within the partition; if there is no such row, it instead returns the default (which must be of the same type as the mentioned value). Both the offset and default are evaluated with respect to the current row. If omitted, the offset defaults to* 1 *and the default becomes* null.

Here are a couple of examples where we can see how it works. In the first example, we will use the `lead` function without any parameters:

```
forumdb=# select x,lead(x) over w from (select generate_series(1,5) as x) V
WINDOW w as (order by x) ;
 x | lead
---+------
 1 | 2
 2 | 3
 3 | 4
 4 | 5
 5 |
(5 rows)
```

As we can see in the `lead` function, the offset starts from the bottom.

Let's now see an example of using the `lead` function with an offset parameter:

```
forumdb=# select x,lead(x,2) over w from (select generate_series(1,5) as x)
V WINDOW w as (order by x) ;
 x | lead
---+------
 1 | 3
 2 | 4
 3 | 5
 4 |
 5 |
(5 rows)
```

The CUME_DIST function

The CUME_DIST function calculates the cumulative distribution of value within a partition. The function is described in the official documentation as follows:

> *The* CUME_DIST *function computes the fraction of partition rows that are less than or equal to the current row and its peers.*

Let's look at an example:

```
forumdb=# select x,cume_dist() over w from (select generate_series(1,5) as
x) V WINDOW w as (order by x) ;
 x | cume_dist
---+-----------
 1 | 0.2
 2 | 0.4
 3 | 0.6
 4 | 0.8
 5 | 1
(5 rows)
```

As the function is mathematically defined, the cume_dist function can never have a value greater than the current value of the field.

The NTILE function

The PostgreSQL NTILE function groups the rows sorted in the partition. Starting from 1, up to the parameter value passed to the NTILE function, each group is assigned a number of buckets. The parameter passed to the NTILE function determines how many records we want the bucket to be composed of.

Now, let's see an example of how it works by trying to split our result set into two buckets:

```
forumdb=# select x,ntile(2) over w from (select generate_series(1,6) as x)
V WINDOW w as (order by x) ;
 x | ntile
---+-------
 1 | 1
 2 | 1
 3 | 1
 4 | 2
 5 | 2
 6 | 2
(6 rows)
```

If we wanted to divide our result set into three buckets, we would run the following statement:

```
forumdb=# select x,ntile(3) over w from (select generate_series(1,6) as x)
V WINDOW w as (order by x) ;
 x | ntile
---+-------
 1 | 1
 2 | 1
 3 | 2
 4 | 2
 5 | 3
 6 | 3
(6 rows)
```

The NTILE() function accepts an integer and tries to divide the window into a number of balanced buckets, specifying to which bucket each row belongs.

In this section, we have introduced some features that allow you to do some basic data mining. For example, lag and lead could be used to compare different lines of a table, and therefore compare the salaries of different employees, or compare collections from different days.

In the next section, we will go into even more detail and explore some more advanced features of window functions.

Using advanced statement window functions

In this section, we will discuss advanced window functions in more detail, and we will explore some techniques that may be useful for carrying out more detailed data analysis.

```
forumdb=# select distinct category, count(*) over w1
from posts
WINDOW w1 as (partition by category RANGE BETWEEN UNBOUNDED PRECEDING AND
CURRENT ROW)
order by category;
 category | count
----------+-------
       12 | 1
       10 | 1
       11 | 3
(3 rows)
```

What does RANGE BETWEEN UNBOUNDED PRECEDING AND CURRENT ROW mean? They are the default conditions, known as the **frame clause**. This means that the data is partitioned, first by category, and then within the partition, the count is calculated by resetting the count every time the frame is changed.

The frame clause

In this section, we'll talk about the frame clause, which allows us to manage partitions in a different way. The frame clause has two forms:

- Rows between start_point and end_point
- Range between start_point and end_point

It only makes sense to use the frame clause if the order by clause is also present. We will use the ROWS BETWEEN clause when we are going to consider a specific set of records relative to the current row. We will use the RANGE BETWEEN clause when we are going to consider a range of values in a specific column relative to the value in the current row.

ROWS BETWEEN start_point and end_point

Now we will look at some simple examples to try to better explain the frame_set clauses. These are typically used to do in-depth data analysis and data mining, among other tasks. Let's start with some examples, beginning here:

```
forumdb=# select x from (select generate_series(1,5) as x) V WINDOW w as
(order by x) ;
 x
---
 1
```

```
2
3
4
5
(5 rows)
```

Suppose that we want to have an incremental sum row by row, the goal that we want to reach is as follows:

x	sum(x)
1	1
2	3
3	6
4	10
5	15

This can be achieved using the following query:

```
forumdb=# SELECT x, SUM(x) OVER w
FROM (select generate_series(1,5) as x) V
WINDOW w AS (ORDER BY x ROWS BETWEEN UNBOUNDED PRECEDING AND CURRENT ROW);
 x | sum
---+-----
 1 | 1
 2 | 3
 3 | 6
 4 | 10
 5 | 15
(5 rows)
```

Now, let's imagine that the query was executed in successive steps, one for each row of the table. In the following diagrams, we will simulate the internal behavior of PostgreSQL, to better understand how the clause ROWS BETWEEN UNBOUNDED PRECEDING AND CURRENT ROW works:

1. First, PostgreSQL uses the order_by_clause condition to order the data inside the window, as seen in the following diagram:

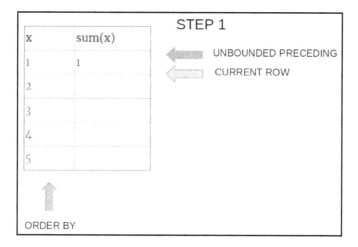

As we can see in the diagram, we have two pointers: the green one for the **UNBOUNDED PRECEDING** clause and the orange pointer for the **CURRENT ROW** clause. The result is **1**, so in the first step both point to the first row. Now, let's see what happens in the next steps.

2. In the second step, the **UNBOUNDED PRECEDING** pointer still points to the first row, whereas the **CURRENT ROW** pointer now points to the second row, and the result of the sum is 1+2 = **3**:

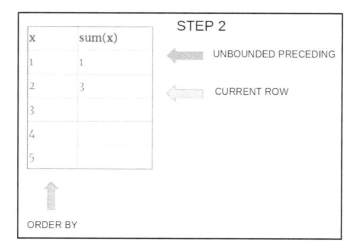

3. The third step is very similar to step 2: the **UNBOUNDED PRECEDING** pointer still points to the first row, whereas the **CURRENT ROW** pointer now points to the third row, and the result of the sum is 1+2+3 = **6**:

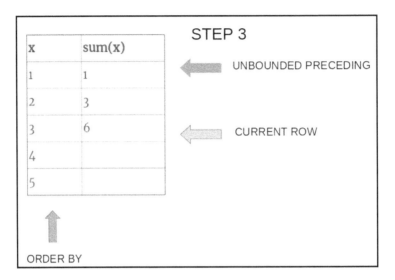

4. The fourth step is almost identical to step 3: the **UNBOUNDED PRECEDING** pointer still points to the first row, whereas the **CURRENT ROW** pointer now points to the fourth row, and the result of the sum is 1+2+3+4 = 10:

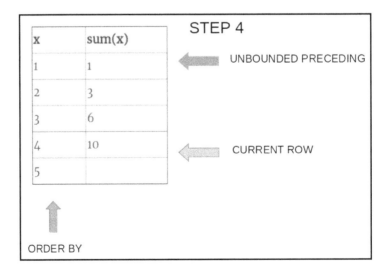

5. And in the fifth and final step, we have the desired result:

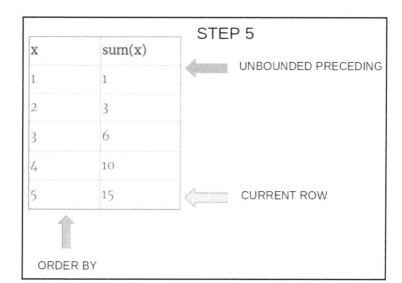

That is how a frameset clause works!

Let's look at some more examples of how the frame clause works using different options.

If for each row of the table we want to find the sum of the current row with the preceding row, we would start from the following:

x
1
2
3
4
5

We want to end up with the following result:

x	sum(x)
1	1
2	3
3	5
4	7
5	9

The query that we have to perform is described in the following example:

```
forumdb=# SELECT x, SUM(x) OVER w
 FROM (select generate_series(1,5) as x) V
 WINDOW w AS (ORDER BY x RANGE BETWEEN 1 PRECEDING AND CURRENT ROW);
 x | sum
---+-----
 1 | 1
 2 | 3
 3 | 5
 4 | 7
 5 | 9
(5 rows)
```

The preceding query works similarly to what we saw before. The only difference is that now the calculation range is between the first row and the current row of the partition, as written in the statement BETWEEN 1 PRECEDING AND CURRENT ROW. In this example, only two lines are used to calculate the sum. The same mechanism can be used to perform an incremental sum, as we can see in the preceding example:

```
forumdb=# SELECT x, SUM(x) OVER w
 FROM (select generate_series(1,5) as x) V
 WINDOW w AS (ORDER by x ROWS UNBOUNDED PRECEDING);
 x | sum
---+-----
 1 | 1
 2 | 3
 3 | 6
 4 | 10
 5 | 15
(5 rows)
```

Now the only difference is that the calculation range is by ROWS UNBOUNDED PRECEDING and not BETWEEN 1 PRECEDING AND CURRENT ROW.

Let's look at another example where window functions simplify our work. Always starting from the series that we've seen before, we know that the total sum is 1+2+3+4+5 = 15, so now suppose that we want to make a reverse sum starting from the max value of the table, that is, 5.

In this example, we want the result to be as follows:

x	sum(x)
1	15
2	14
3	12
4	9
5	5

The query that makes this possible is the following:

```
forumdb=# SELECT x, SUM(x) OVER w
FROM (select generate_series(1,5) as x) V
WINDOW w AS (ORDER BY X ROWS BETWEEN CURRENT ROW AND UNBOUNDED FOLLOWING);
 x | sum
---+-----
 1 | 15
 2 | 14
 3 | 12
 4 | 9
 5 | 5
(5 rows)
```

What makes this possible is the **UNBOUNDED FOLLOWING** clause, which works the opposite way to **UNBOUNDED PRECEDING**. This happens because of the following:

- In the first row, all values are added: 1+2+3+4+5 =15.
- In the second row, these values are added: 2+3+4+5 = 14.
- In the third row, these values are added: 3+4+5 = 12.

RANGE BETWEEN start_point and end_point

As discussed earlier, when we use RANGE BETWEEN, we will consider a RANGE of values with respect to the value in the current row. The difference when it comes to the ROWS clause is that if the field that we use for ORDER BY does not contain unique values for each row, then RANGE will combine all the rows it comes across with non-unique values, rather than processing them one at a time.

In contrast, ROWS will include all of the rows in the non-unique bunch but processes each of them separately.

1. First of all, let's create a simple dataset with duplicate data:

```
forumdb=# select generate_series(1,10) % 5 as x order by 1;
 x
---
 0
 0
 1
 1
 2
 2
 3
 3
 4
 4
(10 rows)
```

2. Now let's do some tests to observe the differences between the ROWS and RANGE clauses. Let's start with the ROWS clause:

```
forumdb=# SELECT x, row_number() OVER w, SUM(x) OVER w FROM (select
generate_series(1,10) % 5 as x) V
WINDOW w AS (ORDER BY x ROWS BETWEEN 1 PRECEDING AND CURRENT ROW);
 x | row_number | sum
---+------------+-----
 0 | 1          | 0
 0 | 2          | 0
 1 | 3          | 1
 1 | 4          | 2
 2 | 5          | 3
 2 | 6          | 4
 3 | 7          | 5_
 3 | 8          | 6
 4 | 9          | 7
 4 | 10         | 8
(10 rows)
```

The preceding query works exactly as we've seen before, it sums the previous row with the current row.

3. Let's now see what happens if we use the RANGE clause instead of the ROWS clause:

```
forumdb=# SELECT x, row_number() OVER w, SUM(x) OVER w
FROM (select generate_series(1,10) % 5 as x) V
WINDOW w AS (ORDER BY x RANGE BETWEEN 1 PRECEDING AND CURRENT ROW);
 x | row_number | sum
---+------------+-----
 0 | 1          | 0
 0 | 2          | 0
 1 | 3          | 2
 1 | 4          | 2
 2 | 5          | 6
 2 | 6          | 6
 3 | 7          | 10
 3 | 8          | 10
 4 | 9          | 14
 4 | 10         | 14
(10 rows)
```

Let's take this result:

x	row_number	sum
0	1	0
0	2	0
1	3	2
1	4	2
2	5	6
2	6	6
3	7	10
3	8	10
4	9	14
4	10	14

Now let's look at the result from the frame point of view:

x	row_number	sum	Frame Number
0	1	0	1
0	2	0	1
1	3	2	2
1	4	2	2
2	5	6	3
2	6	6	3

3	7	10	4
3	8	10	4
4	9	14	5
5	10	14	5

As we can see, there are four frames in the table before, so internally PostgreSQL works in this way: first, PostgreSQL splits the window function into frames using the order by clause and then aggregates the data among the frames; for example:

- The sum of row number 3 is the result of the sum of row number 1 + row number 2 + row number 3 + row number 4: 0+0+1+1=2.
- The sum of row number 4 is the result of the sum of row number 1 + row number 2 + row number 3 + row number 4: 0+0+1+1=2.
- The sum of row number 5 is the result of the sum of row number 3 + row number 4 + row number 5 + row number 6:1+1+2+2=6.
- The sum of row number 6 is the result of the sum of row number 3 + row number 4+ row number 5 + row number 6:1+1+2+2=6.

In the preceding example, we have considered a partition ordered in an ascending way. In the next example, the partition is sorted in a descending way and we will see the difference between ROWS and RANGE in this scenario.

This is the query for the RANGE clause:

```
forumdb=# SELECT x,row_number() OVER w, dense_rank() OVER w,sum(x) OVER w
FROM (select generate_series(1,10) % 5 as x) V
WINDOW w AS (ORDER BY x desc RANGE BETWEEN 1 PRECEDING AND CURRENT ROW);
 x | row_number | dense_rank | sum
---+------------+------------+-----
 4 | 1          | 1          | 8
 4 | 2          | 1          | 8
 3 | 3          | 2          | 14
 3 | 4          | 2          | 14
 2 | 5          | 3          | 10
 2 | 6          | 3          | 10
 1 | 7          | 4          | 6
 1 | 8          | 4          | 6
 0 | 9          | 5          | 2
 0 | 10         | 5          | 2
(10 rows)
```

And this is the query for the ROWS clause. As we can see, things work exactly as in the previous example without the ORDER BY DESC option:

```
forumdb=# SELECT x,row_number() OVER w, dense_rank() OVER w,sum(x) OVER w
FROM (select generate_series(1,10) % 5 as x) V
WINDOW w AS (ORDER BY x desc ROWS BETWEEN 1 PRECEDING AND CURRENT ROW);
 x | row_number | dense_rank | sum
---+------------+------------+-----
 4 | 1          | 1          | 4
 4 | 2          | 1          | 8
 3 | 3          | 2          | 7
 3 | 4          | 2          | 6
 2 | 5          | 3          | 5
 2 | 6          | 3          | 4
 1 | 7          | 4          | 3
 1 | 8          | 4          | 2
 0 | 9          | 5          | 1
 0 | 10         | 5          | 0
(10 rows)
```

In this example, using the sum function, we can better understand the difference between the RANGE and ROWS options. As we can see, the RANGE option aggregates data by frame (RANGE) while the ROWS option aggregates data by rows. The main difference between the ROWS clause and the RANGE clause is that ROWS operates on individual rows, while RANGE operates on groups. That concludes our chapter on window functions.

Summary

In this chapter, we explored how to use window functions. We have seen that by using window functions we can create more complex aggregates compared to those made with the GROUP BY statement, which we saw in Chapter 5, *Advanced Statements*. We learned how to use the ROW_NUMBER (), FIRST_VALUE (), LAST_VALUE (), RANK DENSE_RANK(), LAG (), LEAD (), CUME_DIST (), and NTILE () functions. We have also seen the difference between creating aggregates with the ROWS BETWEEN and RANGE BETWEEN clauses. You can use what you have learned in this chapter in data mining operations to make your work much easier.

For more information on window functions, you can consult the official documentation: https://www.postgresql.org/docs/12/functions-window.html.

In the next chapter, we will talk about server-side programming, we will look at how to create functions to be used on the server side and, if necessary, where to use window functions.

References

- PostgreSQL 12 - window functions official documentation: https://www.postgresql.org/docs/12/functions-window.html

7
Server-Side Programming

In previous chapters, we learned how to execute SQL queries. We started by writing simple queries, then moved on to writing more complex queries; we learned how to use aggregates in the traditional way, and in `Chapter 5`, *Advanced Statements*, we talked about window functions, which are another way to write aggregates. In this chapter, we will add server-side programming to this list of skills. Server-side programming can be useful in many cases as it moves the programming logic from the client side to the database side. For example, we could use it to take a function that has been written many times at different points of the application program and move it inside the server so that it is written only once, meaning that in case of modification, we only have to modify one function. In this chapter, we will also look at how PostgreSQL can manage different server-side programming languages, and we will see that server-side programming can be very useful if you need to process a large amount of data that has been extracted from tables. We will address the fact that all the functions we will write can be called in any SQL statement. We will also see that in some cases, for certain types of functions, it is also possible to create indices on the functions.

Another feature of server-side programming is the chance to define customized data. In this chapter, we will look at some examples of this.

In simple terms, this chapter will discuss the following:

- Exploring data types
- Exploring functions and languages

Exploring data types

As users, we have already had the opportunity to experience the power and versatility of server-side functions – for example, in `Chapter 5`, *Advanced Statements*, we used a query similar to the following:

```
forumdb=# select * from categories where upper(title) like 'A%';
 pk | title   | description
----+---------+-------------
 10 | apple   | fruits
 14 | apricot | fruits
(2 rows)
```

In this piece of code, the `upper` function is a server-side function; this function turns all the characters of a string into uppercase. In this chapter, we will acquire the knowledge to be able to write functions such as the `upper` functions that we called in the preceding query.

In this section, we'll talk about data types. We will briefly mention the standard types managed by PostgreSQL and how to create new ones.

The concept of extensibility

What is extensibility? Extensibility is PostgreSQL's ability to extend its functionality and its data types. Extensibility is an extremely useful PostgreSQL feature because it enables us to have data types, functions, and functional indexes that are not present in the base system. In this chapter, we will cover the extension at the data type level, as well as the addition of new functions.

Standard data types

In previous chapters, even if not explicitly obvious, we have already used standard data types. This happened when we learned how to use **Data Definition Language** (**DDL**) commands. However, we will now be looking more deeply into this topic. The following is a short list of the most used data types:

- Boolean type
- Numeric types
- Character types
- Date/time
- NoSQL data types : `hstore`, `xml`, `json`, and `jsonb`

For each data type, we will show an example operation followed by a brief explanation. For further information on the standard data types supported by PostgreSQL, please refer to the official documentation at `https://www.postgresql.org/docs/12/extend-type-system.html`.

Boolean data type

First, we will introduce the Boolean data type. PostgreSQL supports Boolean data types. The Boolean type (identified by `BOOLEAN` or `BOOL`), like all data types supported by PostgreSQL, can assume the `NULL` value. Therefore, a Boolean data type can take the `NULL`, `FALSE`, and `TRUE` values. The data type input function for the Boolean type accepts the following representations for the `TRUE` state:

State
true
yes
on
1

For the `false` state, we have the following:

State
false
no
off
0

Let's look at some examples, starting with the `users` table:

1. Let's first display the contents of the `users` table:

```
forumdb=# select * from users;
 pk | username    | gecos         | email
----+-------------+---------------+--------------
 1  | myusername  | mygecos       | myemail
 2  | scotty      | scotty_gecos  | scotty_email
(2 rows)
```

2. Now let's add a Boolean data type to the `users` table:

```
forumdb=# alter table users add user_on_line boolean;
ALTER TABLE
```

3. Let's update some values:

```
forumdb=# update users set user_on_line = true where pk=1;
UPDATE 1
```

4. Now, if we want to search for all the records that have the `user_on_line` field set to `true`, we have to perform the following:

```
forumdb=# select * from users where user_on_line = true;
 pk | username   | gecos   | email   | user_on_line
----+------------+---------+---------+--------------
 1  | myusername | mygecos | myemail | t
(1 row)
```

5. If we want the search for all the records who have the `user_on_line` field set to `NULL`, as we saw in `Chapter 4`, *Basic Statements*, we have to perform the following:

```
forumdb=# select * from users where user_on_line is NULL;
 pk | username | gecos        | email        | user_on_line
----+----------+--------------+--------------+--------------
 2  | scotty   | scotty_gecos | scotty_email |
(1 row)
```

Thus, we have explored the Boolean data type.

Numeric data type

PostgreSQL supports several types of numeric data types; the most used ones are as follows:

- `integer` or `int4` (4-byte integer number).
- `bigint` or `int8` (8-byte integer number).
- `real` (4-byte variable-precision, inexact with 6 decimal digit precision).
- `double precision` (8-byte variable precision, inexact with 15 decimal digits precision).
- `numeric` (precision, scale), where the precision of a numeric is the total count of significant digits in the whole number, and the scale of a numeric is the count of decimal digits in the fractional part. For example, 5.827 has a precision of 4 and a scale of 3.

Now, we will look at some brief examples of each type in the upcoming sections.

Integer types

As we can see here, if we cast a number to an integer type such as `integer` or `bigint`, PostgreSQL will make a `trunc` value of the input number:

```
forumdb=# select 1.123456789::integer as my_field;
 my_field
-----
    1
forumdb=# select 1.123456789::int4 as my_field;
 my_field
-----
    1
forumdb=# select 1.123456789::bigint as my_field;
 my_field
-----
    1
forumdb=# select 1.123456789::int8 as my_field;
 my_field
-----
    1
```

Numbers with a fixed precision data type

In the following example, we'll see the same query that we have seen previously, but this time, we'll make a cast to `real` and to `double precision`:

```
forumdb=# select 1.123456789::real as my_field;
 my_field
-----------
 1.1234568
forumdb=# select 1.123456789::double precision as my_field;
 my_field
------------
 1.123456789
```

As can be seen here, in the first query, the result was cut to the sixth digit; this happened because the real type has 6 decimal digit precision.

Numbers with an arbitrary precision data type

In this last section about numeric data types, we'll make the same query that we saw earlier, but we'll make a cast to arbitrary precision:

```
forumdb=# select 1.123456789::numeric(10,1) as my_field;
 my_field
-----
```

```
1.1
forumdb=# select 1.123456789::numeric(10,5) as my_field;
 my_field
---------
 1.12346
forumdb=# select 1.123456789::numeric(10,9) as my_field;
  my_field
-------------
 1.123456789
```

As we can see from the examples shown here, we decide how many digits the scale should be.

But what about if we perform something like the following?

```
forumdb=# select 1.123456789::numeric(10,11) as my_field;
ERROR: NUMERIC scale 11 must be between 0 and precision 10
ROW 1: select 1.123456789::numeric(10,11) as my_field;
```

The result is an error. This is because the data type was defined as a numeric type with a precision value equal to 10, so we can't have a scale parameter equal to or greater than the precision value.

Similarly, the next example will also produce an error:

```
forumdb=# select 1.123456789::numeric(10,10) as my_field;
 ERROR: numeric field overflow
 DETAILS: A field with precision 10, scale 10 must round to an absolute
value less than 1.
```

In the preceding example, the query generates an error because the scale was 10, meaning we should have 10 digits, but we have 11 digits in total:

Digits	1	2	3	4	5	6	7	8	9	10	11
	1	.	1	2	3	4	5	6	7	8	9

However, if in our number we don't have the first digit, the query will work:

```
forumdb=# select 0.123456789::numeric(10,10) as my_field;
    my_field
--------------
 0.1234567890
```

Thus, we have learned all about the various numeric data types.

Character data type

The most used character data types in PostgreSQL are the following:

- `character(n)/char(n)` (fixed-length, blank-padded)
- `character varying(n)/varchar(n)` (variable-length with a limit)
- `text` (variable unlimited length)

Now, we will look at some examples to see how PostgreSQL manages these kinds of data types.

Chars with fixed-length data types

We will check out how they work using the following example:

1. Let's start by creating a new test table:

```
create table new_tags (
pk integer not null primary key,
tag char(10)
);
```

In the previous code, we created a new table named `new_tags` with a `char(10)` field name tag.

2. Now, let's add some records and see how PostgreSQL behaves:

```
forumdb=# insert into new_tags values (1,'first tag');
INSERT 0 1
forumdb=# insert into new_tags values (2,'tag');
INSERT 0 1
```

In order to continue with our analysis, we must introduce two new functions:

- `length (p)`: This counts the number of characters, where `p` is an input parameter and a string.
- `octet_length(p)`: This counts the number of bytes, where `p` is an input parameter and a string.

3. Let's execute the following query:

```
forumdb=# select
pk,tag,length(tag),octet_length(tag),char_length(tag);
 pk | tag        | length | octet_length | char_length
----+------------+--------+--------------+------------
  1 | first tag  | 9      | 10           | 9
  2 | tag        | 3      | 10           | 3
(2 rows)
```

As we can see, the overall length of the space occupied internally by the field is always 10; this is true even if the number of characters entered is different. This happens because we have defined the field as `char(10)`, with a fixed length of 10, so even if we insert a string with a shorter length, the difference between 10 and the number of real characters of the string will be filled with blank characters.

Chars with variable length with a limit data types

In this section, we are going to repeat the same example that we used in the previous section, but this time we'll use the `varchar(10)` data type for the `tag` field:

1. Let's recreate the `new_tags` table:

```
forumdb=# drop table if exists new_tags;
DROP TABLE

forumdb=# create table new_tags (
pk integer not null primary key,
tag varchar(10)
);
CREATE TABLE
```

2. Then, let's insert some data:

```
forumdb=# insert into new_tags values (1,'first tag');
INSERT 0 1
forumdb=# insert into new_tags values (2,'tag');
INSERT 0 1
```

3. Now, if we repeat the same query as before, we obtain the following:

```
forumdb=# select pk,tag,length(tag),octet_length(tag) from new_tags
;
 pk | tag        | length | octet_length
----+------------+--------+--------------
  1 | first tag  | 9      | 9
  2 | tag        | 3      | 3
(2 rows)
```

As we can see, this time, the real internal size and the number of characters in the string are the same.

4. Now, let's try to insert a string longer than 10 characters and see what happens:

```
forumdb=# insert into new_tags values (3,'this sentence has more
than 10 characters');
ERROR: value too long for type character varying(10)
```

PostgreSQL answers correctly with an error because the input string exceeds the dimension of the field.

Chars with a variable length without a limit data types

In this section, we will again use the same example as before, but this time we'll use a text data type for the tag field.

Let's recreate the new_tags table and let's re-insert the same data that we inserted previously:

```
forumdb# drop table if exists new_tags;
DROP TABLE
forumdb# create table new_tags (
pk integer not null primary key,
tag text
);
CREATE TABLE
forumdb# insert into new_tags values (1,'first tag'), (2,'tag'),(3,'this
sentence has more than 10 characters');
INSERT 0 3
```

This time, PostgreSQL correctly inserts all three records. This is because the text data type is a char data type with unlimited length, as we can see in the following query:

```
forumdb# select pk,substring(tag from 0 for
20),length(tag),octet_length(tag) from new_tags ;
 pk | substring          | length | octet_length
```

```
----+--------------------+--------+--------------
 1  | first tag          | 9      | 9
 2  | tag                | 3      | 3
 3  | this sentence has m | 41    | 41
(3 rows)
```

In the preceding example, we can see that the `text` data type behaves exactly like the `varchar(n)` data type we saw earlier. The only difference between `text` and `varchar(n)` is that the `text` type has no size limit. It is important to note that on the preceding query, we used the `substring` function. The `substring` function takes a piece of the string starting from the `from` parameter for *n* characters; for example, if we write `substring(tag from 0 for 20)`, it means that we want the first 20 characters of the `tag` string as output.

With this, we have covered all the `char` data types.

Date/timestamp data types

In this section, we will talk about how to store dates and times in PostgreSQL. PostgreSQL supports both dates and times and the combination of date and time (timestamp). PostgreSQL manages hours both with time zone settings and without time zone settings, as described in the official documentation (`https://www.postgresql.org/docs/12/datatype-datetime.html`):

> *PostgreSQL supports the full set of SQL date and time types. Dates are counted according to the Gregorian calendar.*

Date data types

Managing dates often becomes a puzzle for the developer. This happens because dates are represented differently depending on the country for which we have to store the data – for example, the American way is month/day/year, whereas the Italian format is day/month/year. PostgreSQL helps us by providing the necessary tools to best solve this problem, as seen here:

1. The first thing we have to do is to see how PostgreSQL internally stores dates. To do this, we have to perform the following query:

   ```
   forumdb=# \x
   forumdb=# select * from pg_settings where name ='DateStyle';
   -[ RECORD 1 ]---+-----------------------------------------------------
   -----
   name            | DateStyle
   setting         | ISO, MDY
   ```

```
unit              |
category          | Client Connection Defaults / Locale and
Formatting
short_desc        | Sets the display format for date and time values.
extra_desc        | Also controls interpretation of ambiguous date
inputs.
context           | user
vartype           | string
source            | configuration file
min_val           |
max_val           |
enumvals          |
boot_val          | ISO, MDY
reset_val         | ISO, MDY
sourcefile        | /etc/postgresql/12/main/postgresql.conf
sourceline        | 649
pending_restart   | f
```

First of all, let's take a look at the pg_settings view. Using the pg_settings view, we can view the parameters set in the postgresql.conf configuration file. In the preceding result, we can see that the configuration for displaying the date is MDY (month/day/year). If we want to change this parameter globally, we have to edit the postgresql.conf file.

2. On a Debian or Debian-based distribution, we can edit the file as follows:

   ```
   root@pgdev:/# vim /etc/postgresql/12/main/postgresql.conf
   ```

3. Then, we have to modify the following section:

   ```
   #Locale and Formatting

   datestyle = 'iso, mdy'
   ```

4. After changing this parameter, in the query on pg_settings, the context parameter is 'user'; we just need to do a reload of the server. In this case, a restart is not necessary:

   ```
   root@pgdev:/# /etc/init.d/postgresql reload
   [ ok ] Reloading postgresql configuration (via systemctl):
   postgresql.service.
   ```

For further information about the pg_settings view, we suggest visiting https://www.postgresql.org/docs/12/view-pg-settings.html.

5. We have learned what the internal parameters for date display are, so now, let's look at how to insert, update, and display dates. If we know the value of the date-style parameter, the PostgreSQL way of converting a string into a date is as follows:

```
forumdb=# select '12-31-2020'::date;
    date
------------
 2020-12-31
(1 row)
```

This way is simple but not particularly user-friendly. The best way to manage dates is by using some functions that PostgreSQL provides for us.

6. The first function that we'll talk about is the `to_date()` function. The `to_date()` function converts a given string into a date. The syntax of the `to_date()` function is as follows:

```
forumdb=# select to_date('31/12/2020','dd/mm/yyyy') ;
  to_date
------------
 2020-12-31
(1 row)
```

The `to_date()` function accepts two string parameters. The first parameter contains the value that we want to convert into a date. The second parameter is the pattern of the date. The `to_date()` function returns a date value.

7. Now, let's go back to the `posts` table and execute this query:

```
forumdb=# select pk,title,created_on from posts;
 pk | title          | created_on
----+----------------+-----------------------------
  4 |  Re:my orange  | 2020-01-03 18:46:06.436248+01
  5 |  my tomato     | 2020-01-03 18:47:39.603937+01
  2 |  my orange     | 2020-01-03 18:44:13.266102+01
  6 |  my new orange | 2020-01-05 18:05:10.860354+01
  3 |  my new apple  | 2020-01-03 18:44:32.459516+01
(5 rows)
```

How is it possible that we have date/time combinations (timestamps) if nobody has ever entered these values into the table? It is possible because the `posts` table has been created as follows:

```
forumdb=# \d posts;
 Table "public.posts"
 Column          | Type                       |[...]| Default
-----------------+----------------------------+[...]+-----------------
-------------
 pk              | integer                    |     | [..]
 title           | text                       |     |
 [......]
 created_on      | timestamp with time zone |     |
CURRENT_TIMESTAMP
```

As we can see, the `created_on` field has `CURRENT_TIMESTAMP` as the default value, which means that if no value has been inserted, the current timestamp of the server will be inserted. Suppose now that we want to display the date in a different format – for example, in the Italian format, `created_on: 03-01-2020`.

8. To reach this goal, we have to use another built-in function, the `to_char` function:

```
forumdb=# select pk,title,to_char(created_on,'dd-mm-yyyy') as
created_on
from posts;
 pk | title          | created_on
----+----------------+------------
  4 | Re:my orange   | 03-01-2020
  5 | my tomato      | 03-01-2020
  2 | my orange      | 03-01-2020
  6 | my new orange  | 05-01-2020
  3 | my new apple   | 03-01-2020
(5 rows)
```

As shown here, the `to_char()` function is the inverse of the `to_date()` function. It converts a date into a string using a specific pattern.

Timestamp data types

PostgreSQL can manage dates and times with a time zone and without a time zone. We can store both date and time using the timestamp data type. In PostgreSQL, there is a data type called `timestamp with time zone` to display date and time with a time zone, and a data type called `timestamp without time zone` to store date and time without a time zone.

Let's now create some examples. First of all, let's create a new table:

```
forumdb=# create table new_posts as select pk,title,created_on::timestamp
with time zone as created_on_t, created_on::timestamp without time zone as
create_on_nt from posts;
SELECT 5
```

We have just created a new table called `new_posts` with the following structure:

```
forumdb=# \d new_posts;
 Table "public.new_posts"
 Column        | Type                        | Collation | Nullable |
Default
--------------+-----------------------------+-----------+----------+-------
--
 pk           | integer                     |           |          |
 title        | text                        |           |          |
 created_on_t | timestamp with time zone    |           |          |
 create_on_nt | timestamp without time zone |           |          |
```

This table now has the same values for the `create_on_t` (`timestemp with time zone`) field and for the `created_on_nt` (`timestamp without time zone`) field, as we can see here:

```
forumdb=# select * from new_posts ;
 pk | title          | created_on_t                    | create_on_nt
----+----------------+---------------------------------+---------------------
------
  4 | Re:my orange   | 2020-01-03 18:46:06.436248+01   | 2020-01-03
18:46:06.436248
  5 | my tomato      | 2020-01-03 18:47:39.603937+01   | 2020-01-03
18:47:39.603937
  2 | my orange      | 2020-01-03 18:44:13.266102+01   | 2020-01-03
18:44:13.266102
  6 | my new orange  | 2020-01-05 18:05:10.860354+01   | 2020-01-05
18:05:10.860354
  3 | my new apple   | 2020-01-03 18:44:32.459516+01   | 2020-01-03
18:44:32.459516
(5 rows)
```

Now, let's introduce a PostgreSQL environment variable called the `timezone` variable. This variable tells us the current value of the time zone:

```
forumdb=# show timezone;
 TimeZone
---------------
 Europe/Berlin
(1 row)
```

In this server, the time zone is set to CET; if we want to modify this value only on this session, we have to perform the following query:

```
forumdb=# set timezone='GMT';
SET
```

Now, the time zone is set to GMT:

```
forumdb=# show timezone;
 TimeZone
----------
 GMT
(1 row)
```

Now, if we execute the query that we performed previously again, we will see that the field with the time zone has changed its value:

```
forumdb=# select * from new_posts ;
 pk | title          | created_on_t                   | create_on_nt
----+----------------+--------------------------------+---------------------
------
  4 | Re:my orange   | 2020-01-03 17:46:06.436248+00 | 2020-01-03
18:46:06.436248
  5 | my tomato      | 2020-01-03 17:47:39.603937+00 | 2020-01-03
18:47:39.603937
  2 | my orange      | 2020-01-03 17:44:13.266102+00 | 2020-01-03
18:44:13.266102
  6 | my new orange  | 2020-01-05 17:05:10.860354+00 | 2020-01-05
18:05:10.860354
  3 | my new apple   | 2020-01-03 17:44:32.459516+00 | 2020-01-03
18:44:32.459516
(5 rows)
```

This shows the difference between a timestamp with a time zone and a timestamp without a time zone. For further information on the topic of date and time, please refer to the official documentation at https://www.postgresql.org/docs/12/datatype-datetime.html.

The NoSQL data type

In this section, we will approach the NoSQL data types that are present in PostgreSQL. We will take just a quick look because the NoSQL world is not specific to this book.

PostgreSQL handles the following NoSQL data types:

- hstore
- xml
- json

We will now talk about hstore and json.

The hstore data type

hstore was the first NoSQL data type that was implemented in PostgreSQL. This data type is used for storing key-value pairs in a single value. Before working with the hstore data type, we need to enable the hstore extension on our server:

```
forumdb=# create extension hstore ;
CREATE EXTENSION
```

Let's look at how we can use the hstore data type with an example. Suppose that we want to show all posts with their usernames and with their categories:

```
forumdb=# select p.pk,p.title,u.username,c.title as category
from posts p
inner join users u on p.author=u.pk
left join categories c on p.category=c.pk
order by 1;

 pk | title          | username     | category
----+----------------+--------------+----------
  2 | my orange      | myusername   | orange
  3 | my new apple   | myusername   | apple
  4 | Re:my orange   | scotty       | orange
  5 | my tomato      | scotty       | tomato
  6 | my new orange  | myusername   | orange
(5 rows
```

Suppose now that the table's posts, users, and categories are huge tables and we would like to store all the information about usernames and categories in a single field stored inside the `posts` table. If we could do this, we would no longer need to join three huge tables. In this case, `hstore` can help us:

```
forumdb# select
p.pk,p.title,hstore(ARRAY['username',u.username,'category',c.title]) as
options
from posts p
inner join users u on p.author=u.pk
left join categories c on p.category=c.pk
order by 1;

 pk | title         | options
----+---------------+---------------------------------------------------
  2 | my orange     | "category"=>"orange", "username"=>"myusername"
  3 | my new apple  | "category"=>"apple", "username"=>"myusername"
  4 | Re:my orange  | "category"=>"orange", "username"=>"scotty"
  5 | my tomato     | "category"=>"tomato", "username"=>"scotty"
  6 | my new orange | "category"=>"orange", "username"=>"myusername"
(5 rows)
```

The preceding query first puts in an array the values of the username and category fields, and then transforms them into `hstore`. Now, if we want to store the data in a new table called `posts_options`, we have to perform something like the following:

```
forumdb# create table posts_options as
select p.pk,p.title,hstore(ARRAY['username',u.username,'category',c.title])
as options
from posts p
inner join users u on p.author=u.pk
left join categories c on p.category=c.pk
order by 1;

SELECT 5
```

We now have a new table with the following structure:

```
forumdb=# \d posts_options
 Table "public.posts_options"
 Column  | Type    | Collation | Nullable | Default
---------+---------+-----------+----------+---------
 pk      | integer |           |          |
 title   | text    |           |          |
 options | hstore  |           |          |
```

Next, suppose that we want to search for all the records that have category = 'orange'. We would have to execute the following:

```
forumdb=# select * from posts_options where options->'category' = 'orange';
 pk | title         | options
----+---------------+---------------------------------------------------
  2 | my orange     | "category"=>"orange", "username"=>"myusername"
  4 | Re:my orange  | "category"=>"orange", "username"=>"scotty"
  6 | my new orange | "category"=>"orange", "username"=>"myusername"
(3 rows)
```

Since hstore, as well as the json/jsonb data types, is not a structured data type, we can insert any other key value without defining it first – for example, we can do this:

```
insert into posts_options (pk,title,options) values (7,'my last
post','"enabled"=>"false"') ;
```

The result of the selection on the whole table will be the following:

```
forumdb=# select * from posts_options;
 pk | title         | options
----+---------------+---------------------------------------------------
  2 | my orange     | "category"=>"orange", "username"=>"myusername"
  3 | my new apple  | "category"=>"apple", "username"=>"myusername"
  4 | Re:my orange  | "category"=>"orange", "username"=>"scotty"
  5 | my tomato     | "category"=>"tomato", "username"=>"scotty"
  6 | my new orange | "category"=>"orange", "username"=>"myusername"
  7 | my last post  | "enabled"=>"false"
(6 rows)
```

As we said at the beginning of this section, NoSQL is not the subject of this book, but it is worth briefly going over it. For further information about the stored data type, please refer to the official documentation at https://www.postgresql.org/docs/12/hstore.html.

The JSON data type

In this section, we'll take a brief look at the JSON data type. **JSON** stands for **JavaScript Object Notation**. JSON is an open standard format, and it is formed of key-value pairs. PostgreSQL supports the JSON data type natively. It provides many functions and operators used for manipulating JSON data. PostgreSQL, in addition to the json data type, also supports the jsonb data type. The difference between these two data types is that the first is internally represented as text whereas the second is internally represented in a binary and indexable manner. Let's look at how we can use the json/jsonb data types with an example.

Suppose that we want to show all the posts and tags that we have in our `forumdb` database. Working in a classic relational SQL way, we should write something like the following:

```
forumdb=# select p.pk,p.title,t.tag
from posts p
left join j_posts_tags jpt on p.pk=jpt.post_pk
left join tags t on jpt.tag_pk=t.pk
order by 1;
 pk |      title       |    tag
----+------------------+------------
  2 | my orange        | vegetables
  2 | my orange        | fruits
  3 | my new apple     | fruits
  4 | Re:my orange     |
  5 | my tomato        |
  6 | my new orange    | fruits
(6 rows)
or if we want as result something like:
```

Suppose now that we want to have a result like the following:

pk	title	tag
2	my orange	vegetables,fruits
3	my new apple	fruits
4	Re:my orange	
5	my tomato	
6	my new orange	fruits

In a relational way, we have to aggregate data using the first two fields and perform something like the following:

```
forumdb=# select p.pk,p.title,string_agg(t.tag,',') as tag
from posts p
left join j_posts_tags jpt on p.pk=jpt.post_pk
left join tags t on jpt.tag_pk=t.pk
group by 1,2
order by 1;

 pk | title         | tag
----+---------------+------------------
  2 | my orange     | vegetables,fruits
  3 | my new apple  | fruits
  4 | Re:my orange  |
  5 | my tomato     |
  6 | my new orange | fruits
(5 rows)
```

Now, imagine that we want to generate a simple JSON structure; we would execute the following query:

```
forumdb# select row_to_json(q) as json_data from (
 select p.pk,p.title,string_agg(t.tag,',') as tag
 from posts p
 left join j_posts_tags jpt on p.pk=jpt.post_pk
 left join tags t on jpt.tag_pk=t.pk
group by 1,2 order by 1) Q;
                          json_data
---------------------------------------------------------
 {"pk":2,"title":"my orange","tag":"vegetables,fruits"}
 {"pk":3,"title":"my new apple","tag":"fruits"}
 {"pk":4,"title":"Re:my orange","tag":null}
 {"pk":5,"title":"my tomato","tag":null}
 {"pk":6,"title":"my new orange","tag":"fruits"}
(5 rows)
```

As we can see, with a simple query, it is possible to switch from a classic SQL representation to a NoSQL representation. Now, let's create a new table called `post_json`. This table will have only one `jsonb` field, called `jsondata`:

```
forumdb=# create table post_json (jsondata jsonb);
CREATE TABLE
forumdb=# \d post_json
 Table "public.post_json"
 Column   | Type  | Collation | Nullable | Default
----------+-------+-----------+----------+---------
 jsondata | jsonb |           |          |
```

Now, let's insert some data into the `post_json` table:

```
forumdb# insert into post_json(jsondata)
select row_to_json(q) as json_data from (
  select p.pk,p.title,string_agg(t.tag,',') as tag
  from posts p
  left join j_posts_tags jpt on p.pk=jpt.post_pk
  left join tags t on jpt.tag_pk=t.pk
group by 1,2 order by 1) Q;
INSERT 0 5
```

Now, the `post_json` table has the following records:

```
forumdb=# select jsonb_pretty(jsondata) from post_json;
           jsonb_pretty
--------------------------------
 {                            +
```

```
        "pk": 2,                      +
        "tag": "vegetables,fruits",+
        "title": "my orange"         +
    }
    {                                 +
        "pk": 3,                      +
        "tag": "fruits",              +
        "title": "my new apple"       +
    }
    {                                 +
        "pk": 4,                      +
        "tag": null,                  +
        "title": "Re:my orange"       +
    }
    {                                 +
        "pk": 5,                      +
        "tag": null,                  +
        "title": "my tomato"          +
    }
    {                                 +
        "pk": 6,                      +
        "tag": "fruits",              +
        "title": "my new orange"      +
    }
(5 rows)
```

If we wanted to search for all data that has `tag = "fruits"`, we could use the `@>` jsonb operator. This operator checks whether the left JSON value contains the right JSON path/value entries at the top level; the following query makes this search possible:

```
forumdb=# select jsonb_pretty(jsondata) from post_json where jsondata @>
'{"tag":"fruits"}';
 jsonb_pretty
------------------------------
 {  +
 "pk": 3, +
 "tag": "fruits", +
 "title": "my new apple" +
 }
 {  +
 "pk": 6, +
 "tag": "fruits", +
 "title": "my new orange"+
 }
(2 rows)
```

What we have just written is just a small taste of what can be done through the NoSQL data model. JSON is widely used when working with large tables and when a data structure is needed that minimizes the number of joins to be done during the research phase. A detailed discussion of the NoSQL world is beyond the scope of this book, but we wanted to describe briefly how powerful PostgreSQL is in the approach to unstructured data as well. For more information, please look at the official documentation at `https://www.` `postgresql.org/docs/12/functions-json.html`.

After understanding what data types are and which data types can be used in PostgreSQL, in the next section, we will see how to use data types within functions.

Exploring functions and languages

PostgreSQL is capable of executing server-side code. There are many ways to provide PostgreSQL with the code to be executed. For example, the user can create functions in different programming languages. The main languages supported by PostgreSQL are as follows:

- SQL
- PL/pgSQL
- C

These listed languages are the built-in languages; there are also other languages that PostgreSQL can manage, but before using them, we need to install them on our system. Some of these other supported languages are as follows:

- PL/Python
- PL/Perl
- PL/tcl
- PL/Java

In this section, we'll talk about SQL and PL/pgSQL functions.

Functions

The command structure with which a function is defined is as follows:

```
CREATE FUNCTION function_name(p1 type, p2 type,p3 type, ....., pn type)
 RETURNS type AS
BEGIN
```

```
   -- function logic
END;
LANGUAGE language_name
```

The following steps always apply for any type of function we want to create:

1. Specify the name of the function after the CREATE FUNCTION keywords.
2. Make a list of parameters separated by commas.
3. Specify the return data type after the RETURNS keyword.
4. For the PL/Pgsql language, put some code between the BEGIN and END block.
5. For the PL/Pgsql language, the function has to end with the END keyword followed by a semicolon.
6. Define the language in which the function was written – for example, sql or plpgsql, plperl, plpython, and so on.

This is the basic scheme to which we will refer later in the chapter; this scheme may have small variations in some specific cases.

SQL functions

SQL functions are the easiest way to write functions in PostgreSQL, and we can use any SQL command inside them.

Basic functions

This section will show how to take your first steps into the SQL functions world. For example, the following function makes a sum between two numbers:

```
forumdb=#CREATE OR REPLACE FUNCTION my_sum(x integer, y integer) RETURNS
integer AS $$
  SELECT x + y;
$ LANGUAGE SQL;
CREATE FUNCTION


forumdb=# select my_sum(1,2);
 my_sum
--------
      3
(1 row)
```

As we can see in the preceding example, the code function is placed between $$; we can consider $$ as labels. The function can be called using the SELECT statement without using any FROM clauses. The arguments of a SQL function can be referenced in the function body using either numbers (the old way) or their names (the new way). For example, we could write the same function in this way:

```
CREATE OR REPLACE FUNCTION my_sum(integer, integer) RETURNS integer AS $$
  SELECT $1 + $2;
$$ LANGUAGE SQL;
```

In the preceding function, we can see the old way to reference the parameter inside the function. In the old way, the parameters were referenced positionally, so the value $1 corresponds to the first parameter of the function, $2 to the second, and so on. In the code of the SQL functions, we can use all the SQL commands, including those seen in previous chapters.

SQL functions returning a set of elements

In this section, we will look at how to make a SQL function that returns a result set of a data type. For example, suppose that we want to write a function that takes p_title as a parameter and delete all the records that have title=p_title, as well as returning all the keys of the deleted records. The following function would make this possible:

```
forumdb=# CREATE OR REPLACE FUNCTION delete_posts(p_title text) returns
setof integer as $$
delete from posts where title=p_title returning pk;
$$
LANGUAGE SQL
CREATE FUNCTION
```

This is the situation before we called the delete_posts function:

```
forumdb=# select pk,title from posts order by pk;
 pk | title
----+----------------
  2 | my orange
  3 | my new apple
  4 | Re:my orange
  5 | my tomato
  6 | my new orange
(5 rows)
```

This is the situation after we called the `delete_posts` function:

```
forumdb=# select delete_posts('my tomato');
 delete_posts
--------------
 5
(1 row)

forumdb=# select pk,title from posts order by pk;
 pk | title
----+---------------
 2  | my orange
 3  | my new apple
 4  | Re:my orange
 6  | my new orange
(4 rows)
```

In this function, we've introduced a new kind of data type – the `setof` data type. The `setof` directive simply defines a result set of a data type. For example, the `delete_posts` function is defined to return a set of integers, so its result will be an integer dataset. We can use the `setof` directive with any type of data.

SQL functions returning a table

In the previous section, we saw how to write a function that returns a result set of a single data type; however, it is possible that there will be cases where we need our function to return a result set of multiple fields. For example, let's consider the same function as before, but this time we want the `pk`, `title` pair to be returned as a result, so our function becomes the following:

```
create or replace function delete_posts (p_title text) returns table
(ret_key integer,ret_title text) AS $$
delete from posts where title=p_title returning pk,title;
$$
language SQL;
```

The only difference between this and the previous function is that now the function returns a table type; inside the table type, we have to specify the name and the type of the fields. As we have seen before, this is the situation before calling the function:

```
forumdb=# select pk,title from posts order by pk;
 pk | title
----+---------------
 2  | my orange
 3  | my new apple
```

```
4   | Re:my orange
5   | my tomato
6   | my new orange
(5 rows)
```

This is the correct way to call the function:

```
forumdb=# select * from delete_posts('my tomato');
 ret_key | ret_title
---------+-----------
    5    | my tomato
(1 row)
```

This is the situation after calling the function:

```
forumdb=# select pk,title from posts order by pk;
 pk | title
----+---------------
 2  | my orange
 3  | my new apple
 4  | Re:my orange
 6  | my new orange
(4 rows)
```

The functions that return a table can be treated as real tables, in the sense that we can use them with the `ofin`, `exists`, `join`, and so on options.

Polymorphic SQL functions

In this section, we will briefly talk about polymorphic SQL functions.

Polymorphic functions are useful for DBAs when we need to write a function that has to work with different types of data. To better understand polymorphic functions, let's start with an example. Suppose we want to recreate something that looks like the Oracle NVL function – in other words, we want to create a function that accepts two parameters and replaces the first parameter with the second one if the first parameter is null. The problem is that we want to write a single function that is valid for all types of data (integer, real, text, and so on).

The following function makes this possible:

```
create or replace function nvl ( anyelement,anyelement) returns anyelement
as $$
select coalesce($1,$2);
$$
language SQL;
```

This is how to call it:

```
forumdb=# select nvl(NULL::int,1);
 nvl
-----
 1
(1 row)

forumdb=# select nvl(''::text,'n'::text);
 nvl
-----

(1 row)

forumdb=# select nvl('a'::text,'n'::text);
 nvl
-----
 a
(1 row)
```

For further information, see the official documentation at `https://www.postgresql.org/docs/12/extend-type-system.html`.

PL/pgSQL functions

In this section, we'll talk about the PL/pgsql language. The PL/pgSQL language is the default built-in procedural language for PostgreSQL. As described in the official documentation, the design goals with PL/pgSQL were to create a loadable procedural language that can do the following:

- Can be used to create functions and trigger procedures (we'll talk about triggers in the next chapter).
- Add new control structures.
- Add new data types to the SQL language.

It is very similar to Oracle PL/SQL and it supports the following:

- Variable declarations
- Expressions
- Control structures as conditional structures or loop structures
- Cursors

First overview

As we saw at the beginning of the *SQL functions* section, the prototype for writing functions in PostgreSQL is as follows:

```
CREATE FUNCTION function_name(p1 type, p2 type,p3 type, ....., pn type)
 RETURNS type AS
BEGIN
 -- function logic
END;
LANGUAGE language_name
```

Now, suppose that we want to recreate the `my_sum` function using the PL/pgsql language:

```
forumdb# CREATE OR REPLACE FUNCTION my_sum(x integer, y integer) RETURNS
integer AS
$BODY$
DECLARE
 ret integer;
BEGIN
 ret := x + y;
 return ret;
END;
$BODY$
language 'plpgsql';

forumdb=# select my_sum(2,3);
 my_sum
--------
      5
```

The preceding query provides the same results as the query seen at the beginning of the chapter. Now, let's examine it in more detail:

1. The following is the function header; here, you define the name of the function, the input parameters, and the return value:

   ```
   CREATE OR REPLACE FUNCTION my_sum(x integer, y integer) RETURNS
   integer AS
   ```

2. The following is a label indicating the beginning of the code. We can put any string inside the $$ characters, the important thing is that the same label is present at the end of the function:

   ```
   $BODY$
   ```

3. In the following section, we can define our variables; it is important that each declaration or statement ends with a semicolon:

```
DECLARE
   ret integer;
```

4. With the BEGIN statement, we tell PostgreSQL that we want to start to write our logic:

```
BEGIN
   ret := x + y;
   return ret;
```

 Caveat: do not write a semicolon after BEGIN – it's not correct and it will generate a syntax error.

5. Between the BEGIN statement and the END statement, we can put our own code:

```
END;
```

6. The END instruction indicates that our code has ended:

```
$BODY$
```

7. This label closes the first label and at last, the language statement specifies PostgreSQL, in which language the function is written:

```
language 'plpgsql';
```

Declaring function parameters

After learning about how to write a simple PL/pgsql function, let's go into a little more detail about the single aspects seen in the preceding section. Let's start with the declaration of the parameters. In the next two examples, we'll see how to define, in two different ways, the my_sum function that we have seen before.

The first example is as follows:

```
CREATE OR REPLACE FUNCTION my_sum(integer, integer) RETURNS integer AS
$BODY$
DECLARE
 x alias for $1;
 y alias for $2;
 ret integer;
```

```
BEGIN
 ret := x + y;
 return ret;
END;
$BODY$
language 'plpgsql';
```

The second example is as follows:

```
CREATE OR REPLACE FUNCTION my_sum(integer, integer) RETURNS integer AS
$BODY$
DECLARE
 ret integer;
BEGIN
 ret := $1 + $2;
 return ret;
END;
$BODY$
language 'plpgsql';
```

In example 1, we used `alias`; the syntax of `alias` is, in general, the following:

```
newname ALIAS FOR oldname;
```

In our specific case, we used the positional variable `$1` as the `oldname` value. In the second example, we used the positional approach exactly as we did in the case of SQL functions.

IN/OUT parameters

In the preceding example, we used the `RETURNS` clause in the first row of the function definition; however, there is another way to reach the same goal. In PL/pgSQL, we can define all parameters as input parameters, output parameters, or input/output parameters. For example, say we write the following:

```
CREATE OR REPLACE FUNCTION my_sum_3_params(IN x integer,IN y integer, OUT z
integer) AS
$BODY$
BEGIN
 z := x+y;
END;
$BODY$
language 'plpgsql';
```

We have defined a new function called `my_sum_3_params`, which accepts two input parameters (x and y) and has an output of parameter z. As there are two input parameters, the function will be called with only two parameters, exactly as in the last function:

```
forumdb=# select my_sum_3_params(2,3);
 my_sum_3_params
-----------------
       5
(1 row)
```

With this kind of parameter definition, we can have functions that have multiple variables as a result. For example, if we want a function that, given two integer values, computes their sum and their product, we can write something like this:

```
CREATE OR REPLACE FUNCTION my_sum_mul(IN x integer,IN y integer,OUT w
integer, OUT z integer) AS
$BODY$
BEGIN
  z := x+y;
  w := x*y;
END;
$BODY$
language 'plpgsql';
```

The strange thing is that if we invoke the function as we did before, we will have the following result:

```
forumdb=# select my_sum_mul(2,3);
 my_sum_mul
------------
   (6,5)
(1 row)
```

This result seems to be a little bit strange because the result is not a scalar value but is a record, which is a custom type. To cause the output to be separated as columns, we have to use the following syntax:

```
forumdb=# select * from my_sum_mul(2,3);
 w | z
---+---
 6 | 5
(1 row)
```

We can use the result of the function exactly as if it were a result of a table and write, for example, the following:

```
forumdb=# select * from my_sum_mul(2,3) where w=6;
 w | z
---+---
 6 | 5
(1 row)
```

We can define the parameters as follows:

- `IN`: Input parameters (if omitted, this is the default option)
- `OUT`: Output parameters
- `INOUT`: Input/output parameters

Function volatility categories

In PostgreSQL, each function can be defined as `VOLATILE`, `STABLE`, or `IMMUTABLE`. If we do not specify anything, the default value is `VOLATILE`. The difference between these three possible definitions is well described in the official documentation (`https://www.postgresql.org/docs/12/xfunc-volatility.html`):

*A **VOLATILE** function can do anything, including modifying the database. It can return different results on successive calls with the same arguments. The optimizer makes no assumptions about the behavior of such functions. A query using a volatile function will reevaluate the function at every row where its value is needed. If a function is marked as VOLATILE, it can return different results if we call it multiple times using the same input parameters.*

*A **STABLE** function cannot modify the database and is guaranteed to return the same results given the same arguments for all rows within a single statement. This category allows the optimizer to optimize multiple calls of the function to a single call. In particular, it is safe to use an expression containing such a function in an index scan condition. If a function is marked as STABLE the function will return the same result given the same parameters within the same transaction.*

*An **IMMUTABLE** function cannot modify the database and is guaranteed to return the same results given the same arguments forever. This category allows the optimizer to pre-evaluate the function when a query calls it with constant arguments.*

In the following pages of this chapter, we will only be focusing on examples of volatile functions; however, here we will briefly look at one example of a stable function and one example of an immutable function:

1. Let's start with a stable function – for example, the `now()` function is a stable function. The `now()` function returns the current date and time that we have at the beginning of the transaction, as we can see here:

```
forumdb=# begin ;
BEGIN

forumdb=# select now();
now
------------------------------
2020-02-19 16:33:35.322562+01
(1 row)

forumdb=# select now();
now
------------------------------
2020-02-19 16:33:35.322562+01
(1 row)

forumdb=# commit;
COMMIT

forumdb=# begin ;
BEGIN
forumdb=# select now();
now
------------------------------
2020-02-19 16:33:51.394306+01
(1 row)
forumdb=# commit ;
COMMIT
```

2. Now, let's look at an immutable function – for example, the `lower(string_expression)` function. The `lower` function accepts a string and converts it into a lowercase format. As we can see, if the input parameters are the same, the `lower` function always returns the same result, even if it is performed in different transactions:

```
forumdb=# begin;
BEGIN

forumdb=# select now();
```

```
 now
-------------------------------
 2020-02-19 16:43:40.109944+01
(1 row)

forumdb=# select lower('MICKY MOUSE');
 lower
-------------
 micky mouse
(1 row)

forumdb=# commit ;
COMMIT

forumdb=# begin;
BEGIN

forumdb=# select now();
 now
-------------------------------
 2020-02-19 16:43:52.797172+01
(1 row)

forumdb=# select lower('MICKY MOUSE');
 lower
-------------
 micky mouse
(1 row)

forumdb=# commit;
COMMIT
```

Control structure

PL/pgSQL has the ability to manage control structures such as the following:

- Conditional statements
- Loop statements
- Exception handler statements

Conditional statements

The PL/pgSQL language can manage IF-type conditional statements and CASE-type conditional statements.

IF statements

In PL/pgSQL, the syntax of an IF statement is as follows:

```
IF boolean-expression THEN
 statements
[ ELSIF boolean-expression THEN
 statements
[ ELSIF boolean-expression THEN
 statements
 ...
]
]
[ ELSE
 statements ]
END IF;
```

For example, say we want to write a function that, when given the two input values x and y, returns the following:

- 'first parameter is higher than second parameter if x > y'
- 'second paramater is higher than first parameter if x < y'
- 'the 2 parameters are equals if x = y'

We have to write the following function:

```
CREATE OR REPLACE FUNCTION my_check(x integer default 0, y integer default
0) RETURNS text AS
$BODY$
BEGIN
 IF x > y THEN
 return 'first parameter is higher than second parameter';
 ELSIF x < y THEN
 return 'second paramater is higher than first parameter';
 ELSE
 return 'the 2 parameters are equals';
 END IF;
END;
$BODY$
language 'plpgsql';
```

In this example, we have seen the IF construct in its largest form: IF [...] THEN[...] ELSIF [...] ELSE[...] ENDIF;.

However, its shorter form also exists, as follows:

- `IF [...] THEN[...] ELSE[...] ENDIF;`
- `IF [...] THEN[...] ENDIF;`

Some examples of the results provided by the previously defined function are as follows:

```
forumdb=# select my_check(1,2);
 my_check
-------------------------------------------------
 second paramater is higher than first parameter
(1 row)

forumdb=# select my_check(2,1);
 my_check
-------------------------------------------------
 first parameter is higher than second parameter
(1 row)

forumdb=# select my_check(1,1);
 my_check
---------------------------
 the 2 parameters are equals
(1 row)
```

CASE statements

In PL/pgSQL, it is also possible to use the CASE statement. The CASE statement can have the following two syntaxes.

The following is a simple CASE statement:

```
CASE search-expression
  WHEN expression [, expression [ ... ]] THEN
  statements
  [ WHEN expression [, expression [ ... ]] THEN
  statements
  ... ]
  [ ELSE
  statements ]
END CASE;
```

The following is a searched CASE **statement:**

```
CASE
 WHEN boolean-expression THEN
 statements
 [ WHEN boolean-expression THEN
 statements
 ... ]
 [ ELSE
 statements ]
END CASE;
```

Now, we will perform the following operations:

- We will use the first one, the simple CASE syntax, if we have to make a choice from a list of values.
- We will use the second one when we have to choose from a range of values.

Let's start with the first syntax:

```
CREATE OR REPLACE FUNCTION my_check_value(x integer default 0) RETURNS text
AS
$BODY$
BEGIN
 CASE x
 WHEN 1 THEN return 'value = 1';
 WHEN 2 THEN return 'value = 2';
 ELSE return 'value >= 3 ';
 END CASE;
END;
$BODY$
language 'plpgsql';
```

The preceding my_check_value function returns the following:

- value = 1 if x = 1
- value =2 if x = 2
- value >= 3 if x >= 3

We can see this to be true here:

```
forumdb=# select my_check_value(1);
 my_check_value
----------------
 value = 1
(1 row)
```

```
forumdb=# select my_check_value(2);
 my_check_value
----------------
 value = 2
(1 row)

forumdb=# select my_check_value(3);
 my_check_value
----------------
 value >= 3
(1 row)
```

Now, let's see an example of the searched CASE syntax:

```
CREATE OR REPLACE FUNCTION my_check_case(x integer default 0, y integer
default 0) RETURNS text AS
 $BODY$
 BEGIN
   CASE
     WHEN x > y THEN return 'first parameter is higher than second
parameter';
     WHEN x < y THEN return 'second paramater is higher than first
parameter';
 ELSE return 'the 2 parameters are equals';
 END CASE;
 END;
 $BODY$
 language 'plpgsql';
```

The my_check_case function returns the same data as the my_check function that we
wrote before:

```
forumdb=# select my_check_case(2,1);
 my_check_case
-----------------------------------------------
 first parameter is higher than second parameter
(1 row)

forumdb=# select my_check_case(1,2);
 my_check_case
-----------------------------------------------
 second paramater is higher than first parameter
(1 row)

forumdb=# select my_check_case(1,1);
 my_check_case
-----------------------------
 the 2 parameters are equals
```

```
(1 row)

forumdb=# select my_check_case();
 my_check_case
---------------------------
 the 2 parameters are equals
(1 row)
```

Loop statements

PL/pgSQL can handle loops in many ways. We will look at some examples of how to make a loop next. For further details, we suggest referring to the official documentation at `https://www.postgresql.org/docs/12/plpgsql.html`. What makes PL/pgsql particularly useful is the fact that it allows us to process data from queries through procedural language. We are going to see now how this is possible.

Suppose that we want to build a PL/pgSQL function that, when given an integer as parameters, returns a result set of a composite data type. The composite data type that we want it to return is as follows:

ID	pk field	Integer data type
TITLE	Title field	text data type
RECORD_DATA	Title field + content field	hstore data type

The right way to build a composite data type is as follows:

```
create type my_ret_type as (
  id integer,
  title text,
  record_data hstore
);
```

The preceding statement creates a new data type, a composite data type, which is composed of an integer data type + a `text` data type + an `hstore` data type. Now, if we want to write a function that returns a result set of the `my_ret_type` data type, our first attempt might be as follows:

```
CREATE OR REPLACE FUNCTION my_first_fun (p_id integer) returns setof
my_ret_type as
$$
DECLARE
 rw posts%ROWTYPE; -- declare a rowtype;
 ret my_ret_type;
BEGIN
    for rw in select * from posts where pk=p_id loop
```

```
        ret.id := rw.pk;
        ret.title := rw.title;
        ret.record_data := hstore(ARRAY['title',rw.title,'Title and Content'
                            ,format('%s %s',rw.title,rw.content)]);
      return next ret;
      end loop;
  return;
END;
$$
language 'plpgsql';
```

As we can see, many things are concentrated in these few lines of PL/pgSQL code:

1. `rw posts%ROWTYPE`: With this statement, the `rw` variable is defined as a container of a single row of the `posts` table.

2. `for rw in select * from posts where pk=p_id loop`: With this statement, we cycle within the result of the selection, assigning the value returned by the `select` command each time to the `rw` variable.The next three steps assign the values to the `ret` variable.

3. `return next ret;`: This statement returns the value of the `ret` variable and goes to the next record of the `for` cycle.

4. `end loop;`: This statement tells PostgreSQL that the `for` cycle ends here.

5. `return;`: This is the return instruction of the function.

An important thing to remember is that the PL/pgSQL language is inside the PostgreSQL transaction system. This means that the functions are executed atomically and that the function returns the results not at the execution of the RETURN NEXT command but at the execution of the RETURN command placed at the end of the function. This may mean that, for very large datasets, the PL/pgsql functions can take a long time before returning results.

The record type

In an example that we used previously, we introduced the %ROWTYPE data type. In the PL/pgSQL language, it is possible to generalize this concept. There is a data type called record that generalizes the concept of %ROWTYPE. For example, we can rewrite the my_first_fun in the following way:

```
CREATE OR REPLACE FUNCTION my_second_fun (p_id integer) returns setof
my_ret_type as
$$
DECLARE
```

```
    rw record; -- declare a record variable
    ret my_ret_type;
BEGIN
    for rw in select * from posts where pk=p_id loop
    ret.id := rw.pk;
    ret.title := rw.title;
    ret.record_data := hstore(ARRAY['title',rw.title
                    ,'Title and Content',format('%s
%s',rw.title,rw.content)]);
    return next ret;
 end loop;
 return;
END;
$$
language 'plpgsql';
```

The only difference between `my_first_fun` and `my_second_fun` is in this definition:

```
    rw record; -- declare a record variable
```

This time, the `rw` variable is defined as a record data type. This means that the `rw` variable is an object that can be associated with any records of any table. The result of the two functions, `my_first_fun` and `my_second_fun`, is the same:

```
forumdb=# select * from my_first_fun(3);
-[ RECORD 1 ]------------------------------------------------------------
--------------------
id | 3
title | my new apple
record_data | "title"=>"my new apple", "Title and Content"=>"my new apple
my apple is the best orange in the world"

forumdb=# select * from my_second_fun(3);
-[ RECORD 1 ]------------------------------------------------------------
--------------------
id | 3
title | my new apple
record_data | "title"=>"my new apple", "Title and Content"=>"my new apple
my apple is the best orange in the world"
```

Exception handling statements

PL/pgSQL can also handle exceptions. The BEGIN...END block of a function allows the EXCEPTION option, which works as a catch for exceptions. For example, if we write a function to divide two numbers, we could have a problem with a division by 0:

```
CREATE OR REPLACE FUNCTION my_first_except (x real, y real ) returns real
as
$$
DECLARE
 ret real;
BEGIN
 ret := x / y;
 return ret;
END;
$$
language 'plpgsql';
```

This function works well if $y <> 0$, as we can see here:

```
forumdb=# select my_first_except(4,2);
 my_first_except
-----------------
 2
(1 row)
```

However, if y assumes a zero value, we have a problem:

```
forumdb=# select my_first_except(4,0);
ERROR: division by zero
CONTEXT: PL/pgSQL function my_first_except(real,real) line 5 at assignment
```

To solve this problem, we have to handle the exception. To do this, we have to rewrite our function in the following way:

```
CREATE OR REPLACE FUNCTION my_second_except (x real, y real ) returns real
as
$$
DECLARE
  ret real;
BEGIN
  ret := x / y;
  return ret;
EXCEPTION
  WHEN division_by_zero THEN
      RAISE INFO 'DIVISION BY ZERO';
      RAISE INFO 'Error % %', SQLSTATE, SQLERRM;
      RETURN 0;
```

```
END;
$$
language 'plpgsql' ;
```

The `SQLSTATE` and `SQLERRM` variables contain the status and message associated with the generated error. Now, if we execute the second function, we no longer get an error from PostgreSQL:

```
forumdb=# select my_second_except(4,0);
INFO: DIVISION BY ZERO
INFO: Error 22012 division by zero
 my_second_except
------------------
                0
(1 row)
```

The list of errors that PostgreSQL can manage is available at `https://www.postgresql.org/docs/12/errcodes-appendix.html`.

Summary

In this chapter, we introduced the world of server-side programming. The topic is so vast that there are specific books dedicated just to server-side programming. We have tried to give you a better understanding of the main concepts of server-side programming. We talked about the main data types managed by PostgreSQL, then we saw how it is possible to create new ones using composite data types. We also mentioned SQL functions and polymorphic functions, and finally, provided some information about the PL/pgSQL language.

In the next chapter, we will use these concepts to introduce event management in PostgreSQL. We will talk about event management through the use of triggers and the functions associated with them.

References

- Postgresql 12 – data types official documentation: `https://www.postgresql.org/docs/12/datatype.html`
- Postgresql 12 – SQL functions official documentation: `https://www.postgresql.org/docs/12/xfunc-sql.html`
- Postgresql 12 – PL/PGSQL official documentation: `https://www.postgresql.org/docs/12/plpgsql.html`

8
Triggers and Rules

In the previous chapter, we talked about server-side programming. In this chapter, we will use the concepts introduced in the previous chapter to manage the programming of events in PostgreSQL. The first thing we need to address is what an event in PostgreSQL actually is. In PostgreSQL, possible events are given by the `SELECT`/`INSERT`/`UPDATE`, and `DELETE` statements. There are also events related to **data definition language** (**DDL**) operations; we will talk about those events in Chapter 17, *Event Triggers*.

In PostgreSQL, there are two ways to handle events:

- Rules
- Triggers

In this chapter, we will explore both of these ways and address when it is more appropriate to use one of them rather than the other. As a starting point, we can generally say that rules are usually simple event handlers, while triggers are more complex event handlers. Triggers and rules are often used to update accumulators and to modify or delete records that belong to different tables than the one in which we are modifying records. They are very powerful tools that allow us to perform operations in tables other than the one in which we are modifying the data. Triggers and rules will also be used in the next chapter when we talk about partitioning. This is because, in PostgreSQL, there is still a partitioning model based on triggers and rules.

In this chapter, we will talk about the following:

- Exploring rules in PostgreSQL
- Managing triggers in PostgreSQL
- Event triggers

Exploring rules in PostgreSQL

As mentioned earlier, rules are simple event handlers. At the user level, it is possible to manage all the events that perform write operations, which are as follows:

- INSERT
- DELETE
- UPDATE

The fundamental concept behind rules is to modify the flow of an event. If we are given an event, what we can do when certain conditions occur is as follows:

- Do nothing and then undo the action of that event.
- Trigger another event instead of the default one.
- Trigger another event in conjunction with the default.

So, given a write operation, for example, an INSERT operation, we can perform one of these three actions:

- Cancel the operation.
- Perform another operation instead of the INSERT.
- Execute the INSERT and simultaneously perform another operation.

Understanding the OLD and NEW variables

Before we start working with rules and then with triggers, we need to understand the concept of the OLD and NEW variables.

The OLD and NEW variables represent the state of the row in the table before or after the event. OLD and NEW values are cursors that represent the whole record. To better understand this, consider an UPDATE operation; in this case, the OLD variable contains the value of the record already present in the table, while the NEW variable contains the value that the record of the table will have after the UPDATE operation.

For example, we can consider the `tags` table with the following records:

```
forumdb=# select * from tags;
 pk | tag        | parent
----+------------+--------
 1  | fruits     |
 2  | vegetables |
 3  | apple      | 1
(3 rows)
```

Suppose we want to modify the tag with `pk=3` from `'apple'` to `'orange'` with this UPDATE operation:

```
forumdb=# update tags set tag='orange' where pk=3;
UPDATE 1
```

The OLD variable will have these values:

3	apple	1

The NEW variable will have these values:

3	orange	1

It is quite logical that for certain operations both the OLD variable and the NEW variable may exist, but for other operations, only one of them may exist. Here, we can see this expressed in more detail:

Operation/Variable	NEW	OLD
INSERT	present	absent
DELETE	absent	present
UPDATE	present	present

Now that everything is clearer, we can start working with rules.

Rules on INSERT

Let's start by introducing the rules syntax:

```
CREATE [ OR REPLACE ] RULE name AS ON event
    TO table [ WHERE condition ]
    DO [ ALSO | INSTEAD ] { NOTHING | command | ( command ; command ... ) }
```

As we can see, the rule definition is extremely simple. There are three options that we can have when we decide to use a rule:

1. The `ALSO` option
2. The `INSTEAD` option
3. The `INSTEAD NOTHING` option

The ALSO option

Suppose that, from the `tags` table, we want to copy all records with the field `tag` value starting with the letter `a` in the `a_tag` table:

1. First of all, let's create a new table called `a_tags`:

   ```
   create table a_tags (
       pk integer not null primary key,
       tag text,
       parent integer);
   ```

2. Then let's create the new rule as follows:

   ```
   create or replace rule r_tags1
       as on INSERT to tags
       where NEW.tag ilike 'a%' DO ALSO
       insert into a_tags(pk,tag,parent)values
   (NEW.pk,NEW.tag,NEW.parent);
   ```

   In the rule we have just defined, we simply told PostgreSQL that every time a record is inserted with a `tag` value that starts with the letter "a," as well as being inserted into the `tags` table, it must also be inserted into the `a_tags` table.

3. Now we perform the following query:

   ```
   forumdb=# insert into tags (tag) values ('apple');
   INSERT 0 1
   ```

4. Then we check the records in the `tags` table and the `a_tags` records. We will check in the parent table:

```
forumdb=# select * from tags;
 pk | tag        | parent
----+------------+--------
 1  | fruits     |
 2  | vegetables |
 3  | orange     | 1
 11 | apple      |
(4 rows)
```

In the child table, we will see the following:

```
forumdb=# select * from a_tags;
 pk | tag | parent
----+-------+--------
 12 | apple |
(1 row)
```

The record is present in both tables. A question worth asking is whether the rules are executed before the event or after the event. For example, is the newly created rule executed before `INSERT` or after `INSERT`? The answer is that rules in PostgreSQL are always executed before the event.

The INSTEAD OF option

Suppose now that we want to move all records with the field `tag` starting with the letter `b` in the `b_tags` table:

1. First of all, let's create a new table called `b_tags`:

```
create table b_tags (
  pk integer not null primary key ,
  tag text,
  parent integer);
```

2. Then let's create the new rule:

```
create or replace rule r_tags2
    as on INSERT to tags
    where NEW.tag ilike 'b%'
    DO INSTEAD insert into b_tags(pk,tag,parent)values
(NEW.pk,NEW.tag,NEW.parent);
```

This time, in the rule, we simply told PostgreSQL that every time a record is inserted with a `tag` value that starts with the letter "b," it must be moved into the `b_tags` table.

3. Now let's perform this query:

```
forumdb=# insert into tags (tag) values ('banana');
INSERT 0 0
```

Already from the answer, `INSERT 0 0`, we can guess that nothing has been inserted into the `a_tags` table.

4. Now, we will perform this statement:

```
forumdb=# select * from tags;
 pk | tag        | parent
----+------------+--------
 1  | fruits     |
 2  | vegetables |
 3  | orange     | 1
 11 | apple      |
(4 rows)
```

5. As we can see in the preceding snippet, the value `banana` does not appear in the parent table, and in the child table, we will have the following:

```
forumdb=# select * from b_tags ;
 pk | tag    | parent
----+--------+--------
 13 | banana |
(1 row)
```

The rule that we defined made sure that the record was not inserted in the `tags` table but was inserted in the `b_tags` table.

6. As the last example of the `INSERT` rule, suppose we want nothing to be inserted every time a record is inserted with the `tag` field starting with the letter "c." As we have done before, let's perform the rule:

```
create or replace rule r_tags3
    as on INSERT to tags
    where NEW.tag ilike 'c%'
    DO INSTEAD NOTHING;
```

7. This time, we've said to PostgreSQL that every time the `tags` table receives a record with the field tag starting with the letter "c," this record should not be considered. Let's try what we've said:

```
forumdb=# insert into tags (tag) values ('cedro');
INSERT 0 0
```

8. Even now, we have `INSERT 0 0` as the answer from the server, and we can check that the record has not been inserted in any table:

```
forumdb=# select pk,tag,parent,'tags' as tablename
from tags
union all
select pk,tag,parent,'a_tags' as tablename
from a_tags
union all
select pk,tag,parent,'b_tags' as tablename
from b_tags
order by tablename, tag;

 pk | tag        | parent | tablename
----+------------+--------+-----------
 12 | apple      |        | a_tags
 13 | banana     |        | b_tags
 11 | apple      |        | tags
  1 | fruits     |        | tags
  3 | orange     | 1      | tags
  2 | vegetables |        | tags
(6 rows)
```

As we can see, the record does not appear in any table. In the preceding query, we used `UNION ALL`. `UNION ALL` concatenates the results of the three queries. The important thing is that the field types must be compatible with each other.

Rules on DELETE / UPDATE

In the previous section, we looked at how to use rules on `INSERT` events. In this section, we will see how to use rules on `DELETE` and `UPDATE` events. We will now look at a complete example of how to use the rules, starting from the concepts described above.

The goal we want to reach is described in the following steps:

1. Create a table called `new_tags` equal to the `tags` table; this table will help us to have a clean environment where we can do our tests.
2. Create two tables: a table called `new_a_tags` for a copy of all records with the tags that start with the letter "a" and a table called `new_b_tags` for a copy of all records with the tags that start with the letter "b."
3. Create all the `INSERT`/`DELETE`/`UPDATE` rules that make everything work.

Let's begin.

Creating the new_tags table

The first step is to create a new `new_tags` table. We will create this table based on the existing `tags` table:

```
forumdb=# create table new_tags as select * from tags limit 0;
SELECT 0
forumdb=# \d new_tags
              Table "public.new_tags"
 Column | Type    | Collation | Nullable | Default
--------+---------+-----------+----------+---------
 pk     | integer |           |          |
 tag    | text    |           |          |
 parent | integer |           |          |
```

The preceding statement copies the structure of the fields of the `tags` table into the `new_tags` table but does not copy the constraints or any indices. Now we have to create the primary key constraint on the new table:

```
forumdb=# alter table new_tags alter pk set not null ;
ALTER TABLE
forumdb=# alter table new_tags add constraint new_tags_pk primary key (pk);
ALTER TABLE
forumdb=# \d new_tags
              Table "public.new_tags"
 Column | Type    | Collation | Nullable | Default
--------+---------+-----------+----------+---------
 pk     | integer |           | not null |
 tag    | text    |           |          |
 parent | integer |           |          |
Indexes:
    "new_tags_pk" PRIMARY KEY, btree (pk)
```

With this, step 1 is complete.

Creating two tables

In a similar way to what we just did, let's create `new_a_tags` and `new_b_tags` tables. For the `new_a_tags` table, we will have the following:

```
forumdb=# create table new_a_tags as select * from a_tags limit 0;
SELECT 0
forumdb=# alter table new_a_tags alter pk set not null ;
ALTER TABLE
forumdb=# alter table new_a_tags add constraint new_b_tags_pk primary key
(pk);
ALTER TABLE
forumdb=# \d new_a_tags
 Table "public.new_a_tags"
 Column | Type    | Collation | Nullable | Default
--------+---------+-----------+----------+---------
 pk     | integer |           | not null |
 tag    | text    |           |          |
 parent | integer |           |          |
Indexes:
  "new_b_tags_pk" PRIMARY KEY, btree (pk)
```

In the same way, we will create the `new_b_tags` table:

```
forumdb=# create table new_b_tags as select * from a_tags limit 0;
SELECT 0
forumdb=# alter table new_b_tags alter pk set not null ;
ALTER TABLE
forumdb=# alter table new_b_tags add constraint new_a_tags_pk primary key
(pk);
ALTER TABLE
forumdb=# \d new_b_tags
 Table "public.new_b_tags"
 Column | Type    | Collation | Nullable | Default
--------+---------+-----------+----------+---------
 pk     | integer |           | not null |
 tag    | text    |           |          |
 parent | integer |           |          |
Indexes:
  "new_a_tags_pk" PRIMARY KEY, btree (pk)
```

Step 2 is now complete and we have everything we need to start our complete example.

Managing rules on INSERT, DELETE, and UPDATE events

The goal we want to achieve is shown in the following figure:

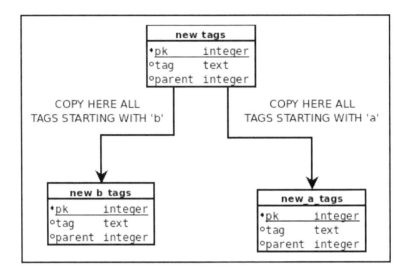

We want all tags starting with the letter "a" to be stored in the new_tags table and also copied to the new_a_tags table, and we want the same for tags that begin with the letter "b."

We have to manage rules for INSERT, DELETE, and UPDATE events in the following ways:

- INSERT rules must recognize all tags starting with the letters "a" or "b" and copy those records into their respective tables – new_a_tags and new_b_tags.
- DELETE rules must recognize all the tags starting with the letters "a" or "b" and delete those records in the respective tables – new_a_tags and new_b_tags.
- Update rules must recognize all the tags that begin with the letters "a" or "b" and, if a record changes its tag, the rule must check whether the record should be copied or deleted in the new_a_tags and new_b_tags tables.

INSERT rules

Let's start by creating two INSERT rules:

```
forumdb=# create or replace rule r_new_tags_insert_a as on INSERT to
new_tags where NEW.tag ilike 'a%' DO ALSO insert into
```

```
new_a_tags(pk,tag,parent)values (NEW.pk,NEW.tag,NEW.parent);
CREATE RULE

forumdb=# create or replace rule r_new_tags_insert_b as on INSERT to
new_tags where NEW.tag ilike 'b%' DO ALSO insert into
new_b_tags(pk,tag,parent)values (NEW.pk,NEW.tag,NEW.parent);
CREATE RULE
```

As we can see, the `new_tags` table now has two new rules:

```
forumdb=# \d new_tags;
 Table "public.new_tags"
 Column | Type    | Collation | Nullable | Default
--------+---------+-----------+----------+---------
 pk     | integer |           | not null |
 tag    | text    |           |          |
 parent | integer |           |          |
Indexes:
 "new_tags_pk" PRIMARY KEY, btree (pk)
Rules:
 r_new_tags_insert_a AS
 ON INSERT TO new_tags
 WHERE new.tag ~~* 'a%'::text DO INSERT INTO new_a_tags (pk, tag, parent)
 VALUES (new.pk, new.tag, new.parent)
 r_new_tags_insert_b AS
 ON INSERT TO new_tags
 WHERE new.tag ~~* 'b%'::text DO INSERT INTO new_b_tags (pk, tag, parent)
 VALUES (new.pk, new.tag, new.parent)
```

To check whether the rules work, let's insert some data:

```
forumdb=# insert into new_tags values(1,'fruits',NULL);
INSERT 0 1
forumdb=# insert into new_tags values(2,'apple',1);
INSERT 0 1
forumdb=# insert into new_tags values(3,'orange',1);
INSERT 0 1
forumdb=# insert into new_tags values(4,'banana',1);
INSERT 0 1
```

Then let's check the parent table:

```
forumdb=# select * from new_tags ;
 pk | tag    | parent
----+--------+--------
 1  | fruits |
 2  | apple  | 1
 3  | orange | 1
```

```
  4  | banana | 1
(4 rows)
```

Now let's see what is in the `table_a` child table:

```
forumdb=# select * from new_a_tags ;
 pk | tag   | parent
----+-------+--------
  2 | apple | 1
(1 row)
```

And what's in the `table_b` child table:

```
forumdb=# select * from new_b_tags ;
 pk | tag    | parent
----+--------+--------
  4 | banana | 1
(1 row)
```

We can see that the two rules work.

DELETE rules

Now let's create the `DELETE` rules. We need rules that, if a record is deleted from the `new_tags` table and it begins with the letter "a" or with the letter "b," its copy in the `new_a_tags` and `new_b_tags` table must also be deleted. For all the records that start with the letter "a," we need this rule:

```
create or replace rule r_new_tags_delete_a as on delete to new_tags where
OLD.tag ilike 'a%' DO ALSO delete from new_a_tags where pk=OLD.pk;
```

Similarly, we need this rule for records beginning with the letter "b":

```
create or replace rule r_new_tags_delete_b as on delete to new_tags where
OLD.tag ilike 'b%' DO ALSO delete from new_b_tags where pk=OLD.pk;
```

The current situation of the `new_tags` table is as follows:

```
forumdb=# \d new_tags
 Table "public.new_tags"
 Column | Type    | Collation | Nullable | Default
--------+---------+-----------+----------+---------
 pk     | integer |           | not null |
 tag    | text    |           |          |
 parent | integer |           |          |
Indexes:
 "new_tags_pk" PRIMARY KEY, btree (pk)
```

```
Rules:
 r_new_tags_delete_a AS
 ON DELETE TO new_tags
 WHERE old.tag ~~* 'a%'::text DO DELETE FROM new_a_tags
 WHERE new_a_tags.pk = old.pk
 r_new_tags_delete_b AS
 ON DELETE TO new_tags
 WHERE old.tag ~~* 'b%'::text DO DELETE FROM new_b_tags
 WHERE new_b_tags.pk = old.pk
 r_new_tags_insert_a AS
 ON INSERT TO new_tags
 WHERE new.tag ~~* 'a%'::text DO INSERT INTO new_a_tags (pk, tag, parent)
 VALUES (new.pk, new.tag, new.parent)
 r_new_tags_insert_b AS
 ON INSERT TO new_tags
 WHERE new.tag ~~* 'b%'::text DO INSERT INTO new_b_tags (pk, tag, parent)
 VALUES (new.pk, new.tag, new.parent)
```

Let's test whether the two new rules work:

```
forumdb=# delete from new_tags where tag = 'apple';
DELETE 1

forumdb=# delete from new_tags where tag = 'banana';
DELETE 1

forumdb=# select * from new_tags ;
 pk | tag    | parent
----+--------+--------
 1  | fruits |
 3  | orange | 1
(2 rows)

forumdb=# select * from new_a_tags ;
 pk | tag | parent
----+-----+--------
(0 rows)

forumdb=# select * from new_b_tags ;
 pk | tag | parent
----+-----+--------
(0 rows)
```

We can see from this that the new rules work.

UPDATE rules

Now we need to introduce a rule that checks whether a tag is updated with a word that starts with "a" or "b." The best way to do this is to first create a function that conducts this check and then create a rule based on that function. Let's start by creating the function:

```
create or replace function move_record (p_pk integer, p_tag text, p_parent
integer,p_old_pk integer,p_old_tag text ) returns void language plpgsql as
$$
BEGIN
   if left(lower(p_tag),1) in ('a','b') THEN
        delete from new_tags where pk = p_old_pk;
        insert into new_tags values(p_pk,p_tag,p_parent);
   end if;
END;
$$;
```

This function takes five parameters as input; the first three parameters are the new values that arrive from the update and the last two parameters are the old values of the record that are present in the record. The function checks these things:

1. If the record in the table starts with the letter "a" or "b."
2. If the old record in the table starts with the letter "a" or "b," it deletes the old record and inserts the new record.

So, finally, the rule is as follows:

```
forumdb=# create or replace rule r_new_tags_update_a as on UPDATE to
new_tags DO ALSO select
move_record(NEW.pk,NEW.tag,NEW.parent,OLD.pk,OLD.tag);
CREATE RULE
```

The rule calls the function in the case of an update. Let's see if this rule works:

```
forumdb=# update new_tags set tag='apple' where tag='orange';
 move_record
-------------

(1 row)
UPDATE 0
forumdb=# select * from new_a_tags ;
 pk |  tag  | parent
----+-------+--------
  3 | apple |      1
(1 row)

forumdb=# select * from new_tags ;
 pk |  tag   | parent
```

```
----+--------+--------
  1 | fruits |
  3 | apple  |       1
(2 rows)
```

Now let's see what happens if a record changes its tag from apple to banana:

```
forumdb=# update new_tags set tag='banana' where tag='apple';
NOTICE: 3 banana 1 3 apple
 move_record
-------------
(1 row)

UPDATE 0
forumdb=# select * from new_tags ;
 pk | tag | parent
----+--------+--------
  1 | fruits |
  3 | banana | 1
(2 rows)

forumdb=# select * from new_a_tags ;
 pk | tag | parent
----+-----+--------
(0 rows)

forumdb=# select * from new_b_tags ;
 pk | tag | parent
----+--------+--------
  3 | banana | 1
(1 row)
```

The rule works! In this short exercise, we have tried to introduce an example of complete rule management. It is a didactic example and there are many other ways to achieve the same goal. In the next section, we will explore another way to manage events in PostgreSQL: triggers.

Managing triggers in PostgreSQL

In the previous section, we talked about rules. In this section, we will talk about triggers, what they are, and how to use them. We need to start by understanding what triggers are; if we have understood what rules are this should be simple. In the previous section, we defined rules as simple event handlers, now we can define triggers as complex event handlers. For triggers, as for rules, there are NEW and OLD records, which assume the same meaning for triggers as they did for rules. For triggers, the manageable events are INSERT / DELETE / UPDATE and TRUNCATE. Another difference between rules and triggers is that with triggers it is possible to handle INSERT / UPDATE / DELETE / and TRUNCATE events before they happen or after they have happened. With triggers, we can also use the INSTEAD OF option, but only on views.

So we can manage the following events:

- BEFORE INSERT/UPDATE/DELETE/TRUNCATE
- AFTER INSERT/UPDATE/DELETE/TRUNCATE
- INSTEAD OF INSERT/UPDATE/DELETE

With rules, it is possible to have only the NEW record for INSERT operations, the NEW and OLD record for UPDATE operations, and the OLD record for DELETE operations. The first two list items can also be used on foreign tables as well as primary tables and the third list item can only be used on views. For further information, see https://www.postgresql.org/docs/12/sql-createtrigger.html.

We will now take the first steps to use triggers and we will find out how to obtain the same results as achieved when using rules. With triggers, we can do everything we can do with rules and much more.

Before continuing, we need to keep two things in mind:

1. If triggers and rules are simultaneously present on the same event in a table, the rules always fire before the triggers.
2. If there are multiple triggers on the same event of a table (for example, BEFORE INSERT), they are executed in alphabetical order.

There is another category of triggers called **event triggers**, which will be covered in the *Event triggers* section.

Trigger syntax

As described in the official document, the syntax for defining a trigger is as follows:

```
CREATE [ CONSTRAINT ] TRIGGER name { BEFORE | AFTER | INSTEAD OF } { event
[ OR ... ] }
 ON table_name
 [ FROM referenced_table_name ]
 [ NOT DEFERRABLE | [ DEFERRABLE ] [ INITIALLY IMMEDIATE | INITIALLY
DEFERRED ] ]
 [ REFERENCING { { OLD | NEW } TABLE [ AS ] transition_relation_name } [
... ] ]
 [ FOR [ EACH ] { ROW | STATEMENT } ]
 [ WHEN ( condition ) ]
 EXECUTE { FUNCTION | PROCEDURE } function_name ( arguments )

where event can be one of:

 INSERT
 UPDATE [ OF column_name [, ... ] ]
 DELETE
 TRUNCATE
```

We will only look at the most used aspects of this syntax; for further information, see `https://www.PostgreSQL.org/docs/12/sql-createtrigger.html`. The key points behind the execution of a trigger are as follows:

1. The event that we want to handle, for example, INSERT, DELETE, or UPDATE.

2. When we want the TRIGGER execution to start (for example, BEFORE INSERT).

3. The trigger calls a function to perform some action.

The function invoked by the trigger must be defined in a particular way, as shown in the prototype here:

```
CREATE OR REPLACE FUNCTION function_name RETURNS trigger as
$$
DECLARE
....
BEGIN

    RETURN
END;
$$
LANGUAGE 'plpgsql';
```

The functions that are called by the triggers are functions that have no input parameters and must return a TRIGGER type; these functions have no input parameters and they take the parameters from the NEW / OLD records. Starting with this prototype of the preceding function, a possible TRIGGER definition on the BEFORE INSERT event can be described as follows:

```
CREATE TRIGGER trigger_name BEFORE INSERT on table_name FOR EACH ROW
EXECUTE PROCEDURE function_name.
```

In the next section, we will try to implement what we wrote with the rules, this time applying triggers.

Triggers on INSERT

In this section, we will see how to make our first triggers:

1. Let's go back to the rule that we wrote in the *The ALSO option* section:

```
create or replace rule r_tags1
 as on INSERT to tags
 where NEW.tag ilike 'a%' DO ALSO
 insert into a_tags(pk,tag,parent)values
(NEW.pk,NEW.tag,NEW.parent);
```

2. Now let's see how we can achieve the same goal using a trigger. First, let's go back to the initial situation:

```
forumdb=# drop table if exists new_tags cascade;
forumdb=# create table new_tags as select * from tags limit 0;
forumdb=# truncate table a_tags;
forumdb=# select * from new_tags ;
 pk | tag | parent
----+-----+--------
(0 rows)

forumdb=# select * from a_tags ;
 pk | tag | parent
----+-----+--------
(0 rows)
```

3. Now we can create the function, which will then be called by the trigger:

```
CREATE OR REPLACE FUNCTION f_tags() RETURNS trigger as
$$
BEGIN
```

```
IF lower(substring(NEW.tag from 1 for 1)) = 'a' THEN
insert into a_tags(pk,tag,parent)values
(NEW.pk,NEW.tag,NEW.parent);
END IF;
RETURN NEW;
END;
$$
LANGUAGE 'plpgsql';
```

Let's take a deeper look at what the code means:

- The statement `lower (substring (NEW.tag from 1 for 1))` takes the first character of a string and converts it into lowercase.
- The `RETURN NEW` statement passes the new record from the table to the `INSERT` in the `new_tags` table.

4. Now let's define the trigger on the `BEFORE INSERT` event of the `t_tags` table:

```
CREATE TRIGGER t_tags BEFORE INSERT on new_tags FOR EACH ROW
EXECUTE PROCEDURE f_tags();
```

5. So when a value is inserted into the `new_tags` table, before executing the `INSERT`, the trigger is executed and returns the `NEW` record to the default action (`INSERT` on the `new_tags` table). Now let's check that it works:

```
forumdb=# insert into new_tags (pk,tag,parent) values
(1,'fruits',NULL);
INSERT 0 1

forumdb=# insert into new_tags (pk,tag,parent) values
(2,'apple',1);
INSERT 0 1

forumdb=# select * from new_tags ;
 pk | tag | parent
----+--------+--------
 1 | fruits |
 2 | apple | 1
(2 rows)

forumdb=# select * from a_tags ;
 pk | tag | parent
----+-------+--------
 2 | apple | 1
(1 row)
```

As we can see here, it works!

6. We will proceed from here, step by step, to better understand the difference between working with rules and working with triggers. The goal we want to achieve with triggers is to receive the same result as we can achieve with the following rule:

```
create or replace rule r_tags2
 as on INSERT to tags
 where NEW.tag ilike 'b%'
 DO INSTEAD insert into b_tags(pk,tag,parent)values
(NEW.pk,NEW.tag,NEW.parent);
```

7. For now, let's use the same procedure we used in the rules, by creating a new function that will then be fired from the trigger:

```
CREATE OR REPLACE FUNCTION f2_tags() RETURNS trigger as
$$
BEGIN
 IF lower(substring(NEW.tag from 1 for 1)) = 'b' THEN
 insert into b_tags(pk,tag,parent)values
(NEW.pk,NEW.tag,NEW.parent);
 RETURN NULL;
 END IF;
 RETURN NEW;
END;
$$
LANGUAGE 'plpgsql';

CREATE TRIGGER t2_tags BEFORE INSERT on new_tags FOR EACH ROW
EXECUTE PROCEDURE f2_tags();
```

8. The lower statement, `(substring (NEW.tag from 1 for 1)) = 'b'`, is practically identical to what we first saw in relation to rules. The difference is the `RETURN NULL`, which means that if the `NEW.tag` value starts with `'b'`, then a `NULL` value is returned to the default action and then the `INSERT` on the `new_tags` table will not insert any value. If, instead, the `IF` condition is not satisfied, then the function returns `NEW` and the record is inserted into the `new_tags` table.

Let's see if it works:

```
forumdb=# insert into new_tags (pk,tag,parent) values
(1,'fruits',NULL);
INSERT 0 1
forumdb=# insert into new_tags (pk,tag,parent) values
(2,'apple',1);
INSERT 0 1
forumdb=# insert into new_tags (pk,tag,parent) values
(3,'banana',1);
INSERT 0 0
forumdb=# select * from new_tags ;
 pk | tag | parent
----+--------+--------
 1 | fruits |
 2 | apple | 1
(2 rows)
forumdb=# select * from a_tags ;
 pk | tag | parent
----+-------+--------
 2 | apple | 1
(1 row)
forumdb=# select * from b_tags ;
 pk | tag | parent
----+--------+--------
 3 | banana | 1
(1 row)
```

As we can see, it works.

9. We will now look at how to write the whole procedure using a single trigger.
First, let's go back to the initial conditions of our environment. As before, we
delete the data in the tables and, using the CASCADE option, we delete the triggers
and the functions associated with them:

```
forumdb=# TRUNCATE new_tags;
TRUNCATE TABLE
forumdb=# TRUNCATE a_tags;
TRUNCATE TABLE
forumdb=# TRUNCATE b_tags;
TRUNCATE TABLE
forumdb=# DROP TRIGGER t_tags ON new_tags CASCADE;
DROP TRIGGER
forumdb=# DROP TRIGGER t2_tags ON new_tags CASCADE;
DROP TRIGGER
```

10. In this last step, we will combine what we have written in the functions `f1_tags ()` and `f2_tags ()` into a single function, `f3_tags ()`, that will be fired from the `t3_tags` trigger:

```
CREATE OR REPLACE FUNCTION f3_tags() RETURNS trigger as
$$
BEGIN
 IF lower(substring(NEW.tag from 1 for 1)) = 'a' THEN
     insert into a_tags(pk,tag,parent)values
(NEW.pk,NEW.tag,NEW.parent);
     RETURN NEW;
 ELSIF lower(substring(NEW.tag from 1 for 1)) = 'b' THEN
     insert into b_tags(pk,tag,parent)values
(NEW.pk,NEW.tag,NEW.parent);
     RETURN NULL;
 ELSE
     RETURN NEW;
 END IF;
END;
$$
LANGUAGE 'plpgsql';

CREATE TRIGGER t3_tags BEFORE INSERT on new_tags FOR EACH ROW
EXECUTE PROCEDURE f3_tags();
```

This function contains the logic of the two functions previously seen. In this way, we can solve the problem in a more elegant way by using a single function and a single trigger. Let's see if it works:

```
forumdb=# insert into new_tags (pk,tag,parent) values
(1,'fruits',NULL);
INSERT 0 1
forumdb=# insert into new_tags (pk,tag,parent) values
(2,'apple',1);
INSERT 0 1
forumdb=# insert into new_tags (pk,tag,parent) values
(3,'banana',1);
INSERT 0 0
forumdb=# select * from new_tags ;
 pk | tag | parent
----+--------+--------
 1 | fruits |
 2 | apple | 1
(2 rows)
forumdb=# select * from a_tags ;
 pk | tag | parent
----+-------+--------
```

```
   2 | apple | 1
(1 row)
forumdb=# select * from b_tags ;
 pk | tag | parent
----+--------+--------
  3 | banana | 1
```

As can be seen, the function works.

One final thing to note about the function used is that the same function can be written in a simpler way as follows:

```
CREATE OR REPLACE FUNCTION f3_tags() RETURNS trigger as
$$
BEGIN
  IF lower(substring(NEW.tag from 1 for 1)) = 'a' THEN
      nsert into a_tags(pk,tag,parent)values
(NEW.pk,NEW.tag,NEW.parent);
  ELSIF lower(substring(NEW.tag from 1 for 1)) = 'b' THEN
      insert into b_tags(pk,tag,parent)values
(NEW.pk,NEW.tag,NEW.parent);
      RETURN NULL;
  END IF;
  RETURN NEW;
END;
$$
LANGUAGE 'plpgsql';
```

This is possible because the RETURN statement returns the control to the function caller.

The TG_OP variable

As shown in the official documentation at https://www.PostgreSQL.org/docs/12/plpgsql-trigger.html, control of the triggers in PostgreSQL is allowed using special variables, two of which we have already seen (the NEW variable and the OLD variable). There is another special variable called TG_OP, which tells us from which event the trigger is fired. The possible values of the TG_OP variable are INSERT, DELETE, UPDATE, and TRUNCATE.

Triggers on UPDATE / DELETE

Let's look at the example we used before when learning about rules and try to do the same with triggers:

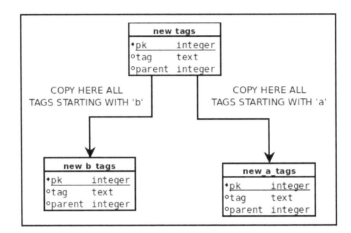

What we want to do is make it so that all tags starting with the letter "a" have to be stored in the new_tags table and also copied to the new_a_tags table, and we want the same for tags that begin with the letter "b." For the management of the INSERT event, we wrote these rules:

```
forumdb=# create or replace rule r_new_tags_insert_a as on INSERT to
new_tags where NEW.tag ilike 'a%' DO ALSO insert into
new_a_tags(pk,tag,parent)values (NEW.pk,NEW.tag,NEW.parent);
CREATE RULE

forumdb=# create or replace rule r_new_tags_insert_b as on INSERT to
new_tags where NEW.tag ilike 'b%' DO ALSO insert into
new_b_tags(pk,tag,parent)values (NEW.pk,NEW.tag,NEW.parent);
CREATE RULE
```

These and subsequent rules that we wrote for handling INSERT, UPDATE, and DELETE events, will be managed by a single function called by the triggers that will start before the INSERT, UPDATE, and DELETE events. First, let's return to the initial conditions in our environment:

```
TRUNCATE new_tags;
TRUNCATE a_tags;
TRUNCATE b_tags;
drop trigger t3_tags ON new_tags cascade;
```

Now, as before, we will proceed step by step. The first step is to write the section of code that will be performed during the INSERT event. Then, we will see how to extend the function to manage the DELETE and UPDATE events. The function that will handle all three events will be the fcopy_ins () function; this function will be invoked by the tcopy_tags_ins, tcopy_tags_upd, and tcopy_tags_del triggers. The function using the TG_OP variable will be able to discriminate between the INSERT, UPDATE, and DELETE events.

Let's start by writing the fcopy_ins () function to handle the INSERT event:

```
CREATE OR REPLACE FUNCTION fcopy_tags() RETURNS trigger as
$$
BEGIN
IF TG_OP = 'INSERT' THEN
    IF lower(substring(NEW.tag from 1 for 1)) = 'a' THEN
        insert into a_tags(pk,tag,parent)values
(NEW.pk,NEW.tag,NEW.parent);
    ELSIF lower(substring(NEW.tag from 1 for 1)) = 'b' THEN
        insert into b_tags(pk,tag,parent)values
(NEW.pk,NEW.tag,NEW.parent);
    END IF;
    RETURN NEW;
END IF;
END;
$$
LANGUAGE 'plpgsql';

CREATE TRIGGER tcopy_tags_ins BEFORE INSERT on new_tags FOR EACH ROW
EXECUTE PROCEDURE fcopy_tags();
```

Now let's see if, for the INSERT event, this code works:

```
forumdb=# insert into new_tags (pk,tag,parent) values (1,'fruits',NULL);
INSERT 0 1
forumdb=# insert into new_tags (pk,tag,parent) values (2,'apple',1);
INSERT 0 1
forumdb=# insert into new_tags (pk,tag,parent) values (3,'banana',1);
INSERT 0 1

forumdb=# select * from a_tags;
 pk | tag | parent
----+-------+--------
  2 | apple | 1
(1 row)

forumdb=# select * from b_tags;
 pk | tag | parent
```

```
----+--------+--------
 3 | banana | 1
(1 row)

forumdb=# select * from new_tags;
 pk | tag | parent
----+--------+--------
 1 | fruits |
 2 | apple | 1
 3 | banana | 1
(3 rows)
```

It is clear that it works!

Next, let's handle the DELETE event. The things we need to do are the following:

- Add some lines of code to the function for managing the DELETE operation.
- Create a new trigger on the DELETE event.

The function becomes as follows:

```
CREATE OR REPLACE FUNCTION fcopy_tags() RETURNS trigger as
$$
BEGIN
IF TG_OP = 'INSERT' THEN
    IF lower(substring(NEW.tag from 1 for 1)) = 'a' THEN
        insert into a_tags(pk,tag,parent)values
(NEW.pk,NEW.tag,NEW.parent);
    ELSIF lower(substring(NEW.tag from 1 for 1)) = 'b' THEN
        insert into b_tags(pk,tag,parent)values
(NEW.pk,NEW.tag,NEW.parent);
    END IF;
    RETURN NEW;
END IF;
IF TG_OP = 'DELETE' THEN
    IF lower(substring(OLD.tag from 1 for 1)) = 'a' THEN
            DELETE FROM a_tags WHERE pk = OLD.pk;
        ELSIF lower(substring(OLD.tag from 1 for 1)) = 'b' THEN
            DELETE FROM b_tags WHERE pk = OLD.pk;
    END IF;
    RETURN OLD;
END IF;
END;
$$
LANGUAGE 'plpgsql';
```

This piece of code was added:

```
IF TG_OP = 'DELETE' THEN
    IF lower(substring(OLD.tag from 1 for 1)) = 'a' THEN
        DELETE FROM a_tags WHERE pk = OLD.pk;
    ELSIF lower(substring(OLD.tag from 1 for 1)) = 'b' THEN
        DELETE FROM b_tags WHERE pk = OLD.pk;
    END IF;
 RETURN OLD;
END IF;
```

This piece of code deletes the data in the a_tags and b_tags tables if the record to be deleted begins with the letter "a" or with the letter "b." The trigger for handling the DELETE event is the following:

```
CREATE TRIGGER tcopy_tags_del
AFTER DELETE on new_tags FOR EACH ROW EXECUTE PROCEDURE fcopy_tags();
```

The trigger is executed AFTER DELETE; in this case it would have made no difference if we created the TRIGGER BEFORE or AFTER INSERT functions. Let's see if this trigger on the DELETE event works:

```
forumdb=# delete from new_tags where pk=2;
DELETE 1

forumdb=# delete from new_tags where pk=3;
DELETE 1

forumdb=# select * from a_tags;
 pk | tag | parent
----+-----+--------
(0 rows)

forumdb=# select * from b_tags;
 pk | tag | parent
----+-----+--------
(0 rows)

forumdb=# select * from new_tags;
 pk | tag | parent
----+--------+--------
 1 | fruits |
(1 row)
```

As we can see, the TRIGGER works.

For the last step, we need to manage the UPDATE event. Let's write the function and the triggers as a full version from scratch. First, let's bring our environment back to the initial conditions:

```
forumdb=# DROP TRIGGER tcopy_tags_ins ON new_tags cascade;
DROP TRIGGER
forumdb=# DROP TRIGGER tcopy_tags_del ON new_tags cascade;
DROP TRIGGER

forumdb=# TRUNCATE new_tags;
TRUNCATE TABLE

forumdb=# TRUNCATE a_tags;
TRUNCATE TABLE

forumdb=# TRUNCATE b_tags;
TRUNCATE TABLE

forumdb=# insert into new_tags (pk,tag,parent) values (1,'fruits',NULL);
INSERT 0 1

forumdb=# insert into new_tags (pk,tag,parent) values (2,'apple',1);
INSERT 0 1

forumdb=# insert into new_tags (pk,tag,parent) values (3,'banana',1);
INSERT 0 1
```

Now we can write the complete function with all the triggers for the INSERT, UPDATE, and DELETE events:

```
CREATE OR REPLACE FUNCTION fcopy_tags() RETURNS trigger as
$$
BEGIN
IF TG_OP = 'INSERT' THEN
    IF lower(substring(NEW.tag from 1 for 1)) = 'a' THEN
        insert into a_tags(pk,tag,parent)values
(NEW.pk,NEW.tag,NEW.parent);
    ELSIF lower(substring(NEW.tag from 1 for 1)) = 'b' THEN
        insert into b_tags(pk,tag,parent)values
(NEW.pk,NEW.tag,NEW.parent);
    END IF;
    RETURN NEW;
 END IF;
IF TG_OP = 'DELETE' THEN
    IF lower(substring(OLD.tag from 1 for 1)) = 'a' THEN
        DELETE FROM a_tags WHERE pk = OLD.pk;
    ELSIF lower(substring(OLD.tag from 1 for 1)) = 'b' THEN
        DELETE FROM b_tags WHERE pk = OLD.pk;
```

```
        END IF;
        RETURN OLD;
END IF;
IF TG_OP = 'UPDATE' THEN
    IF (lower(substring(OLD.tag from 1 for 1)) in( 'a','b') ) THEN
         DELETE FROM a_tags WHERE pk=OLD.pk;
         DELETE FROM b_tags WHERE pk=OLD.pk;
         DELETE FROM new_tags WHERE pk = OLD.pk;
         INSERT into new_tags(pk,tag,parent) values
(NEW.pk,NEW.tag,NEW.parent);
    END IF;
    RETURN NEW;
END IF;
END;
$$
LANGUAGE 'plpgsql';

CREATE TRIGGER tcopy_tags_ins
    BEFORE INSERT on new_tags FOR EACH ROW EXECUTE PROCEDURE fcopy_tags();
CREATE TRIGGER tcopy_tags_del
    AFTER DELETE on new_tags FOR EACH ROW EXECUTE PROCEDURE fcopy_tags();
CREATE TRIGGER tcopy_tags_upd
    AFTER UPDATE on new_tags FOR EACH ROW EXECUTE PROCEDURE fcopy_tags();
```

In this case, the trigger must be defined AFTER UPDATE and not BEFORE UPDATE because in the UPDATE section, we have the instruction DELETE FROM new_tags WHERE pk = OLD.pk; if the trigger had been defined BEFORE UPDATE, we would have had an error because we would have attempted to delete a record reserved for UPDATE.

Let's see if the complete function works:

```
forumdb=# select * from new_tags;
 pk | tag    | parent
----+--------+--------
 1 | fruits |
 2 | apple  | 1
 3 | banana | 1
(3 rows)

forumdb=# select * from a_tags;
 pk | tag   | parent
----+-------+--------
 2 | apple | 1
(1 row)

forumdb=# select * from b_tags;
 pk | tag   | parent
```

```
----+--------+--------
  3 | banana | 1
(1 row)

forumdb=# update new_tags set tag='apricot' where pk=3;
UPDATE 1
forumdb=# select * from b_tags;
 pk | tag | parent
----+-----+--------
(0 rows)

forumdb=# select * from a_tags;
 pk | tag     | parent
----+---------+--------
  2 | apple   | 1
  3 | apricot | 1
(2 rows)

forumdb=# select * from new_tags;
 pk | tag     | parent
----+---------+--------
  1 | fruits  |
  2 | apple   | 1
  3 | apricot | 1
(3 rows)
```

As this shows, the trigger approach works. Now, in this section, we have seen how to modify events that are DML through the use of rules and triggers. In the next section, we will see how it is also possible to intercept and modify events related to DDL operations using event triggers.

Event triggers

Rules and triggers act as **Data Manipulation Level** (**DML**) statements, which means they are triggered by something that changes the data but not the data layout or the table properties. PostgreSQL provides so-called *event triggers*, which are particular triggers that fire on **DDL** (**Data Definition Level**) statements. The purpose of the event trigger is therefore to manage and react to events that are going to change the data structure rather than the data content. Triggers can be used in many ways to enforce specific policies across your databases.

Once fired, an event trigger receives an *event* and a *command tag*, both of which are useful for introspection and providing information about what fired the trigger. In particular, the *command tag* contains a description of the command (for example, CREATE or ALTER), while the *event* contains the category that fired the trigger, in particular, the following:

- ddl_command_start and ddl_command_end indicate respectively the beginning and the completion of the DDL command.
- sql_drop indicates that a DROP command is near to completion.
- table_rewrite indicates that a full table rewrite is about to begin.

As with DML triggers, there are particular commands to create, delete, and modify an event trigger:

- CREATE EVENT TRIGGER to add a new event trigger
- DROP EVENT TRIGGER to delete an existing trigger
- ALTER EVENT TRIGGER to modify an existing trigger

Here is the synopsis for the creation of a new event trigger:

```
CREATE EVENT TRIGGER name
    ON event
    [ WHEN filter_variable IN (filter_value [, ... ]) [ AND ... ] ]
    EXECUTE { FUNCTION | PROCEDURE } function_name ()
```

Similar to their DML counterpart triggers, event triggers are associated with a mnemonic name and a function to execute once they are fired. However, unlike ordinary triggers, event triggers do not specify to which table they are attached; in fact, event triggers are not related to any particular table but rather to DDL commands.

Event triggers must be created by the database administrator and have a database scope, meaning they live and act in the database they have been defined in.

There are a couple of special functions that can help developers to perform introspection within an event trigger in order to understand the exact event that fired the trigger. The most important functions are as follows:

- pg_event_trigger_commands(), which returns a tuple for every command that was executed during the DDL statement.
- pg_event_trigger_dropped_objects(), which reports a tuple for every dropped object within the same DDL statement.

Along with the preceding utility functions, it is important to carefully read the event trigger documentation in order to understand when a command will fire an event trigger or not. Explaining event triggers in further detail is out of the scope of this section; instead, we will look at a practical example in the following section. For more information about event triggers, please refer to the official documentation or to the Packt Book *PostgreSQL 11 Server-Side Programming*.

An example of an event trigger

In order to better understand how event triggers work, let's build a simple example of a trigger that prevents any ALTER TABLE like commands in a database.

The first step is to define a function that will be executed once the trigger has been fired; such a function needs to inspect the DDL statement properties in order to understand whether it has been invoked by means of an ALTER TABLE command. The introspection is done using the pg_event_trigger_ddl_commands() special function, which returns a tuple for every DDL statement executed within the same command. Such tuples contain a field named command_tag, which reports the command group (uppercase), and object_type, which reports the object type (lowercase) that the DDL statement has been executed against. The function must return a trigger type, specifically an event trigger type, therefore the function can be defined as follows:

```
CREATE OR REPLACE FUNCTION
f_avoid_alter_table()
RETURNS EVENT_TRIGGER
AS
$code$
DECLARE
event_tuple record;
BEGIN

   FOR event_tuple IN SELECT *
                      FROM pg_event_trigger_ddl_commands()  LOOP
        IF event_tuple.command_tag = 'ALTER TABLE' AND
event_tuple.object_type = 'table' THEN
            RAISE EXCEPTION 'Cannot execute an ALTER TABLE!';
        END IF;
   END LOOP;
END
$code$
LANGUAGE plpgsql;
```

As you can see, if the function discovers that the executed command has an `'ALTER TABLE'` tag and a `'table'` object type, it raises an exception causing the whole statement to fail.

Once the function is in place, it is possible to attach it to an event trigger:

```
forumdb=# CREATE EVENT TRIGGER tr_avoid_alter_table
          ON ddl_command_end EXECUTE FUNCTION f_avoid_alter_table();
```

At this point, the trigger is active and the function will be fired for every DDL command once the system is approaching the end of a command.

It is now possible to test the trigger and see whether a user is allowed to execute `ALTER TABLE`:

```
forumdb=> ALTER TABLE tags ADD COLUMN thumbs_up int DEFAULT 0;
ERROR:  Cannot execute an ALTER TABLE!
CONTEXT:  PL/pgSQL function f_avoid_alter_table() line 10 at RAISE
```

As we can see, an exception is raised as soon as the `ALTER TABLE` command is executed.

While event triggers can be used, as in the preceding example, to prevent users from executing particular commands, a better strategy is to avoid inappropriate command executions by means of permissions whenever possible. Event triggers are complex and are used to provide support for things such as logical replication, auditing, and other infrastructures.

Summary

In this chapter, we covered the topic of triggers and rules. We explored rules and triggers using some identical examples. We established that rules are simple event handlers and triggers are complex event handlers.

We introduced the concept of trigger variables:

- NEW
- OLD
- TG_OP

As well as data-manipulation-based triggers, we briefly introduced the PostgreSQL event triggers that allow developers and database administrators to have more control over firing and executing functions.

We have come to understand that triggers are extremely complex event handlers. In this chapter, we started to show the power of these tools made available to the PostgreSQL DBA; in the next chapter, we will talk about partitioning and we will utilize the topics covered in this chapter.

In the next chapter, we will return to talking about triggers and we will use triggers to create a certain type of partitioning present in PostgreSQL.

References

- PostgreSQL 12 Rules on INSERT, UPDATE, and DELETE official documentation: https://www.PostgreSQL.org/docs/12/rules-update.html
- PostgreSQL 12 trigger functions official documentation: https://www.PostgreSQL.org/docs/12/plpgsql-trigger.html
- PostgreSQL 12 ALTER TRIGGER official documentation: https://www.PostgreSQL.org/docs/12/sql-altertrigger.html
- PostgreSQL 12 DROP TRIGGER official documentation: https://www.PostgreSQL.org/docs/12/sql-droptrigger.html
- PostgreSQL 12 event trigger official documentation: https://www.postgresql.org/docs/12/functions-event-triggers.html
- *PostgreSQL 11 Server-Side Programming – Quick Start Guide*: https://www.packtpub.com/big-data-and-business-intelligence/postgresql-11-server-side-programming-quick-start-guide

9
Partitioning

In the previous chapter, we talked about rules and triggers. In this chapter, we will talk about partitioning. Partitioning is a technique that allows us to split a huge table into smaller tables, to make queries more efficient. In this chapter, we will see how we can partition data, and, in some cases, we will see how to use the rules and triggers seen in the previous chapter to make partitioning possible. We will start by introducing the basic concepts of partitioning, and then we will see what possibilities PostgreSQL offers to implement partitioning.

This chapter will cover the following topics:

- Basic concepts
- Partitioning using table inheritance
- Declarative partitioning

Basic concepts

First of all, let's try to understand why we have to partition data. We should start by saying that a common constant of all databases is that their size always grows. It is, therefore, possible that a database, after a few months of growth, can reach a size of gigabytes, terabytes, or petabytes.

Another thing we must always keep in mind is that not all tables grow at the same rate or to the same level; there are tables that will be bigger than other tables and there will be indexes too that will be bigger than other indexes.

We also need to know that there is a part of our server's RAM memory shared among all the Postgres processes that is used to manage the data that is present in tables. This part of the server's RAM is called `shared_buffers`.

The way PostgreSQL works is as follows:

1. Data is taken from disks.
2. Data is placed in shared buffers.
3. Data is processed in shared buffers.
4. Data is downloaded to disks.

Typically, in a dedicated server only for PostgreSQL, the size of `shared_buffers` is about one-third or one-quarter of the total server RAM memory. A useful link to set some PostgreSQL configuration parameters is `https://pgtune.leopard.in.ua`.

When a table grows excessively compared to the `shared_buffers` size, there is a possibility that performance will decrease. In this case, partitioning data can help us. Partitioning data means splitting a very large table into smaller tables in a way that is transparent to the client program. The client program will think that the server still has only one single table. Data partitioning can be done in two ways:

• Using table inheritance
• Using declarative partitioning

After figuring out when it is recommended to partition data, let's see what types of table partitioning are possible. PostgreSQL 12 manages the following types of table partitioning:

• Range partitioning
• List partitioning
• Hash partitioning

We will now describe these three methods in detail.

Range partitioning

Range partitioning is where the table is divided into "intervals." The intervals must not overlap and the range is defined through the use of a field or a set of fields. For further information, see `https://www.postgresql.org/docs/12/ddl-partitioning.html`.

Let's look at an example of the definition of range partitioning. Suppose we have this table:

field date	field_value
2020-03-01	1
2020-03-02	10
2020-04-01	12
2020-04-15	1

Now consider that we want to split this table into two tables. The first table (**TABLE A**) will contain all the records with a `field_date` value between 2020-03-01 and 2020-03-31, and the second table (**TABLE B**) will contain all the records with a `field_date` value between 2020-04-01 and 2020-04-30. So, our goal is to have two tables as follows:

field date	field_value
2020-03-01	1
2020-03-02	10

This is TABLE B:

field date	field_value
2020-04-01	12
2020-04-15	1

What we have seen is an example of partitioning by range.

List partitioning

In list partitioning, the table will be partitioned using a list of values. For further information, see `https://www.postgresql.org/docs/12/ddl-partitioning.html`.

Let's look at an example of the definition of list partitioning. Suppose we have this table:

field_state	field_city
United States	Washington
United States	San Francisco
Italy	Rome
Japan	Tokio

Suppose now that we want to split this table into *n* tables, with one table for each state. The first table (**TABLE A**) will contain all the records with a `field_state` value equal to United States, the second table (**TABLE B**) will contain all records with a `field_state` value equal to Italy, and the third table (**TABLE C**) will contain records with a `field_state` value equal to Japan. So, our goal is to have three tables as follows:

TABLE A:

field_state	field_city
United States	Washington
United States	San Francisco

TABLE B:

field_state	field_city
Italy	Rome

TABLE C:

field_state	field_city
Japan	Tokyo

This is an example of partitioning by list.

Hash partitioning

Using hash partitioning, the table will be partitioned using a hash value that will be used as the value to split data into different tables. For further information, see `https://www.postgresql.org/docs/12/ddl-partitioning.html`.

Let's look at an example of the definition of list partitioning. Suppose we have this table:

field date	field_value
2020-03-01	1
2020-03-02	10
2020-04-01	12
2020-04-15	1

Suppose now that we have a hash function that transforms a date into a hash value; for example:

- `hash(2020-03-01)` = AAAAAAB
- `hash(2020-03-02)` = AAAAAAC
- `hash(2020-04-01)` = AAAAAAB
- `hash(2020-04-15)` = AAAAAAC

So, after the partitioning process, we will have two tables:

TABLE A :

field date	field_value
2020-03-01	1
2020-04-01	12

TABLE B :

field date	field_value
2020-03-02	10
2020-04-15	1

This is an example of partitioning by hash.

In the following sections, we will see how PostgreSQL implements list, range, and hash partitioning.

Table inheritance

Another introductory topic that we must look at is the inheritance of tables. PostgreSQL takes up the concept of inheritance from databases to objects. The concept is very simple and can be summarized as follows: suppose we have two tables, table A and table B. If we define table A as a parent table and table B as child table, this means that all the records in table B will be accessible from table A. Let's now try to give an example of what we have just described:

1. For example, let's define two tables.

   The first table, the parent table, is defined as follows:

   ```
   create table table_a (
     pk integer not null primary key,
   ```

```
    tag text,
    parent integer);
```

And the second table, the child table, is defined as follows:

```
create table table_b () inherits (table_a);

alter table table_b add constraint table_b_pk primary key(pk);
```

2. The child table inherits all the fields from the parent table. The parent table is described as seen here:

```
forumdb=# \d table_a;
 Table "public.table_a"
 Column | Type     | Collation  | Nullable  | Default
--------+----------+------------+-----------+---------
 pk     | integer  |            | not null  |
 tag    | text     |            |           |
 parent | integer  |            |           |
Indexes:
 "table_a_pkey" PRIMARY KEY, btree (pk)
Number of child tables: 1 (Use \d+ to list them.)
```

And for more details, let's use the \d+ command:

```
forumdb=# \d+ table_a;
 Table "public.table_a"
 Column | Type     | Collation | Nullable | Default | Storage  |
Stats target | Description
--------+----------+-----------+----------+---------+----------+----
-----------+-------------
 pk     | integer  |           | not null |         | plain    |
 |
 tag    | text     |           |          |         | extended |
 |
 parent | integer  |           |          |         | plain    |
 |
Indexes:
 "table_a_pkey" PRIMARY KEY, btree (pk)
Child tables: table_b
Access method: heap
```

In this last table, we can see that table_b is a child table of table_a.

3. Let's do the same for the table called table_b:

```
forumdb=# \d table_b;
 Table "public.table_b"
```

```
 Column | Type     | Collation | Nullable | Default
--------+----------+-----------+----------+---------
 pk     | integer  |           | not null |
 tag    | text     |           |          |
 parent | integer  |           |          |
Indexes:
 "table_b_pk" PRIMARY KEY, btree (pk)
Inherits: table_a
```

Here, we can see that `table_b` is a child table of `table_a`.

4. Now let's see how these two tables behave if we insert, modify, or delete data. For example, let's make some inserts as follows:

```
forumdb=# insert into table_a (pk,tag,parent) values
(1,'fruits',0);
INSERT 0 1

forumdb=# insert into table_b (pk,tag,parent) values
(2,'orange',0);
INSERT 0 1
```

5. Let's see how our data reacts if we execute the `select` command:

```
forumdb=# select * from table_b ;
 pk | tag    | parent
----+--------+--------
 2  | orange | 0
(1 row)
```

We can see that `table_b` has one record.

6. Now we execute the following command:

```
forumdb=# select * from table_a ;
 pk | tag    | parent
----+--------+--------
 1  | fruits | 0
 2  | orange | 0
(2 rows)
```

It seems that `table_a` has two records. This happens because this table inherits the other table's attributes. If we execute a SELECT command on a parent table, we will see all the records that belong to the parent table and all the records that belong to the child table.

7. If we want to see all the records that belong only to `table_a`, we have to use the `ONLY` clause, as seen here:

```
forumdb=# select * from only table_a ;
pk | tag      | parent
----+--------+--------
1  | fruits | 0
(1 row)
```

8. Let's see what happens if we `UPDATE` some records, for example, if we execute the following:

```
forumdb=# update table_a set tag='apple' where pk=2;
UPDATE 1
```

We performed an `update` operation on `table_a`, but this update was physically done on `table_b` by means of the inheritance of the tables, as we can see here:

```
forumdb=# select * from table_b;
 pk | tag   | parent
----+-------+--------
 2  | apple | 0
(1 row)
```

9. The same happens if we use a `delete` statement as follows:

```
forumdb=# delete from table_a where pk=2;
DELETE 1
```

Here, again, the `delete` operation performed on `table_a` has its effect on `table_b`; as we can see here, `table_a` will have these records:

```
forumdb=# select * from table_a;
 pk | tag | parent
----+--------+--------
 1 | fruits | 0
(1 row)
```

And `table_b` will now have no records:

```
forumdb=# select * from table_b;
 pk | tag | parent
----+-----+--------
(0 rows)
```

In PostgreSQL 12, inheritance propagates the operations performed on the parent table to the child tables.

Dropping tables

To conclude the topic of inheritance, we need to address how to delete tables. If we want to delete a child table, for example, to drop `table_b`, we have to run the following statement:

```
forumdb=# drop table table_b;
DROP TABLE
```

If we want to DROP a parent table and all its linked child tables, we have to run the following:

```
forumdb=# drop table table_a cascade;
NOTICE: drop cascades to table table_b
DROP TABLE
```

After showing the concepts behind partitioning, in the next section, we will make a table partition using the concept of inheritance.

Exploring partitioning using inheritance

At this point in the book, we have all the elements necessary to partition data. In the previous chapter, we learned how to use triggers and rules to move data between tables. In this chapter, we just learned that it is possible using inheritance to query a parent table and manage the data of child tables. In the next section, we will do partitioning using the declarative partitioning method, available on PostgreSQL starting from version 10.x.

An example of list partitioning

In this first example, we will try to create a partitioned table using the list partitioning logic. In the next section, we will create a partitioned table using the range partitioning logic, starting from a table such as the following:

pk	tag	level
1	vegetables	0
2	fruits	0

3	orange	1
4	apple	1
5	red apple	2

Our goal is to split our data using the value of the `level` field. The goal that we want to reach is to have 4 levels and we want to have 1 table for each level. We also want all records with `level = 0` to be stored in the `part_level_0` table, all records with `level = 1` to be stored in the `part_level_1` table, and so on, as shown here.

This is table `part_tags_level_0`:

pk	tag	level
1	vegetables	0
2	fruits	0

This is table `part_tags_level_1`:

pk	tag	level
3	orange	1
4	apple	1

This is table `part_tags_level_2`:

pk	tag	level
5	red apple	2

We will now dive into the code in the coming sections.

Creating tables

The first step is to create a parent table and its child tables:

1. The parent table will be defined as follows:

```
CREATE SEQUENCE part_tags_pk_seq;

CREATE TABLE part_tags (
    pk INTEGER NOT NULL DEFAULT nextval('part_tags_pk_seq')
PRIMARY KEY,
    tag VARCHAR(255) NOT NULL,
    level INTEGER DEFAULT 0
);
```

2. And the child tables will be defined as follows:

```
CREATE TABLE part_tags_level_0 (
    CHECK(level = 0 )
) INHERITS (part_tags);

CREATE TABLE part_tags_level_1 (
    CHECK(level = 1 )
) INHERITS (part_tags);

CREATE TABLE part_tags_level_2 (
    CHECK(level = 2 )
) INHERITS (part_tags);

CREATE TABLE part_tags_level_3 (
    CHECK(level = 3 )
) INHERITS (part_tags);
```

The check command on the level field has two functions:

- The first is to avoid incorrect values in the daughter tables.
- The second is to have PostgreSQL perform the correct pruning of the data.

The second function is linked to the constraint_exclusion parameter, as shown in the parameter description:

```
forumdb=# select name,short_desc,extra_desc from pg_settings where name
='constraint_exclusion';
-[ RECORD 1 ]---+---------------------------------------------------------
---------------------
name            | constraint_exclusion
short_desc | Enables the planner to use constraints to optimize queries.
extra_desc | Table scans will be skipped if their constraints guarantee
that no rows match the query.
```

This parameter makes it possible for the query optimizer to exclude some child tables from the search. For example, if we execute select * from part_tags where level = 1, then the query optimizer will check the conditions of the check imposed on the child tables and will only query the table for which check (level = 1) is satisfied. Possible values for the constraint_exclusion parameter are the following:

- on: With this value set, PostgreSQL examines all tables.
- off: With this value set, PostgreSQL doesn't examine any constraints.

- `partition`: With this value, PostgreSQL checks the constraints for the `UNION ALL` subqueries and only for inheritance child tables, and `partition` is the default setting.

For further information, see `https://www.postgresql.org/docs/12/runtime-config-query.html#GUC-CONSTRAINT-EXCLUSION`.

3. After creating child tables, let's now create their primary keys:

```
ALTER TABLE ONLY part_tags_level_0 add constraint
part_tags_level_0_pk primary key (pk);
ALTER TABLE ONLY part_tags_level_1 add constraint
part_tags_level_1_pk primary key (pk);
ALTER TABLE ONLY part_tags_level_2 add constraint
part_tags_level_2_pk primary key (pk);
ALTER TABLE ONLY part_tags_level_3 add constraint
part_tags_level_3_pk primary key (pk);
```

4. The next step is to create all the indexes that we need for our queries. Suppose, for example, that we want to index `like` and `ilike` queries. First of all, we have to create the trigram extension (`https://www.postgresql.org/docs/12/pgtrgm.html`):

```
CREATE EXTENSION pg_trgm ;
```

5. And then we have to create a `GIN` index on the `tag` field:

```
CREATE INDEX part_tags_level_0_tag on part_tags_level_0 using GIN
(tag gin_trgm_ops);
CREATE INDEX part_tags_level_1_tag on part_tags_level_1 using GIN
(tag gin_trgm_ops);
CREATE INDEX part_tags_level_2_tag on part_tags_level_2 using GIN
(tag gin_trgm_ops);
CREATE INDEX part_tags_level_3_tag on part_tags_level_3 using GIN
(tag gin_trgm_ops);
```

We have finished the phase of creating the tables and their indexes.

Creating triggers and functions, and inserting data

Now that we have created the structure on which our data will be stored, we must tell PostgreSQL how to make sure that the data will be inserted not into the parent table but into the child tables. The mechanism is very similar to what we saw in the previous chapter and we can create it using triggers or rules. In this example, we will use triggers:

1. First of all, let's create the function that will be called by the trigger:

```
CREATE OR REPLACE FUNCTION insert_part_tags () RETURNS TRIGGER as
$$
BEGIN
 IF NEW.level = 0 THEN
     INSERT INTO part_tags_level_0 values (NEW.*);
 ELSIF NEW.level = 1 THEN
     INSERT INTO part_tags_level_1 values (NEW.*);
 ELSIF NEW.level = 2 THEN
     INSERT INTO part_tags_level_2 values (NEW.*);
 ELSIF NEW.level = 3 THEN
     INSERT INTO part_tags_level_3 values (NEW.*);
 ELSE
     RAISE EXCEPTION 'Error in part_tags, level out of range';
 END IF;
 RETURN NULL;
END;
$$
language 'plpgsql';
```

2. Then let's create the trigger:

```
CREATE TRIGGER insert_part_tags_trigger BEFORE INSERT ON part_tags
FOR EACH ROW EXECUTE PROCEDURE insert_part_tags();
```

This trigger moves the record into the child tables and returns to the parent table a NULL record. In this way, no one record will be inserted into the parent table.

3. Let's now make some INSERT statements:

```
forumdb=# insert into part_tags (tag,level) values
('vegetables',0);
INSERT 0 0
forumdb=# insert into part_tags (tag,level) values ('fruits',0);
INSERT 0 0
forumdb=# insert into part_tags (tag,level) values ('orange',1);
INSERT 0 0
forumdb=# insert into part_tags (tag,level) values ('apple',1);
INSERT 0 0
forumdb=# insert into part_tags (tag,level) values ('red apple',2);
INSERT 0 0
```

As we can see, if we query the parent table, we receive the following results:

```
forumdb=# select * from part_tags;
 pk | tag | level
----+------------+-------
 6 | vegetables | 0
 7 | fruits | 0
 8 | orange | 1
 9 | apple | 1
 10 | red apple | 2
(5 rows)
```

In the parent table, there is no data:

```
forumdb=# select * from only part_tags;
 pk | tag | level
----+-----+-------
(0 rows)
```

4. But as we can see, all the data is stored in the child tables part_tags_level_0, part_tags_level_1, and part_tags_level_2. We can see the data present in the part_tags_level_0 table here:

```
forumdb=# select * from only part_tags_level_0;
 pk | tag        | level
----+------------+-------
 6 | vegetables | 0
 7 | fruits     | 0
(2 rows)
```

This is the data present in the part_tags_level_1 table:

```
forumdb=# select * from only part_tags_level_1;
 pk | tag    | level
```

```
----+--------+-------
 8  | orange | 1
 9  | apple  | 1
(2 rows)
```

This is the data present in the `part_tags_level_2` table:

```
forumdb=# select * from only part_tags_level_2;
 pk | tag | level
----+-----------+-------
10 | red apple | 2
(1 row)
```

Now, we have finished creating the trigger, and we have added data to it.

Creating triggers and functions and updating data

In terms of the deletion of data, as we saw in the *Exploring partitioning using inheritance* section, the deletion of a record in the parent table is automatically propagated to the child tables. For example, let's execute the following:

```
forumdb=# delete from part_tags where tag='apple';
DELETE 1
forumdb=# select * from only part_tags_level_1;
 pk | tag | level
----+--------+-------
 8 | orange | 1
(1 row)

forumdb=# select * from part_tags;
 pk | tag        | level
----+------------+-------
 6  | vegetables | 0
 7  | fruits     | 0
 8  | orange     | 1
 10 | red apple  | 2
(4 rows)
```

As we can see in the preceding example, the record has been correctly deleted. The same thing happens when we modify a record and the change remains within the same child table. For example, let's run the following:

```
forumdb=# update part_tags set tag='apple' where pk=8;
UPDATE 0
```

As we can see here, the record has been correctly updated:

```
forumdb=# select * from part_tags;
 pk | tag        | level
----+------------+-------
 6  | vegetables | 0
 7  | fruits     | 0
 8  | apple      | 1
 10 | red apple  | 2
(4 rows)

forumdb=# select * from only part_tags_level_1;
 pk | tag   | level
----+-------+-------
 8  | apple | 1
(1 row)
```

A problem occurs when the update has to be moved to another child table, for example, if we run the following:

```
forumdb=# update part_tags set level=1,tag='apple' where pk=10;
ERROR: new row for relation "part_tags_level_2" violates check constraint
"part_tags_level_2_level_check"
```

To make it possible to move data among child tables, we have to add another trigger for each child table. That's because all records are stored in the child tables:

```
CREATE OR REPLACE FUNCTION update_part_tags() RETURNS TRIGGER AS
$$
BEGIN
 IF (NEW.level != OLD.level) THEN
     DELETE FROM part_tags where pk = OLD.PK;
     INSERT INTO part_tags values (NEW.*);
 END IF;
 RETURN NULL;
END;
$$
LANGUAGE 'plpgsql';

CREATE TRIGGER update_part_tags_trigger BEFORE UPDATE ON part_tags_level_0
FOR EACH ROW EXECUTE PROCEDURE update_part_tags();
CREATE TRIGGER update_part_tags_trigger BEFORE UPDATE ON part_tags_level_1
FOR EACH ROW EXECUTE PROCEDURE update_part_tags();
CREATE TRIGGER update_part_tags_trigger BEFORE UPDATE ON part_tags_level_2
FOR EACH ROW EXECUTE PROCEDURE update_part_tags();
CREATE TRIGGER update_part_tags_trigger BEFORE UPDATE ON part_tags_level_3
FOR EACH ROW EXECUTE PROCEDURE update_part_tags();
```

The trigger works in the same way as we saw in the previous chapter. Now let's try again with this update:

```
forumdb=# update part_tags set level=1,tag='apple' where pk=5;
UPDATE 0
forumdb=# select * from part_tags;
 pk | tag        | level
----+------------+-------
 6  | vegetables | 0
 7  | fruits     | 0
 8  | apple      | 1
 10 | apple      | 1
(4 rows)
```

As we can see, it works! In this section, we've talked about how to partition a table using the inheritance method. Starting from PostgreSQL 10.x, it is possible to reach the goal in a much easier way, therefore in the next section, we will talk about declarative partitioning.

Exploring declarative partitioning

In this section, we will talk about declarative partitioning. It is available in PostgreSQL starting from version 10, but it is best in version 12 in terms of features and performance. We will now look at an example of partitioning by range and an example of partitioning by list.

List partitioning

In the first example of declarative partitioning, we will use the same example as we looked at when we introduced partitioning using inheritance. We will see that things become much simpler using the declarative partitioning method:

1. First of all, let's drop the parent table and its child tables that we made previously:

    ```
    DROP TABLE IF EXISTS part_tags cascade;
    ```

2. Now let's recreate the same tables using the declarative method. First, we must define our parent table:

    ```
    CREATE TABLE part_tags (
     pk INTEGER NOT NULL DEFAULT nextval('part_tags_pk_seq') ,
     level INTEGER NOT NULL DEFAULT 0,
     tag VARCHAR (255) NOT NULL,
    ```

```
primary key (pk,level)
)
PARTITION BY LIST (level);
```

As we can see from the preceding example, we have to define what kind of partitioning we want to apply. In this case, it is LIST PARTITIONING. Another important thing to note is that the field used to partition the data must be part of the primary key.

3. Next, let's define the child tables:

```
CREATE TABLE part_tags_level_0 PARTITION OF part_tags FOR VALUES IN
(0);
CREATE TABLE part_tags_level_1 PARTITION OF part_tags FOR VALUES IN
(1);
CREATE TABLE part_tags_level_2 PARTITION OF part_tags FOR VALUES IN
(2);
CREATE TABLE part_tags_level_3 PARTITION OF part_tags FOR VALUES IN
(3);
```

With these SQL statements, we are defining the fact that all records with a level value equal to 0 will be stored in the part_tags_level_0 table, all the records with a level value equal to 1 will be stored in the part_tags_level_1 table, and so on.

4. Now, as we did in the previous section, let's define the indexes for the parent table and for all child tables. We can do this using this simple statement:

```
CREATE INDEX part_tags_tag on part_tags using GIN (tag
gin_trgm_ops);
```

5. As shown here, our partition procedure is finished.

For the parent tables, we have the following:

```
forumdb=# \d part_tags;
 Column | Type                   | Collation | Nullable | Default
--------+------------------------+-----------+----------+----------
--------
 pk     | integer                |           | not null |
nextval('part_tags_pk_seq'::regclass)
 level  | integer                |           | not null | 0
 tag    | character varying(255) |           | not null |
Partition key: LIST (level)
Indexes:
    "part_tags_pkey" PRIMARY KEY, btree (pk, level)
    "part_tags_tag" gin (tag gin_trgm_ops)
Number of partitions: 4 (Use \d+ to list them.)
```

For the child tables, we have the following:

```
forumdb=# \d part_tags_level_0;
 Table "public.part_tags_level_0"
 Column | Type                   | Collation | Nullable | Default
--------+------------------------+-----------+----------+----------
--------
 pk     | integer                |           | not null |
nextval('part_tags_pk_seq'::regclass)
 level  | integer                |           | not null | 0
 tag    | character varying(255) |           | not null |
Partition of: part_tags FOR VALUES IN (0)
Indexes:
  "part_tags_level_0_pkey" PRIMARY KEY, btree (pk, level)
  "part_tags_level_0_tag_idx" gin (tag gin_trgm_ops
```

6. Let's now perform some INSERT operations:

```
insert into part_tags (tag,level) values ('vegetables',0);
insert into part_tags (tag,level) values ('fruits',0);
insert into part_tags (tag,level) values ('orange',1);
insert into part_tags (tag,level) values ('apple',1);
insert into part_tags (tag,level) values ('red apple',2);
```

7. Finally, let's check whether everything is okay:

```
forumdb=# select * from part_tags;
 pk | level | tag
----+-------+-------------
 6  | 0     | vegetables
 7  | 0     | fruits
 8  | 1     | orange
```

```
  9 | 1      | apple
 10 | 2      | red apple

forumdb=# select * from part_tags_level_0;
 pk | level | tag
----+-------+------------
  6 | 0     | vegetables
  7 | 0     | fruits
(2 rows)

forumdb=# select * from part_tags_level_1;
 pk | level | tag
----+-------+--------
  8 | 1     | orange
  9 | 1     | apple
(2 rows)
```

Thus, we have successfully created partitions using lists.

Range partitioning

After having seen how it is possible to partition by list in a very simple way, let's look at how to partition by range:

1. As before, let's DROP the existing part_tags table and its child table:

   ```
   DROP TABLE IF EXISTS part_tags cascade;
   ```

2. Suppose that we want to have a table exactly the same as the previous one, but now we want the part_tags table to have an ins_date field where we will store the day on which the tag was added. What we want to do is partition by range on the ins_date field in order to put all the records entered in January 2020, February 2020, March 2020, and April 2020 into different tables. Here, we have all the statements that make this possible; they are very similar to the statements that we saw in the previous section:

   ```
   CREATE TABLE part_tags (
       pk INTEGER NOT NULL DEFAULT nextval('part_tags_pk_seq'),
       ins_date date not null default now()::date,
       tag VARCHAR (255) NOT NULL,
       level INTEGER NOT NULL DEFAULT 0,
       primary key (pk,ins_date)
   )
   PARTITION BY RANGE (ins_date);
   ```

```
CREATE TABLE part_tags_date_01_2020 PARTITION OF part_tags FOR
VALUES FROM ('2020-01-01') TO ('2020-01-31');
CREATE TABLE part_tags_date_02_2020 PARTITION OF part_tags FOR
VALUES FROM ('2020-02-01') TO ('2020-02-28');
CREATE TABLE part_tags_date_03_2020 PARTITION OF part_tags FOR
VALUES FROM ('2020-03-01') TO ('2020-03-31');
CREATE TABLE part_tags_date_04_2020 PARTITION OF part_tags FOR
VALUES FROM ('2020-04-01') TO ('2020-04-30')

CREATE INDEX part_tags_tag on part_tags using GIN (tag
gin_trgm_ops);
```

As we can see, the only two differences are PARTITION BY RANGE and FOR VALUES FROM .. TO ...

3. In this example, as in the previous example about list partitioning, we have obtained the parent table and all the child tables in a simple way:

```
forumdb=# \d part_tags;
 Partitioned table "public.part_tags"
 Column  | Type                  | Collation | Nullable | Default
---------+-----------------------+-----------+----------+--------
---------
 pk      | integer               |           | not null |
nextval('part_tags_pk_seq'::regclass)
 ins_date | date                 |           | not null |
now()::date
 tag     | character varying(255) |          | not null |
 level   | integer               |           | not null | 0
Partition key: RANGE (ins_date)
Indexes:
 "part_tags_pkey" PRIMARY KEY, btree (pk, ins_date)
 "part_tags_tag" gin (tag gin_trgm_ops)
Number of partitions: 4 (Use \d+ to list them.)

forumdb=# \d part_tags_date_01_2020;
                          Table "public.part_tags_date_01_2020"
  Column  | Type                  | Collation | Nullable | Default
---------+-----------------------+-----------+----------+--------
---------
 pk      | integer               |           | not null |
nextval('part_tags_pk_seq'::regclass)
 ins_date | date                 |           | not null |
now()::date
 tag     | character varying(255) |          | not null |
 level   | integer               |           | not null | 0
Partition of: part_tags FOR VALUES FROM ('2020-01-01') TO
('2020-01-31')
```

```
Indexes:
    "part_tags_date_01_2020_pkey" PRIMARY KEY, btree (pk, ins_date)
    "part_tags_date_01_2020_tag_idx" gin (tag gin_trgm_ops)
```

4. As we did earlier, let's do some INSERT operations:

```
insert into part_tags (tag,ins_date,level) values
('vegetables','2020-01-01',0);
insert into part_tags (tag,ins_date,level) values
('fruits','2020-01-01',0);
insert into part_tags (tag,ins_date,level) values
('orange','2020-02-01',1);
insert into part_tags (tag,ins_date,level) values
('apple','2020-03-01',1);
insert into part_tags (tag,ins_date,level) values ('red
apple','2020-04-01',2);
```

5. And let's check now whether everything is okay:

```
forumdb=# select * from part_tags;
 pk | ins_date   | tag        | level
----+------------+------------+-------
 11 | 2020-01-01 | vegetables | 0
 12 | 2020-01-01 | fruits     | 0
 13 | 2020-02-01 | orange     | 1
 14 | 2020-03-01 | apple      | 1
 15 | 2020-04-01 | red apple  | 2
(5 rows)

forumdb=# select * from part_tags_date_01_2020;
 pk | ins_date   | tag        | level
----+------------+------------+-------
 11 | 2020-01-01 | vegetables | 0
 12 | 2020-01-01 | fruits     | 0
(2 rows)

forumdb=# select * from part_tags_date_02_2020;
 pk | ins_date   | tag    | level
----+------------+--------+-------
 13 | 2020-02-01 | orange | 1
(1 row)

forumdb=# select * from part_tags_date_03_2020;
 pk | ins_date   | tag   | level
----+------------+-------+-------
 14 | 2020-03-01 | apple | 1
(1 row)
```

```
forumdb=# select * from part_tags_date_04_2020;
 pk | ins_date   | tag       | level
----+------------+-----------+-------
 15 | 2020-04-01 | red apple | 2
(1 row)
```

As we can see, all the data has been partitioned correctly.

Partition maintenance

In this section, we will look at how to do the following:

- **Attach a new partition**: If we want to attach a new partition to the parent table, we have to execute the following:

  ```
  CREATE TABLE part_tags_date_05_2020 PARTITION OF part_tags FOR
  VALUES FROM ('2020-05-01') TO ('2020-05-30');
  ```

- As we can see here, a new partition called part_tags_date_05_2020 has been added to the parent table part_tags:

  ```
  forumdb=# \d part_tags;
   Partitioned table "public.part_tags"
   Column    | Type                  | Collation | Nullable | Default
  -----------+-----------------------+-----------+----------+--------
  -----------
   pk        | integer               |           | not null |
  nextval('part_tags_pk_seq'::regclass)
   ins_date  | date                  |           | not null |
  now()::date
   tag       | character varying(255) |          | not null |
   level     | integer               |           | not null | 0
  Partition key: RANGE (ins_date)
  Indexes:
    "part_tags_pkey" PRIMARY KEY, btree (pk, ins_date)
    "part_tags_tag" gin (tag gin_trgm_ops)
  Number of partitions: 5 (Use \d+ to list them.)

  forumdb=# \d part_tags_date_05_2020
   Table "public.part_tags_date_05_2020"
   Column    | Type                  | Collation | Nullable | Default
  -----------+-----------------------+-----------+----------+--------
  -----------
   pk        | integer               |           | not null |
  nextval('part_tags_pk_seq'::regclass)
   ins_date  | date                  |           | not null |
  ```

```
now()::date
 tag       | character varying(255) |           | not null |
 level     | integer                |           | not null | 0
Partition of: part_tags FOR VALUES FROM ('2020-05-01') TO
('2020-05-30')
Indexes:
 "part_tags_date_05_2020_pkey" PRIMARY KEY, btree (pk, ins_date)
 "part_tags_date_05_2020_tag_idx" gin (tag gin_trgm_ops
```

- **Detach an existing partition**: If we want to detach an existing partition from the parent table, we have to execute the following:

```
ALTER TABLE part_tags DETACH PARTITION part_tags_date_05_2020 ;
```

As we can see here, the partition called part_tags_date_05_2020 has been detached from the parent table part_tags:

```
forumdb=# \d part_tags;
 Partitioned table "public.part_tags"
 Column   | Type                   | Collation | Nullable | Default
----------+------------------------+-----------+----------+--------
-----------
 pk       | integer                |           | not null |
nextval('part_tags_pk_seq'::regclass)
 ins_date | date                   |           | not null |
now()::date
 tag      | character varying(255) |           | not null |
 level    | integer                |           | not null | 0
Partition key: RANGE (ins_date)
Indexes:
 "part_tags_pkey" PRIMARY KEY, btree (pk, ins_date)
 "part_tags_tag" gin (tag gin_trgm_ops)
Number of partitions: 4 (Use \d+ to list them.)

forumdb=# \d+ part_tags;
[.. some informtions about fields....]
Partition key: RANGE (ins_date)
Indexes:
    "part_tags_pkey" PRIMARY KEY, btree (pk, ins_date)
    "part_tags_tag" gin (tag gin_trgm_ops)
Partitions: part_tags_date_01_2020 FOR VALUES FROM ('2020-01-01')
TO ('2020-01-31'),
           part_tags_date_02_2020 FOR VALUES FROM ('2020-02-01')
TO ('2020-02-28'),
           part_tags_date_03_2020 FOR VALUES FROM ('2020-03-01')
TO ('2020-03-31'),
           part_tags_date_04_2020 FOR VALUES FROM ('2020-04-01')
TO ('2020-04-30')
```

- **Attach an already existing table to the parent table**: Suppose we have a table called `part_tags_already_exists` already present in our database containing all the tags with a date entered prior to 2019-12-31. This table has the following structure:

```
forumdb=# \d part_tags_already_exists;
 Table "public.part_tags_already_exists"
 Column   | Type                   | Collation | Nullable | Default
----------+------------------------+-----------+----------+--------
------------------
 pk       | integer                |           | not null |
nextval('part_tags_pk_seq'::regclass)
 ins_date | date                   |           | not null |
now()::date
 tag      | character varying(255) |           | not null |
 level    | integer                |           | not null | 0
Indexes:
 "part_tags_already_exists_pkey" PRIMARY KEY, btree (pk, ins_date)
 "part_tags_already_exists_tag_idx" gin (tag gin_trgm_ops)
```

If we want to attach this table containing all the tags with a date entered prior to 2019-12-31 to the parent table, we have to run this statement:

```
ALTER TABLE part_tags ATTACH PARTITION part_tags_already_exists FOR
VALUES FROM ('1970-01-01') TO ('2019-12-31');
```

In this way, the `part_tags_already_exists` table becomes a child table for the parent table, `part_tags`.

Summary

In this chapter, we introduced the topic of table partitioning in PostgreSQL. Partitioning tables is useful when we have tables that become bigger and bigger, making queries slower and slower. We started by introducing the basic concepts of partitioning. We talked about range partitioning, list partitioning, and hash partitioning. We went through some examples of list partitioning and range partitioning. We also saw the two ways in which PostgreSQL solves the problem of partitioning tables: partitioning using inheritance and declarative partitioning.

In the following part of this book, we will return to talking about partitioning in Chapter 13, *Indexes and Performance Optimization*. In the next chapter, we will talk about how PostgreSQL manages users, roles, and in general, the security of our database.

References

- PostgreSQL official documentation about table partitioning: `https://www.postgresql.org/docs/12/ddl-partitioning.html`
- PostgreSQL official documentation about inherintance: `https://www.postgresql.org/docs/12/tutorial-inheritance.html`
- PostgreSQL official documentation about rules: `https://www.postgresql.org/docs/12/rules-update.html`
- PostgreSQL official documentation about triggers: `https://www.postgresql.org/docs/12/sql-createtrigger.html`
- PostgreSQL official documentation about `CREATE TABLE`: `https://www.postgresql.org/docs/12/sql-createtable.html`
- PostgreSQL tuning: `https://pgtune.leopard.in.ua`
- PostgreSQL official documentation about `CONSTRAINT EXCLUSION`: `https://www.postgresql.org/docs/12/runtime-config-query.html#GUC-CONSTRAINT-EXCLUSION`
- PostgreSQL official documentation about trigrams: `https://www.postgresql.org/docs/12/pgtrgm.html`

Section 3: Administering the Cluster

3

In this section, you will learn how to interact with the cluster in order to keep it under control, to tune it, and search for problems or bottlenecks. You will learn about various aspects such as user management, access control, transaction control, and various extensions that can be used with clusters.

This section contains the following chapters:

- *Chapter 10, Users, Roles, and Database Security*
- *Chapter 11, Transactions, MVCC, WALs, and Checkpoints*
- *Chapter 12, Extending the database: the extension ecosystem*
- *Chapter 13, Indexes and Performance Optimization*
- *Chapter 14, Logging and Auditing*
- *Chapter 15, Backup and Restore*
- *Chapter 16, Configuration and Monitoring*

10
Users, Roles, and Database Security

PostgreSQL is a rock solid database, and it pays great attention to security, providing a very rich infrastructure to handle permissions, privileges, and security policies. This chapter builds on the basic concepts introduced in `Chapter 3`, *Managing Users and Connections*, revisiting the role concept and extending knowledge with a particular focus on security and privileges attached to roles (both users and groups). You will learn how to configure every aspect of a role to carefully manage security, from connection to accessing the data within a database.

However, PostgreSQL goes far beyond this and provides a strong mechanism known as **Role Level Security**, which allows a fine-grain definition of policies to mask out part of the data to certain users.

In this chapter, you will also learn about the **Access Control List** (**ACL**) and the way PostgreSQL handles permissions internally, which is the result of granting or revoking privileges. Finally, you will look briefly at the password encryption algorithms that PostgreSQL provides to store role passwords safely.

This chapter covers the following topics:

- Understanding roles
- Access control lists
- Granting and revoking permissions
- Row-level security
- Role password encryption
- SSL connections

Understanding roles

In Chapter 3, *Managing Users and Connections*, you have seen how to create new roles, a stereotype that can act both as a single user or a group of users. The CREATE ROLE statement was used to create the role, and you learned about the main properties a role can be associated with.

This section extends the concepts you have read about in Chapter 3, *Managing Users and Connections*, introducing more interesting and security-related properties of a role.

Just as a quick reminder, the synopsis for creating a new role is the following:

```
CREATE ROLE name [ [ WITH ] option [ ... ] ]
```

Here, an option can be indicated in a positive form, that is, associating the property with the role, or in a negative form with the NO prefix, which removes the property from the role. Some properties are not assigned to new roles by default, so you should take your time and consult the documentation of the CREATE ROLE statement in order to see what the default value is for every property. If you are in doubt, associate explicitly the properties you need and negate those you absolutely don't want your roles to have.

Properties related to new objects

There are two main capabilities that a role can acquire in order to create new objects, and both should be given to trusted parties:

- CREATEROLE allows a role to create other roles (and therefore database accounts and groups).
- CREATEDB allows a role to create other databases within the cluster.

By default, if not specified explicitly, a new role is created without such capabilities, hence:

```
template1=# CREATE ROLE luca;
```

Is wholly equivalent to the following command:

```
template1=# CREATE ROLE luca
            WITH NOCREATEROLE
                 NOCREATEDB;
```

Properties related to superusers

With the SUPERUSER property, the role is created as a cluster administrator, that is, a role that has every right on every object within the cluster, including the capability to change the PostgreSQL configuration, terminate user connections, and halt the cluster.

It is possible to have as many superusers as you need in a cluster. However, being a class of users without any particular restriction, it is a good habit to avoid giving all the permissions to untrusted users and void giving all the permissions to untrusted users unless it is not strictly necessary.

Properties related to replication

The REPLICATION property is used to specify that the new role will be able to use the replication protocol, a particular message-based protocol that PostgreSQL uses to replicate data from one cluster to another.

REPLICATION is an option that allows a role to access all the data within the cluster without any particular restriction. Therefore, it is usually granted to just those roles used for replication. You may think of granting the REPLICATION option like granting the SUPERUSER, and therefore you have to grant it carefully.

Due to its security implications, if not specified otherwise, the NOREPLICATION option is set.

Properties related to row-level security

Role Level Security (**RLS**) is a policy enforcement mechanism that prevents certain roles from gaining access to specific tuples within specific tables. In other words, it applies security constraints at the level of table rows, hence we can also use the name **row-level security**.

There is a single option that drives RLS: BYPASSRLS. If the role has such an option, the role **bypasses** (that is, is not subjected to) all security constraints on every row within the cluster. The default for this option, as you can imagine, is to negate it (that is, NOBYPASSRLS), so that roles are subjected to security enforcement whenever possible.

It is important to note that cluster superusers are always able to bypass row-level security policies.

You will learn more about RLS in the *Row-level security* section of this chapter.

Changing properties of existing roles: the ALTER ROLE statement

As you can imagine, once they have been created, roles are not immutable: you can add or remove properties to a role by means of the ALTER ROLE statement. The synopsis for the statement is very similar to the one used to create a role, and is as follows:

```
ALTER ROLE name [ [ WITH ] option [ ... ] ]
```

Here, name is the unique role name and the options are specified in the exact same manner as in the CREATE ROLE statement.

As an example, imagine you want to provide the luca role with the capabilities to create databases and new roles. You can issue two ALTER ROLE statements or combine the options as follows:

```
forumdb=# ALTER ROLE luca WITH CREATEDB;
ALTER ROLE

forumdb=# ALTER ROLE luca WITH CREATEROLE;
ALTER ROLE

-- equivalent to the above two statements
forumdb=# ALTER ROLE luca CREATEROLE CREATEDB;
ALTER ROLE
```

And if you, later on, change your mind, you can remove one or both options by assigning the negated form:

```
forumdb=# ALTER ROLE luca NOCREATEROLE, NOCREATEDB;
ALTER ROLE
```

The ALTER ROLE statement is always executable by a cluster superuser, but can also be executed by a non-superuser role that has the CREATEROLE option (that is, can create, and therefore manipulate, other roles), but only if the statement is applied to a non-superuser role.

Renaming an existing role

The `ALTER ROLE` statement also allows for a change in the name of the role: the `RENAME` clause allows for a role to be substituted by another unique role name. As an example, let's rename a role's *short* username to a longer one:

```
forumdb=# ALTER ROLE enrico RENAME TO enrico_pirozzi;
ALTER ROLE
```

In case the system is using MD5 passwords, `ALTER ROLE` will issue a warning. Since MD5 passwords are computed using the role name as a "salt", changing the role name means the salt is no longer valid and therefore the password is reset. In other words, if the system is using MD5 passwords, the command will have produced the following warning:

```
forumdb=# ALTER ROLE enrico RENAME TO enrico_pirozzi;
NOTICE:  MD5 password cleared because of role rename
ALTER ROLE
```

This simply reminds you to change the password for the renamed role. This is not a problem for SCRAM-SHA-256 passwords since they don't use the role name as a salt (further details can be found in the *Role Password Encryption* section later in this chapter).

SESSION_USER versus CURRENT_USER

The `ALTER ROLE` statement operates on an existing role, specified by its role name. It is, however, possible to refer to the current role with two particular keywords: `SESSION_USER` and `CURRENT_USER`.

 Mind the usage of `user` in the `SESSION_USER` and `CURRENT_USER` special keywords. They still refer to the concept of *role*, but for backward compatibility, they use the `user` nomenclature. While there is a `CURRENT_ROLE` keyword, there is not the equivalent `SESSION_ROLE` one.

`SESSION_USER` is the role name of the role that is connected to the database.

`CURRENT_USER` is the role name of the role that has been explicitly set by a `SET ROLE` statement.

Once a connection is established, the two keywords refer to the very same role that is the one that opened the connection (that is, the one specified in the connection parameters or in the connection string). If the role performs an explicit `SET ROLE` operation, `SESSION_USER` remains unchanged, while `CURRENT_USER` reflects the last specified role.

Let's see this in action. Suppose the user `luca` opens a connection to the database. In the beginning, both `SESSION_USER` and `CURRENT_USER` hold the same value:

```
$ psql  -U luca forumdb

forumdb=> SELECT current_user, session_user;
 current_user | session_user
--------------+--------------
 luca         | luca
(1 row)
```

Assume the `luca` role is a member of the `forum_stats` group, so that it is possible to perform an explicit *transformation* to such role:

```
forumdb=> SET ROLE forum_stats;
SET
forumdb=> SELECT current_user, session_user;
 current_user | session_user
--------------+--------------
 forum_stats  | luca
(1 row)
```

As you can see, after the `SET ROLE` statement, `CURRENT_USER` changed its value to reflect the role the user is actually playing, while `SESSION_USER` holds the *original* value by which the user connected to the database.

Per-role configuration parameters

Along with role properties and granted permissions, roles can also be attached with some configuration parameters that can document the usage. Essentially, it is possible to attach a list of `SET` commands to a role so that, every time the role connects to a database, such commands are implicitly executed.

Let's say the user `luca` executes a `SET` command for the `client_min_messages` value every time they connect to the database:

```
$ psql  -U luca forumdb

forumdb=> SET client_min_messages TO 'DEBUG';
SET
```

This can be annoying and, most notably, risky. The user could forget to execute the
SET command that they require for proper functioning of the connection. It is possible to
change the role so that they will be executing the SET command automatically as soon as a
connection is established:

```
forumdb=# ALTER ROLE luca
        IN DATABASE forumdb
        SET client_min_messages TO 'DEBUG';
ALTER ROLE
```

And now, every time the luca role connects to the forumdb database, the SET command is
automatically executed:

```
$ psql -U luca forumdb

forumdb=> SHOW client_min_messages;

 client_min_messages
---------------------
 debug
(1 row)
```

The general syntax for changing runtime parameters for a role is as follows:

```
ALTER ROLE name IN DATABASE dbname SET parameter_name TO parameter_value
```

Here, you have to specify the role name or the special keyword ALL for every existing role,
the database name, and the name and value of the parameter you want to change.

It is also possible to discard any per-role configuration with the RESET ALL clause, as in the
following example:

```
forumdb=# ALTER ROLE luca
        IN DATABASE forumdb
        RESET ALL;
ALTER ROLE
```

Inspecting roles

There are different ways to inspect existing roles and get information about their properties. One quick approach, as already seen in Chapter 3, *Managing Users and Connections*, is to use the \du command in psql:

```
forumdb=> \du
                            List of roles
   Role name    |                   Attributes                    |    Member of
 ---------------+-------------------------------------------------+--------------------
   book_authors | Cannot login                                    | {}
   enrico       |                                                 | {book_authors,
                |                                                 |    forum_admins}
   forum_admins | Cannot login                                    | {}
   forum_stats  | Cannot login                                    | {}
   luca         | 1 connection                                    | {book_authors,
                |                                                 |    forum_stats}
   postgres     | Superuser, Create role,                         |
                | Create DB, Replication, Bypass RLS | {}
   test         |                                                 | {forum_stats}
```

The **Attributes** column provides a mnemonic description of the role properties, while the **Member of** column states all the groups a role is within. As an example, the luca role is limited to a single connection and belongs to the book_authors and forum_stats groups.

Besides the special commands of psql, you can always query the system catalog to get information about the existing roles. The main entry point is the table pg_authid, which contains one row per existing role with a column that reflects every property of the role (that is, what you defined via the CREATE ROLE or ALTER ROLE statements), for example:

```
forumdb=# \x
Expanded display is on.
forumdb=# SELECT * FROM pg_authid WHERE rolname = 'luca';
-[ RECORD 1 ]--+------------------------------------
oid            | 16390
rolname        | luca
rolsuper       | f
rolinherit     | t
rolcreaterole  | f
rolcreatedb    | f
rolcanlogin    | t
rolreplication | f
rolbypassrls   | f
rolconnlimit   | 1
rolpassword    | md5bd18b4163ec8a3ad833d867a5933c8ec
rolvaliduntil  |
```

Every role has a unique name and also an `OID` value, which represents the role as a numerical value. This is similar to how users are represented in the Unix system (and many others) where the numerical value of a role is used only internally.

Many of the role properties have a Boolean value, where 'f' means false (that is, NO-option) and 't' means true (that is, with-option). For instance, in the preceding example, you can see that `rolcreatedb` is false, which means that the role has been created (or altered) with the `NOCREATEDB` option.

The role password (`rolpassword` field) is expressed as a hash. In this case, this tells us that the password has been encrypted with MD5 since the hash starts with the string 'md5'. On the other hand, if the password has been encrypted with SCRAM-SHA-256, the hash starts with 'SCRAM-SHA-256'.

There is another possible catalog, named `pg_roles`, which displays the very same information about `pg_authid` with the exception of the `rolpassword` field, which is always masked out:

```
forumdb=> SELECT * FROM pg_roles WHERE rolname = 'luca';
-[ RECORD 1 ]--+---------
rolname        | luca
rolsuper       | f
rolinherit     | t
rolcreaterole  | f
rolcreatedb    | f
rolcanlogin    | t
rolreplication | f
rolconnlimit   | 1
rolpassword    | ********
rolvaliduntil  |
rolbypassrls   | f
rolconfig      |
oid            | 16390
```

Why two similar views of the same data? In order to query `pg_authid`, you must be a cluster superuser, while every user can query `pg_roles` since there is no hint regarding the role password.

What about group membership? You can query the special `pg_auth_members` catalog to get information about what roles are members of what other roles. As an example, the following query provides a list of groups:

```
forumdb=> SELECT r.rolname, g.rolname AS group,
                m.admin_option AS is_admin
         FROM pg_auth_members m
```

```
            JOIN pg_roles r ON r.oid = m.member
            JOIN pg_roles g ON g.oid = m.roleid
       ORDER BY r.rolname;
  rolname    |         group         | is_admin
-------------+-----------------------+----------
 enrico      | book_authors          | f
 enrico      | forum_admins          | f
 luca        | forum_stats           | f
 luca        | book_authors          | f
 pg_monitor  | pg_read_all_settings  | f
 pg_monitor  | pg_read_all_stats     | f
 pg_monitor  | pg_stat_scan_tables   | f
 test        | forum_stats           | f
(8 rows)
```

Roles that inherit from other roles

We have already seen in `Chapter 3`, *Managing Users and Connections*, that a role can contain other roles, therefore behaving as a group.

When a role becomes a member of another role, it gets all the permissions of the group role. However, there are cases where such privileges are dynamically granted, that is, the member role will have the privileges transparently, and cases where the privileges will be granted statically, that is, the member role needs to explicitly become the group role in order to use its privileges. The INHERIT property of a role discriminates how roles can use the group privileges. If a role has the INHERIT property, it will propagate its permissions to its members, while if it has the NOINHERIT property, it will not propagate.

In order to understand the difference and the implication, let's build a couple of groups and assign members to them:

```
forumdb=# CREATE ROLE forum_admins WITH NOLOGIN;
CREATE ROLE

forumdb=# CREATE ROLE forum_stats WITH NOLOGIN;
CREATE ROLE

forumdb=# REVOKE ALL ON users FROM forum_stats;
REVOKE

forumdb=# GRANT SELECT (username, gecos) ON users TO forum_stats;
GRANT

forumdb=# GRANT forum_admins TO enrico;
GRANT ROLE
```

```
forumdb=# GRANT forum_stats  TO luca;
GRANT ROLE
```

It is quite simple to see how the `enrico` role can perform what the `forum_admins` role allows him to do on the `users` table: being a member of the `forum_admins` group, the `enrico` role can perform any action against the `users` table. This can be demonstrated by a couple of simple instructions:

```
$ psql  -U enrico forumdb

forumdb=> SELECT * FROM users;
 pk | username  |     gecos      |        email
----+-----------+----------------+---------------------
  1 | fluca1978 | Luca Ferrari   | fluca1978@gmail.com
  2 | sscotty71 | Enrico Pirozzi | sscotty71@gmail.com
(2 rows)

forumdb=> UPDATE users SET gecos = upper( gecos );
UPDATE 2

forumdb=> SELECT * FROM users;
 pk | username  |     gecos      |        email
----+-----------+----------------+---------------------
  1 | fluca1978 | LUCA FERRARI   | fluca1978@gmail.com
  2 | sscotty71 | ENRICO PIROZZI | sscotty71@gmail.com
(2 rows)
```

As you can see, the user `enrico` has actually changed the name and surname of the existing users to a full uppercase string. Let's now see what the other user can do:

```
$ psql  -U luca forumdb

forumdb=> SELECT * FROM users;
ERROR:  permission denied for table users
forumdb=> SELECT username, gecos FROM users;
 username  |     gecos
-----------+----------------
 fluca1978 | LUCA FERRARI
 sscotty71 | ENRICO PIROZZI
(2 rows)

forumdb=> UPDATE users SET gecos = lower( gecos );
ERROR:  permission denied for table users
```

As you can see, the user `luca` cannot perform anything other than what has been granted to the `forum_stats` role, that is, a group he belongs to.

It is possible to change the privileges of the user `luca` by either assigning to the role the new grants or by adding another group with more privileges. For instance, if we want all users in the `forum_stats` group to not be able to read anything other than the columns `username` and `gecos`, while providing `luca` a special grant even if he belongs to such group, it is possible to explicitly set the permission to `luca`, and to him alone:

```
forumdb=# GRANT SELECT ON users TO luca;
GRANT
```

Once the permission has been granted (by a superuser), the `luca` role can use it:

```
% psql -h miguel -U luca forumdb

forumdb=> SELECT * FROM users;
 pk | username  |      gecos      |         email
----+-----------+-----------------+---------------------
  1 | fluca1978 | LUCA FERRARI    | fluca1978@gmail.com
  2 | sscotty71 | ENRICO PIROZZI  | sscotty71@gmail.com
(2 rows)
```

As you can see, the special permission granted to `luca` wins out against the more restrictive one granted to the `forum_stats` group, of which `luca` is a member.

In order to be able to configure your users and groups, you need to understand the privilege chain.

Understanding how privileges are resolved

When a role performs a SQL statement, PostgreSQL checks whether such a role is allowed to perform the task against the object. For example, when the user `luca` performs `SELECT` against the table users, PostgreSQL verifies whether the role has been granted to do so.

In case the role has not been granted explicitly, PostgreSQL searches for all the groups the role belongs to. In the event that one of the group has the permission requested, the operation is allowed. In the event that no group has the requested permission, and the permission has not been set for the `PUBLIC` catch-all special role, the operation is rejected.

However, this is only a part of the story: when the system checks the groups a role belongs to, it does stop searching for a permission in case a group has the NOINHERIT property. When you define a role, you can define the INHERIT (and its counterpart NOINHERIT) property, which dictates how the permission resolution should happen. In the case of INHERIT (the default), permissions are dynamically inherited, otherwise they are not.

In the previous section, you have seen the INHERITS default behavior in action. The luca role did get the inherited permissions that allowed him to perform SELECT of only two columns on the users table. Even if luca is not granted that permission, the system checks all the groups he belongs to in order find one, and it finds it in the forums_stats group. Since the permissions are dynamically inherited from a role to all its contained ones, that is, from a group to its members, this is akin to luca having such permission set on his own role, and so the operation is allowed.

In order to have a better understanding of this, let's introduce another group, named forum_emails, that can read the emails on the users table, and assign such a group to forum_stats. We would expect that forum_stats, being a member of forum_emails, can read the email column, but since the forum_emails group has been created with the NOINHERIT property, it cannot:

```
-- remove any explicit SELECT permission
-- so luca will have only those from its group
forumdb=# REVOKE SELECT ON users FROM luca;
REVOKE

-- create the new group
forumdb=# CREATE ROLE forum_emails WITH NOLOGIN NOINHERIT;
CREATE ROLE

-- assign permissions
forumdb=# GRANT SELECT (email) ON users TO forum_emails;
GRANT

-- assign the role to the group
forumdb=# GRANT forum_emails TO forum_stats;
GRANT ROLE
```

Now, luca is a member of forum_stats and forum_emails, but since the latter "does not export" its permissions to its members, luca cannot get the permissions to read the email column:

```
% psql  -U luca forumdb

forumdb=> SELECT username, gecos, email FROM users;
ERROR:  permission denied for table users
```

Being a member of a role means the role can always explicitly *become* the group:

```
forumdb=> SELECT current_role;
 current_role
--------------
 luca
(1 row)

forumdb=> SET ROLE TO forum_emails;
SET
forumdb=> SELECT current_role;
 current_role
--------------
 forum_emails
(1 row)

forumdb=> SELECT email FROM users;
         email
---------------------
 fluca1978@gmail.com
 sscotty71@gmail.com
(2 rows)

forumdb=> SELECT gecos FROM users;
ERROR:  permission denied for table users
```

Let's now change the INHERIT property of the forum_emails role so that it does have such a property and therefore propagates it to its members:

```
forumdb=# ALTER ROLE forum_emails WITH INHERIT;
ALTER ROLE
```

And now let's see whether the luca role can use both privileges of the forum_stats and forum_emails groups simultaneously:

```
$ psql -U luca forumdb

forumdb=> SELECT gecos, username, email FROM users;
     gecos      | username |        email
----------------+----------+---------------------
 LUCA FERRARI   | fluca1978 | fluca1978@gmail.com
 ENRICO PIROZZI | sscotty71 | sscotty71@gmail.com
(2 rows)
```

Great! Now the role can use both group privileges at the very same time without having to explicitly change its current role.

Role inheritance overview

When a role is a member of one or more other roles, the privileges resolution goes like this:

- If the role has the privilege requested, nothing more is checked and the operation is allowed (for example, you granted permission to this role).
- If the role does not have the privilege, the latter is searched for in all the parents, that is, all the groups of which the role is a direct member. If the privilege is found in one of the parent groups, the operation is allowed.
- If neither the role nor parent groups have the requested privileged, the latter is searched for in the grandparents of the role. If it is found in any of the grandparents, and there is the INHERIT property set on such a grandparent, the operation is allowed, otherwise it is rejected.
- If the privilege has not been found, the search continues within the parent of the grandparents that have the INHERIT property set.

In other words, a certain privilege is automatically inherited if a role is a direct member of a group, or is dynamically inherited if the role is a member of a group that has the INHERIT property set.

In any case, the role can always exploit the privilege via an explicit SET ROLE statement, which means the INHERT property is used only to prevent the role from changing into the group.

You can think of this as a way to force a role to explicitly declare it is going to perform some important task, and so it must explicitly SET the role it is going to use for that task.

Access control lists

PostgreSQL stores permissions assigned to roles and objects as **Access Control Lists** (**ACLs**), and, when needed, it examines the ACLs for a specific role and a database object in order to understand whether the command or query can be performed. In this section, you will learn what ACLs are, how they are stored, and how to interpret them to understand what permissions an ACL provides.

An ACL is a representation of a group of permissions with the following structure:

```
grantee=flags/grantor
```

Here, we see the following:

- `grantee` is the role name of the role to which the permissions are applied.
- `flags` is the string representing the permissions.
- `grantor` is the user who granted the permissions.

Whenever the granted and grantee results in the same name, the role is the owner of the database object.

The flags that can be used in an ACL are those reported in the following table. As you can see, not all the flags apply to all the objects: for example it does not make sense to have a "delete" permission on a function, and it does not make sense to have an "execute" permission on a table:

Flag	Description	Statements	Applies to
a	append, insert new data	INSERT	tables, columns
r	read, get data	SELECT	tables, columns, and sequences
w	write, update data	UPDATE	tables
d	delete data	DELETE	tables
D	delete all data	TRUNCATE	tables

C	create a new object	`CREATE`	databases, schemas, and table spaces
c	connect to a database		database
t	trigger, react to data changes	`CREATE TRIGGER`	tables
T	crate temporary objects	`CREATE TEMP`	tables
x	cross reference between data	`FOREIGN KEY`	tables
X	execute runnable code	`CALL, SELECT`	functions, routines, and procedures

U	use of various objects		sequences, schemas, foreign objects, types, and languages

With the list of possible flags in mind, it now becomes easy to decode an ACL such as the following, which is related to a table object:

```
luca=arw/enrico
```

First of all, identify the roles involed: `luca` and `enrico`. `luca` is the role before the equals sign, hence it is the role the ACL refers to, which means this ACL describes what permissions the `luca` role has. The other role, `enrico`, is after the slash sign and therefore is the role that granted `luca` permissions. Now, with respect to the flags, the ACL provides an append (a), read (r), and write (w) permissions. The above reads as "`enrico` granted `luca` to perform `INSERT`, `UPDATE`, and `SELECT` on the table."

Let's now see an example of ACLs from a table in the database: you can use the special `\dp` psql command to get information about a table:

```
forumdb=> \dp categories
                            Access privileges
 Schema |    Name     | Type  |  Access privileges   | Column privileges  |
 Policies
--------+-------------+-------+----------------------+--------------------
+----------
 public | categories  | table | enrico=arwdDxt/enrico+|                    |
        |             |       | luca=arw/enrico      +|                    |
        |             |       | =d/enrico            |                    |
```

The ACLs are clearly reported in the `Access privileges` column of the command output. The first line of the ACLs makes a statement regarding the owner of the `categories` table: since the grantee and the grantor are the same role (`enrico`), this is the table owner. Moreover, `enrico` has the append (a), read (r), write (w), delete (d), truncate (D), trigger (t), and cross reference (x) permissions. If you think carefully, this means that the role can do everything possible in relation to a table object. Therefore, it is possible to read this as "*a table owner can do everything on that table.*"

The second line of the ACL is the one decoded above, and reads as "luca can INSERT, UPDATE, and SELECT data." The third line of the ACL is a little more obscure: the grantor is still the enrico role, but there is no grantee before the equals sign. This means that ACL refers to every role. Since the ACL includes only the delete (d) permission, this means that every role in the database can delete rows from the table, as enrico desires.

ACLs are processed to find a match. Imagine that the luca role wants to delete a row from the table, and therefore issues a DELETE statement. Is that statement allowed or rejected? Reading the ACL related to the luca role (luca=arw/enrico), it is clear that the role cannot delete anything from the table. However, there is a "catch-all" ACL that allows every role to perform a DELETE operation (=d/enrico), hence even the luca role is allowed to remove tuples.

On the other hand, a different role (for example, forum_stats) is not allowed to perform any INSERT on the table because there is no specific permission either for that role or for any other role.

But how are those ACLs being produced? First of all, they have all been created by the user enrico, so assuming he is the one connected to the database, the sequence of GRANT statements should have been as follows:

```
-- generates ACL: luca=arw/enrico
forumdb=> GRANT SELECT, UPDATE, INSERT
          ON categories
          TO luca;
GRANT

-- generates ACL: =d/enrico
forumdb=> GRANT DELETE ON categories
          TO PUBLIC;
GRANT
```

Now that you have seen how PostgreSQL manages ACLs and how it translates GRANT and REVOKE commands into ACLs, it is time to see what the default permissions are that are granted to a role.

Default ACLs

What happens if an object is created and neither any GRANT or REVOKE is applied to it? The system does not store any ACL for such an object, as you can see by creating a simple empty table and inspecting privileges on it:

```
forumdb=> CREATE TABLE foo();
CREATE TABLE
forumdb=> \dp foo
                                Access privileges
 Schema | Name | Type  | Access privileges | Column privileges | Policies
--------+------+-------+-------------------+-------------------+----------
 public | foo  | table |                   |                   |
(1 row)
```

Since there is no ACL associated with the table, how can PostgreSQL know what roles are permitted to do what on the object? The answer lies in the default privileges: PostgreSQL applies a set of default privileges to the object and checks against its default list.

Most notably, if the role is the owner of the object, this has all the available privileges for such an object. If the role is not the owner, the PUBLIC permissions are inspected, that is, all permissions assigned to the special PUBLIC role for that kind of object are used.

The list of PUBLIC associated privileges is quite short, for security reasons, and can be summarized as:

- Execute permission (X) on routines
- Connect and create temporary objects on databases (cT)
- Use of languages, types, and domain (U)

As you can see, by default, the PUBLIC set of privileges does not allow a role to do anything really dangerous, and therefore the only way to authorize a role to perform actions against objects is to GRANT and REVOKE permissions carefully.

The first time GRANT is performed against an object, PostgreSQL also introduces the default ACL for the owner of that object. In the case of the preceding table, foo, the owner will have an ACL such as luca=arwdDxt/luca (assuming the luca role is the owner), so suppose we give permissions to manipulate data to enrico:

```
forumdb=> \dp foo
                                Access privileges
 Schema | Name | Type  | Access privileges | Column privileges | Policies
--------+------+-------+-------------------+-------------------+----------
 public | foo  | table |                   |                   |
(1 row)
```

```
forumdb=> GRANT SELECT, INSERT,
              UPDATE, DELETE
        ON foo TO enrico;
GRANT

forumdb=> \dp foo
                        Access privileges
 Schema | Name | Type  | Access privileges | Column privileges | Policies
--------+------+-------+-------------------+-------------------+----------
 public | foo  | table | luca=arwdDxt/luca+|                   |
        |      |       | enrico=arwd/luca  |                   |
(1 row)
```

As you can see, after GRANT, the ACL is made by two entries, the one we just granted, and the one that was implicitly applied to the owner.

It is also important to note that ACLs store what a role can do, not what it cannot do. Everything not listed in the ACLs is rejected. To better understand this, consider revoking a permission to the enrico role:

```
forumdb=> REVOKE TRUNCATE ON foo FROM enrico;
REVOKE

forumdb=> \dp foo
                        Access privileges
 Schema | Name | Type  | Access privileges | Column privileges | Policies
--------+------+-------+-------------------+-------------------+----------
 public | foo  | table | luca=arwdDxt/luca+|                   |
        |      |       | enrico=arwd/luca  |                   |
(1 row)
```

As you can see, the revoke did not change the ACL line for the enrico role. The role did not have the permission revoked; that is, it was already implicitly revoked, hence no change at all is requested.

Similarly, revoking permissions from PUBLIC does not affect already existing ACLs. If we remove the insert permission from every user, enrico will still retain his own permission because ACLs are stored in additively:

```
forumdb=> REVOKE INSERT ON foo FROM PUBLIC;
REVOKE

forumdb=> \dp foo
                        Access privileges
 Schema | Name | Type  | Access privileges | Column privileges | Policies
--------+------+-------+-------------------+-------------------+----------
 public | foo  | table | luca=arwdDxt/luca+|                   |
```

```
            |         |         | enrico=arwd/luca  |                    |
 (1 row)
```

To summarize, ACLs are always empty for a freshly created object. In this situation, the object owner has all available permissions and other roles have default permissions associated with PUBLIC. The first GRANT or REVOKE statement executed against that object will create the explicit owner ACL and, in the case of GRANT, will add another one accordingly.

ACLs are stored as granted privileges. What is not explicitly set in an ACL is implicitly rejected as it has been revoked.

Knowing default ACLs

It is now clear that the owner of an object has all the possible permissions related to such an object. But what about other roles? It is possible to inspect the default ACL provided once an object is instantiated via the special function acldefault.

The function accepts two arguments – a type of object (for example, a relation/table, a function) and the OID value of the role that is supposed to create the object. The function will return the ACLs that will be in place since the creation of the object.

For example, in order to see the permissions provided when your role creates a new table (type 'r'), you can perform the following query:

```
forumdb=> SELECT acldefault( 'r', r.oid )
          FROM pg_roles r
          WHERE r.rolname = CURRENT_ROLE;
     acldefault
---------------------
 {luca=arwdDxt/luca}
(1 row)
```

Nothing new here, but what about the creation of a function (type 'f')? It is now easy to see the following:

```
forumdb=> SELECT acldefault( 'f', r.oid )
          FROM pg_roles r
          WHERE r.rolname = CURRENT_ROLE;
      acldefault
----------------------
 {=X/luca,luca=X/luca}
(1 row)
```

This time, two ACLs are produced: the first grants all users the executable permission, while the latter specifies that the owner is the luca role with executable permissions, too.

You can inspect all the default ACLs for a specific user by means of its OID and the type of object, where the main types are 'r' for tables, 'c' for columns, 'l' for languages, and 'f' for routines and procedures. Other types are available. Please refer to the official documentation. It is now time to see how to manipulate ACLs and permissions in a practical way. In the next section, you will learn how to deal with permission management.

Granting and revoking permissions

As you have seen in Chapter 3, *Managing Users and Connections*, a role contains a collection of permissions that are provided by means of a GRANT statement and removed by means of a REVOKE statement. Permissions are stored internally as ACLs, as you have seen in the previous section.

This section revisits the GRANT and REVOKE statements to better help you understand how to use them, with respect to different database objects.

The GRANT statement has the following synopsis:

```
GRANT <permission, permission, ...> ON <database-object> TO <role>;
```

Here, you list all the permissions you want to associate with the target role for the specified database object. It is also possible to extend the GRANT statement with the WITH GRANT OPTION clause, which will cause the target role to be able to grant the same permissions it has received to another role.

The REVOKE statement has a similar synopsis:

```
REVOKE <permission, permission, ..> ON <database-object> FROM <role>;
```

There is a special role, named PUBLIC, that can be used when dealing with permission management. It is not a concrete role, rather a marker to indicate "all available roles." In other words, if you grant a permission to PUBLIC, you are implicitly granting such permission to all available roles.

But what does "all available roles" mean? It means all existing and future roles. The PUBLIC role represents any role that will ever be present in the system, at the time the permission is managed and in the future.

According to the above, in order to prevent any user from accessing your objects, you should always remove all the permissions from the special PUBLIC role, and then selectively provide the permissions you need to specific roles.

In the following sections, we will detail different permissions to assign or remove grouping, and classify them depending on the database object. As a general rule of thumb, the list of permissions depends on the action you can run against the database object.

In many cases, the special keyword ALL is a substitute for every permission related to the database object.

Permissions related to tables

We have already seen the main permissions related to a database table. They refer to the main statements that can run against a table object, such as SELECT, INSERT, UPDATE, DELETE, and TRUNCATE. Moreover, it is possible to use the special keywords TRIGGER and REFERENCES to create triggers and foreign keys within the table.

Of course, the special keyword ALL does include all the preceding permissions.

As an example, in order to provide the forum_stats role with the permissions to read, update, and insert data into the categories table, without granting permissions to execute the other actions, you can do the following:

```
forumdb=# REVOKE ALL
        ON categories FROM forum_stats;
REVOKE

forumdb=# GRANT SELECT, INSERT, UPDATE
        ON categories TO forum_stats;
GRANT

forumdb=# \dp categories
                              Access privileges
 Schema |    Name    | Type  |     Access privileges      | Column
privileges | Policies
--------+------------+-------+----------------------------+------------------
--+----------
 public | categories | table | postgres=arwdDxt/postgres+|
 |
        |            |       | forum_stats=arw/postgres   |
 |
(1 row)
```

The first REVOKE statement is not mandatory, but it is good practice. Since we want to ensure that the role has precisely the permissions we are going to grant and not one more, removing all the permissions from the role ensures that a previous GRANT statement does not persist.

As you can see, the ACL for the `forum_stats` user reflects the permissions we granted.

Column-based permissions

Since certain statements related to table objects can address columns directly, for example, SELECT and UPDATE, it is also possible to grant or revoke column permissions. The synopsis is the same, but you can list the columns that the permission refers to.

Column permissions can be applied only to SELECT, UPDATE, INSERT, and REFERENCES permissions because those are the ones that can refer to columns explicitly; the special keyword ALL encapsulates the entire list of permissions.

As an example, consider a scenario where the `forum_stats` user can interact with the table users only on the `gecos` and `username` columns, able to read both of them but to update just the first one. The permissions could be assigned as follows:

```
forumdb=# REVOKE ALL ON users
          FROM forum_stats;
REVOKE

forumdb=# GRANT SELECT (username, gecos),
                UPDATE (gecos)
          ON users TO forum_stats;
GRANT
```

As already stressed, the first REVOKE statement is good practice to ensure that the permissions for the role are reset before we assign the ones we want. Then, we grant the SELECT and UPDATE permissions, specifying the columns every statement will be able to interact with.

The side effect of the preceding GRANT statement is that the `forum_stats` role is no longer able to issue SELECT or UPDATE with a column list wider than the one specified in GRANT:

```
-- denied, not all the columns can be read!
forumdb=> SELECT * FROM users;
ERROR:  permission denied for table users

-- allowed
forumdb=> SELECT gecos, username FROM users;
     gecos       | username
-----------------+-----------
 LUCA FERRARI    | fluca1978
 ENRICO PIROZZI  | sscotty71
(2 rows)
```

```
-- denied, the 'username' column cannot be updated!
forumdb=> UPDATE users SET username = upper( username );
ERROR:  permission denied for table users

-- allowed
forumdb=> UPDATE users SET gecos = lower( gecos );
UPDATE 2
```

Let's now inspect the permissions for the `users` table:

```
forumdb=> \dp users
                                          Access privileges
 Schema | Name  | Type  |       Access privileges       |      Column
 privileges      | Policies
--------+-------+-------+-------------------------------+-------------------
---------+----------
 public | users | table | postgres=arwdDxt/postgres    +| username:
+|
        |       |       | forum_admins=arwdDxt/postgres |
forum_stats=r/postgres +|
        |       |       |                               | gecos:
+|
        |       |       |                               |
forum_stats=rw/postgres+|
```

There are two important things here that are different from all the previous examples. First, the `Access privileges` column does not include any entry related to the `forum_stats` role even if we explicitly granted permission. Second, the `Column privileges` column is now full of rows related to the `forum_stats` role.

Every row in `Column privileges` refers to exactly one column of the table and contains an ACL for every allowed role. For instance, the username column has the ACL `forum_stats=r/postgres`, which means that the `forum_stats` role has read permission (that is, `SELECT`) on such a column. The gecos column has the ACL `forum_stats=rw/postgres`, which reads as the `forum_stats` role can both read and write on the column (that is, `SELECT`, `UPDATE`).

To summarize, if the role has been granted one or more permissions on all the columns, the ACL is placed under the **Access privileges** column, and if the permissions are related to specific columns, the ACL is shown under the **Column privileges** column.

You must be careful to not make permissions conflict with one another. For instance, assume we wrongly provide a SELECT permission to the forum_stats role:

```
forumdb=# GRANT SELECT
          ON users TO forum_stats;
GRANT
```

If we inspect the permissions after such a statement, we can see that the ACL has been inserted as an access privilege:

```
forumdb=# \dp users
                                   Access privileges
 Schema | Name  | Type  |        Access privileges        |        Column
 privileges      | Policies
--------+-------+-------+---------------------------------+-------------------
--------+----------
 public | users | table | postgres=arwdDxt/postgres     +| username:
+|
        |       |       | forum_admins=arwdDxt/postgres+|
 forum_stats=r/postgres +|
        |       |       | forum_stats=r/postgres          | gecos:
+|
        |       |       |                                 |
 forum_stats=rw/postgres+|
        |       |       |                                 | email:
+|
        |       |       |                                 |
 forum_emails=r/postgres |
```

Which permission will be considered in the case of a SELECT statement?

It is easy to test and see that PostgreSQL considers the last granted permission more open than the column one. Therefore, the role has been granted to select every column on the table:

```
forumdb=> SELECT * FROM users;
 pk | username  |     gecos      |        email
----+-----------+----------------+---------------------
  1 | fluca1978 | luca ferrari   | fluca1978@gmail.com
  2 | sscotty71 | enrico pirozzi | sscotty71@gmail.com
(2 rows)
```

Fixing the problem may not be as simple as you think. Revoking read permission on the columns you don't want the role to have access to will not do what you may expect:

```
forumdb=# REVOKE SELECT (pk, email)
          ON users FROM forum_stats;
REVOKE
```

If you remember, REVOKE does not store an ACL, but modifies existing ones. In this particular case, since there is nothing related to the preceding pk and email columns, the REVOKE statement does not change anything:

```
forumdb=> \dp users
                                      Access privileges
 Schema | Name  | Type  |           Access privileges         |      Column
 privileges      | Policies
--------+-------+-------+-------------------------------------+------------------
---------+----------
 public | users | table | postgres=arwdDxt/postgres     +| username:
+|
        |       |       | forum_admins=arwdDxt/postgres+|
forum_stats=r/postgres +|
        |       |       | forum_stats=r/postgres              | gecos:
+|
        |       |       |                                     |
forum_stats=rw/postgres+|
        |       |       |                                     | email:
+|
        |       |       |                                     |
forum_emails=r/postgres |
```

The rule of thumb is that every specific GRANT statement is canceled by the counterpart, REVOKE. In this example, since the last GRANT statement was issued without a specific list of columns, we need to issue a REVOKE statement without the list of columns:

```
forumdb=# REVOKE SELECT
          ON users FROM forum_stats;
REVOKE
```

However, this also removes the column-based grant permissions, so after REVOKE, the forum_stats role will no longer be able to perform SELECT against the username and gecos columns. In order to re-enable the role, you must re-issue the GRANT statement for the targeted columns.

The preceding example has shown you that the application of permissions at a fine-grain level requires attention and care, because an overly wide GRANT or REVOKE statement can produce results you would not expect at a glance.

Permissions related to sequences

A sequence is a table-like object that produces a transaction safe stream of new values, usually used for autogenerated (synthetic) keys.

There are three main permissions associated with a sequence: USAGE allows the querying of new values from the sequence; the SELECT privilege allows querying of the last or current value from the sequence (but not to get a new one); and lastly, the UPDATE privilege is another PostgreSQL-specific extension that allows the value of the sequence to be set and/or reset.

Since the USAGE privilege is the only one recognized by the SQL standard, if you grant it to a role, the role will automatically be able to also perform the actions that require the SELECT and UPDATE privileges. The latter two permissions are there only to allow you a finer-grain configuration of permissions against a sequence.

A general synopsis of the GRANT and REVOKE commands is as follows:

```
GRANT <permission> ON SEQUENCE <sequence> TO <role>;
REVOKE <permission> ON SEQUENCE <sequence> FROM <role>;
```

The special keyword ALL encapsulates all the permissions applicable to a sequence.

In order to understand how privileges on a sequence work, let's consider the sequence used to generate the primary keys of the categories table: categories_pk_seq.

First of all, remove all privileges from the luca role so that he can no longer interact with the sequence:

```
forumdb=# REVOKE ALL
          ON SEQUENCE categories_pk_seq
          FROM luca;
REVOKE
```

Now, if the luca role tries to get a new value from the sequence, he gets a permission denied error:

```
forumdb=> SELECT nextval( 'categories_pk_seq' );
ERROR:  permission denied for sequence categories_pk_seq
```

Giving the sequence the USAGE privilege allows the luca role to query the sequence again:

```
forumdb=# GRANT USAGE ON SEQUENCE categories_pk_seq TO luca;
GRANT
```

Now, the role can successfully apply the setval function:

```
forumdb=> SELECT setval( 'categories_pk_seq', 10 );
 setval
--------
     10
```

```
(1 row)

forumdb=> SELECT nextval( 'categories_pk_seq' );
 nextval
---------
      11
(1 row)
```

Remember that the USAGE privilege encapsulates both SELECT and UPDATE privileges, so once you have granted USAGE to a role, the sequence can be queried and set to a specific value.

Permissions related to schemas

A schema is a namespace for various objects, mainly tables and suchlike. There are primarily two permissions that can be applied to a schema: CREATE, to allow the creation of objects within the schema; and USAGE, to allow the role to "use" objects in the schema (assuming it has appropriate permissions on the object). That can look a little confusing at first, since if the role does not have the USAGE permission, it will not be able to access the object even if it is the owner.

The general synopsis for GRANT and REVOKE involves the explicit ON SCHEMA clause (to distinguish them from targeting a table):

```
GRANT <permission> ON SCHEMA <schema> TO <role>;
REVOKE <permission> ON SCHEMA <schema> FROM <role>;
```

As in other similar statements, the keyword ALL encapsulates all the permissions. In order to better understand the two different permissions, let's create a configuration schema and see how to enable access to it:

```
forumdb=# CREATE SCHEMA configuration;
CREATE SCHEMA
```

The schema has been created by a superuser, and therefore the user luca does not have any privileges in it, so he is not able to create a table:

```
forumdb=> CREATE TABLE configuration.conf( param text,
                                            value text,
                                            UNIQUE (param) );
ERROR:  permission denied for schema configuration
LINE 1: CREATE TABLE configuration.conf( param text, value text, UNI...
```

In order to allow the user `luca` to create new objects within the schema, the CREATE permission has to be granted. However, without the USAGE permission, the role will not be able to access anything in the schema, so you need to provide both permissions at the same time:

```
forumdb=# GRANT CREATE ON SCHEMA configuration TO luca;
GRANT

forumdb=# GRANT USAGE ON SCHEMA configuration TO luca;
GRANT
```

Therefore, the `luca` role can now create a new object within the schema:

```
forumdb=> CREATE TABLE configuration.conf( param text,
                                           value text,
                                           UNIQUE (param) );

CREATE TABLE

forumdb=> INSERT INTO configuration.conf
          VALUES( 'posts_per_page', '10' );
INSERT 0 1
```

Without the USAGE permission, the role is no longer able to access any object within the schema, even if it is the owner of the object:

```
forumdb=# REVOKE USAGE ON SCHEMA configuration FROM luca;
REVOKE
```

In fact, the user can no longer read their own data:

```
forumdb=> SELECT * FROM configuration.conf;
ERROR:  permission denied for schema configuration
LINE 1: SELECT * FROM configuration.conf;
```

On the other hand, what you probably want is to let the role be able to handle data within the schema, but not to create new objects, so something like the following:

```
forumdb=# GRANT USAGE ON SCHEMA configuration TO luca;
GRANT

forumdb=# REVOKE CREATE ON SCHEMA configuration FROM luca;
REVOKE
```

You can think of a schema as a container for other database objects. In order to access the container, you must have the USAGE permission, and in order to create new objects, you must have the CREATE permission. Nevertheless, USAGE does not provide you with unlimited access to any object within the schema. Instead, it provides you with access to objects depending on the permissions you have on such objects.

ALL objects in the schema

Since schemas are named containers of database objects, they can be used as a shortcut to apply different privileges to every object contained in the schema by means of the ALL <objects> IN SCHEMA clause.

By way of an example, in order to apply a set of equal permissions to all the tables contained in a schema, you can do the following:

```
forumdb=> REVOKE ALL
          ON ALL TABLES IN SCHEMA configuration
          FROM luca;
REVOKE

forumdb=> GRANT SELECT, INSERT, UPDATE
          ON ALL TABLES IN SCHEMA configuration
          TO luca;
GRANT
```

This can greatly simplify the management of large schemas.

At the moment, you can use the clause for the following:

- Tables, as in ON ALL TABLES IN SCHEMA
- Sequences, as in ON ALL SEQUENCES IN SCHEMA
- Routines, as in ON ALL ROUTINES IN SCHEMA (with the variants ON ALL PROCEDURES IN SCHEMA and ON ALL FUNCTIONS IN SCHEMA)

Permissions related to languages

There is a single permission that applies to a language: USAGE. Such permission allows a role to use the language. The special keyword ALL, which exists for compatibility with other GRANT and REVOKE statements, simply applies just that one permission.

As an example, in order to deny any role to execute any snippet of PL/Perl code, you need to revoke the permission from the special group `PUBLIC`:

```
forumdb=# REVOKE USAGE ON LANGUAGE plperl FROM PUBLIC;
REVOKE
```

In this way, even a trusted user such as `luca` cannot execute a PL/Perl snippet:

```
forumdb=> DO LANGUAGE plperl $$ elog( INFO, "Hello World" ); $$;
ERROR:  permission denied for language plperl
```

If you want to allow the `luca` role to execute PL/Perl code, you need to grant it explicitly:

```
forumdb=# GRANT USAGE ON LANGUAGE plperl TO luca;
GRANT
```

Permissions related to routines

The special keyword `ROUTINES` includes both `FUNCTIONS` and `PROCEDURES`. There is a single permission associated with `ROUTINES`, that is, the `EXECUTE` permission in order to be able to run (execute) the code in the routine.

In order to demonstrate the permission, let's create a very simple routine, `get_max`, that returns the maximum between two integers:

```
forumdb=> CREATE FUNCTION get_max( a int, b int )
RETURNS int AS $$
BEGIN
  IF a > b THEN
    RETURN a;
  ELSE
    RETURN b;
  END IF;
END $$ LANGUAGE plpgsql;
```

This now prevents any role apart from `luca` from executing such a routine:

```
forumdb=# REVOKE EXECUTE ON ROUTINE get_max FROM PUBLIC;
REVOKE
forumdb=# GRANT EXECUTE ON ROUTINE get_max TO luca;
GRANT
```

Any role other than `luca` will receive a `permission denied` error if invoking the function:

```
-- executing as enrico
forumdb=> SELECT get_max( 10, 20 );
ERROR:  permission denied for function get_max
```

Since `get_max` is a function, we could have written the `GRANT` and `REVOKE` permission with the `FUNCTION` keyword instead of the catch all, `ROUTINE`. This is a matter of preference.

In particular, the `ROUTINE` keyword becomes handy when you want to apply permissions to all functions and procedures within a schema, something like the following:

```
forumdb=# GRANT EXECUTE ON ALL ROUTINES IN SCHEMA my_schema;
```

Permissions related to databases

There are a bunch of permissions related to databases: `CONNECT` allows or rejects incoming connections without any regard to host-based access control; `TEMP` allows the creation of temporary objects (for example, tables) in the database; and `CREATE` allows the creation of new objects within the database.

The general synopsis is as follows:

```
GRANT <permission> ON DATABASE <database> TO <role>;
REVOKE <permission> ON DATABASE <database> FROM <role>;
```

For instance, if you need to lock out every user from the database, for instance, because you have to do maintenance work, you can issue the following `REVOKE` command:

```
forumdb=# REVOKE CONNECT ON DATABASE forumdb FROM PUBLIC;
REVOKE
```

New incoming connections will be rejected for a `permission denied` error:

```
$ psql  -U luca forumdb
psql: error: could not connect to server: FATAL:  permission denied for
database "forumdb"
DETAIL:  User does not have CONNECT privilege.
```

Now, if you want the `luca` role to be the only one able to connect to the database and to create objects but not temporary ones, you need to issue the following command:

```
forumdb=# REVOKE ALL ON DATABASE forumdb FROM public;
REVOKE

forumdb=# GRANT CONNECT, CREATE ON DATABASE forumdb TO luca;
GRANT
```

Other GRANT and REVOKE statements

There are other GRANT and REVOKE groups to control permission on table spaces, types, and foreign data wrappers. They will not be discussed here, but you should now have quite a clear workflow for applying permissions to different objects within the cluster and the databases.

Assigning the object owner

You have seen that the owner of an object has all the available permissions on such objects. During this time, you may wish to change the ownership of an object to another role, which, in turn, gets all the permissions. Usually, the change of ownership is done through a special ALTER statement such as the following:

```
ALTER <object> OWNER TO <role>;
```

For instance, to change the ownership of a table, you can issue the following command:

```
forumdb=# ALTER TABLE categories OWNER TO luca;
ALTER TABLE
```

Whereas to change the ownership of a function, you can issue the following command:

```
-- equivalent to: ALTER FUNCTION get_max OWNER TO luca;
forumdb=# ALTER ROUTINE get_max OWNER TO luca;
ALTER ROUTINE
```

Similar statements exist for all the other kinds of objects.

Inspecting ACLs

In order to see which permissions have been granted to roles and objects, you can use the already mentioned psql special command \dp (describe permissions), which reports the ACLs configured for a specific object (a table, for instance). The command performs a query against the special catalog pg_class, which contains a specific field named relacl that is an array of ACLs. You can see this as follows:

```
forumdb=> \dp categories
                                   Access privileges
 Schema |    Name     | Type  |    Access privileges    | Column privileges |
Policies
--------+-------------+-------+-------------------------+-------------------+--
--------
 public | categories  | table | luca=arwdDxt/luca     +|                   |
        |             |       | forum_stats=arw/luca  |                   |
(1 row)

forumdb=> SELECT relname, relacl
          FROM pg_class WHERE relname = 'categories';
  relname    |                     relacl
-------------+-------------------------------------------
 categories  | {luca=arwdDxt/luca,forum_stats=arw/luca}
(1 row)
```

As you can see, the output from the \dp command and from the query is the same, except for the formatting of the output.

You can also use the special function aclexplode to get more descriptive information about what the ACL means. The function returns a set of records, each one with the OID of the grantor, of the grantee, and a textual description of the permission granted. It is, therefore, possible to build a query like the following:

```
forumdb=> WITH acl AS (
            SELECT relname,
                   (aclexplode(relacl)).grantor,
                   (aclexplode(relacl)).grantee,
                   (aclexplode(relacl)).privilege_type
          FROM pg_class )
        SELECT g.rolname AS grantee,
               acl.privilege_type AS permission,
               gg.rolname AS grantor
        FROM acl
        JOIN pg_roles g ON g.oid = acl.grantee
        JOIN pg_roles gg ON gg.oid = acl.grantor
        WHERE acl.relname = 'categories';
```

This returns all the individual permissions assigned to the table categories, as shown here:

```
    grantee     | permission | grantor
----------------+------------+---------
 luca           | INSERT     | luca
 luca           | SELECT     | luca
 luca           | UPDATE     | luca
 luca           | DELETE     | luca
 luca           | TRUNCATE   | luca
 luca           | REFERENCES | luca
 luca           | TRIGGER    | luca
 forum_stats    | INSERT     | luca
 forum_stats    | SELECT     | luca
 forum_stats    | UPDATE     | luca
```

Row-level security

In the previous part of the chapter, you have seen the permission mechanism by which PostgreSQL allows roles (both users and groups) to access different objects within the database and data contained in the objects. In particular, with regard to tables, you have learned how to restrict access to just a specific column list within the tabular data.

PostgreSQL provides another interesting mechanism to restrict access to tabular data: row-level security. The idea is that row-level security can decide which tuples the role can gain access to, either in read or write mode. Therefore, if the column-based permissions provides a way of limiting the vertical shape of the tabular data, the RLS provides a way to restrict the horizontal shape of the data itself.

When is it appropriate to use RLS? Imagine you have a table that contains data related to users, and you don't want your users to be able to tamper with other users' data. In such a case, restricting the access of every user to just their own tuples could provide good isolation that prevents the data from being tampered with. Another fairly common scenario is a multi-homed system, where you store the same data, but for different companies in the very same tables. You don't want a company to be able to spy on or inspect thw data of another company, and so again RLS can prove useful.

Of course, RLS is not a silver bullet, and many of the solutions you could come up with involving RLS could have been realized with other techniques, but being aware of this important feature could make your data much more resistant to misuse.

The RLS framework works on so-called policies. A policy decision is a set of rules according to which certain tuples should be available. Depending on the policies you apply, your roles (that is, users) will be able to read and/or write certain tuples.

Applying RLS to a table is usually a two-step process: first, you have to define a policy (or more than one), and then you have to enable the policy against the table. Please be aware that superusers, owners, and roles with the special BYPASSRLS property will not be subject to any RLS.

 ATTENTION: In the case of a database backup, for example, via pg_dump, the user who executes the backup must be able to bypass row-level security policies; that is, it must have the BYPASSRLS property, or the backup will fail.

A policy defines the availability of tuples according to a logic criterion, that is, a filtering condition. A tuple can be available only for reading, or only for writing, or for both. The general synopsis for a policy is as follows:

```
CREATE POLICY <name>
ON <table>
FOR <statement>
TO <role>
USING <filtering condition>
WITH CHECK <writing condition>
```

Here, the following applies:

- name is the name of the policy; this is used to find it within the system.
- table is the table you want to apply the policy to.
- statement is any of SELECT, UPDATE, DELETE, INSERT, or the special keyword ALL to indicate all of the available statements.
- filtering condition is a condition used to restrict the result set of available tuples, typically, the tuples you want the role to be able to retrieve from your table.
- writing condition is an option clause that provides a restriction on writing down tuples.

A policy can be removed with the DROP POLICY command and can be rewritten with a specific ALTER POLICY command.

Let's now look at a couple of examples to better understand how a policy can be built. Assume we want to allow a database user to see only the tuples in the posts table that belong to them. Therefore, the condition is to match the user themselves against a SELECT statement. The policy could look like the following:

```
forumdb=> CREATE POLICY show_only_my_posts
          ON posts
```

```
        FOR SELECT
        USING ( author = ( SELECT pk FROM users
                            WHERE username = CURRENT_ROLE ) );
CREATE POLICY
```

The policy has been named `show_only_my_posts` and acts against the `posts` table for every `SELECT` statement. A tuple will be returned in the final result set only if there is a match of the `USING` clause, which means only if the author is found in the `users` table and is the current database user.

Having created the policy does not mean that the latter is active; you need to enable the policy on the table it refers to with a specific `ALTER TABLE` command:

```
forumdb=> ALTER TABLE posts ENABLE ROW LEVEL SECURITY;
ALTER TABLE
```

The preceding `ALTER TABLE` will enable all the policies created for such a table, in our case just one, but you have to be aware that if other policies are there, they will be activated too.

> You must be the owner of the table in order to enable or disable row-level security.

Now the role has been restricted to "see" just their own posts, but what about creating new posts? Since there is no particular restriction about writes in the policy, the user is able to create every tuple in the `posts` table. We can limit the user write ability, for instance, making it clear that they can only modify posts that belong to them and within a certain period of time, let's say 1 day. This results in a policy such as the following:

```
forumdb=> CREATE POLICY manage_only_my_posts
        ON posts
        FOR ALL
        USING ( author = ( SELECT pk FROM users
                            WHERE username = CURRENT_ROLE ) )
        WITH CHECK ( author = ( SELECT pk FROM users
                            WHERE username = CURRENT_ROLE )
                    AND
                    last_edited_on + '1 day'::interval >=
CURRENT_TIMESTAMP );
CREATE POLICY
```

Since the row-level security has already been activated for the `posts` table, the freshly created policy will be immediately active.

In this case, whatever statement the user is going to execute against the table, they will only see their own posts (the `USING` clause) and will not be able to write (that is, `INSERT`, `UPDATE`, `DELETE`) any tuple that does not belong to them and is not in a time range of 1 day (the `CHECK` clause).

What is happening under the hood? PostgreSQL silently applies the `USING` and `CHECK` clauses at every query you issue against the table to filter the possible tuples. For example, if you observe the query plan of a non-filtering `SELECT` command, you will see that the `CURRENT_ROLE` filter is applied as in the `USING` clause:

```
forumdb=> EXPLAIN SELECT * FROM posts;
                                    QUERY PLAN
-----------------------------------------------------------------------
--------------
 Seq Scan on posts  (cost=8.17..76.17 rows=1000 width=74)
   Filter: (author = $0)
   InitPlan 1 (returns $0)
     -> Index Scan using users_username_key on users  (cost=0.15..8.17
rows=1 width=4)
         Index Cond: (username = (CURRENT_ROLE)::text)
```

The filter has been applied by PostgreSQL even if the query does not mention it. This means that PostgreSQL is always "forced" to execute the query and filter the results for you, so you cannot expect any performance gain in using RLS. After all, the tuples must be excluded somewhere!

Now, if you try to modify the tuples in a way that violates the `CHECK` condition, PostgreSQL will claim and will not allow you to perform the changes:

```
forumdb=> UPDATE posts
          SET last_edited_on = last_edited_on - '2 weeks'::interval;
   ERROR:  new row violates row-level security policy for table "posts"
```

You can always inspect RLS via the special \dp command in `psql` (the following output has been trimmed to fit the page boundaries):

```
forumdb=> \dp posts
Access privileges
|                                                         Policies
+----------------------------------------------------------------------
------------------------------------------
| show_only_my_posts (r):
+
|   (u): (author = ( SELECT users.pk
+
|     FROM users
+
|    WHERE (users.username = (CURRENT_ROLE)::text)))
+
| manage_only_my_posts:
+
|   (u): (author = ( SELECT users.pk
+
|     FROM users
+
|    WHERE (users.username = (CURRENT_ROLE)::text)))
+
|   (c): ((author = ( SELECT users.pk
+
|     FROM users
+
|    WHERE (users.username = (CURRENT_ROLE)::text))) AND ((last_edited_on +
'1 day'::interval) >= CURRENT_TIMESTAMP))
(1 row)
```

Lastly, you can disable or enable back policies on a table by issuing a specific ALTER TABLE command, such as the following:

```
forumdb=> ALTER TABLE posts DISABLE ROW LEVEL SECURITY;
ALTER TABLE

-- to enable the RLS again
forumdb=> ALTER TABLE posts ENABLE ROW LEVEL SECURITY;
ALTER TABLE
```

We have now learned all about role level security, and can move on to working on password encryption for the roles.

Role password encryption

The passwords associated with roles are always stored in an encrypted form, even if the role is created without the ENCRYPTED PASSWORD property. PostgreSQL determines the algorithm to use in order to encrypt the password via the password_encryption option in the postgresql.conf configuration file. By default, the value of the option is set to md5, which means that the password is computed as MD5 hashes. The only other option available since PostgreSQL 10 is scram-sha-256, which will make the encryption much more robust.

You can quickly check the configuration from the operating system command line:

```
$ sudo -u postgres grep password_encryption $PGDATA/postgresql.conf
password_encryption = scram-sha-256    # md5 or scram-sha-256
```

Alternatively, you can inspect the pg_settings system catalog:

```
forumdb=# SELECT name, setting, enumvals
        FROM pg_settings
        WHERE name = 'password_encryption';
        name          |    setting    |      enumvals
---------------------+---------------+--------------------
 password_encryption | scram-sha-256 | {md5,scram-sha-256}
(1 row)
```

It is important to note that you cannot change the password encryption algorithm of a live system without resetting all the passwords of the active roles. In other words, if you decide to migrate from md5 to scram-sha-256, you need to issue appropriate ALTER ROLE statements to insert a new password for every role you have defined in the database.

 Since the pg_authid.rolpassword field starts with the encryption algorithm, either 'md5' or 'SCRAM-SHA-256', it is simple to inspect the system catalog and find out roles that have not been updated to a new encryption algorithm.

SSL connections

The **Secure Socket Layer** (**SSL**) allows PostgreSQL to accept encrypted network connections, which means every single piece of data in every packet is encrypted and therefore protected against network spoofing, as long as you handle your keys and certificates appropriately.

In order to enable the SSL extension, you first need to configure the server, then accept incoming SSL connections, and finally instrument the clients to connect in SSL mode.

Configuring the cluster for SSL

In order to let SSL do the encryption, the server must have private and public certificates. Creating and managing certificates is beyond the scope of this book, and is a complex topic. If you or your organization already have certificates, the only thing you have to do is to import the certificate and key files into your PostgreSQL server.

Assuming your certificate and key files are named `server.crt` and `server.key`, respectively, you have to configure the following parameters in the `postgresql.conf` configuration file:

```
ssl = on
ssl_key_file = '/postgres/pgdata/ssl/server.key'
ssl_cert_file = '/postgres/pgdata/ssl/server.crt'
```

This is done, of course, with the absolute path to your files. The first line tells PostgreSQL to enable SSL, while the other two lines tell the server where to find the files required to establish an encrypted connection. Of course, those files must be readable by the user who runs the PostgreSQL cluster (usually the `postgres` operating system user).

Once you have enabled SSL, you need to adjust the `pg_hba.conf` file to allow the host-based access machinery to handle SSL-based connections. In particular, if you don't want to accept plain connections, you need to substitute every `host` entry with `hostssl`, for instance:

```
hostssl    all       luca     carmensita        scram-sha-256
hostssl    all       test     192.168.222.1/32 scram-sha-256
```

If you want to accept both plain and encrypted connections, you can leave `host` as the connection method.

Connecting to the cluster via SSL

When connecting to PostgreSQL, the client will switch automatically to an SSL connection in case the host-based access has a `hostssl` entry, otherwise it will default to the standard plain connection.

In case `pg_hba.conf` has a `host` line, this means that it can accept both SSL and plain connections. Therefore, you need to force the connection to be SSL when you are initiating it. In `psql`, this can only be achieved by using a connection string and specifying the `sslmode=require` parameter to enable it. The server, if accepting the connection, will report the SSL protocol in use:

```
$ psql "postgresql://luca@localhost:5432/forumdb?sslmode=require"
psql (12.1)
SSL connection (protocol: TLSv1.3, cipher: TLS_AES_256_GCM_SHA384, bits:
256, compression: off)
Type "help" for help.

forumdb=>
```

If you omit the `sslmode` parameter, or use the standard `psql` connection parameters, the connection will be turned into `SSL` if the `pg_hba.conf` file has a `hostssl` line that matches. For instance, the following three connections produce the same result (an encrypted connection):

```
$ psql -h localhost -U luca forumdb
psql (12.1)
SSL connection (protocol: TLSv1.3, cipher: TLS_AES_256_GCM_SHA384, bits:
256, compression: off)
Type "help" for help.

forumdb=> \q
$ psql "postgresql://luca@localhost:5432/forumdb"
psql (12.1)
SSL connection (protocol: TLSv1.3, cipher: TLS_AES_256_GCM_SHA384, bits:
256, compression: off)
Type "help" for help.

forumdb=> \q
$ psql "postgresql://luca@localhost:5432/forumdb?sslmode=require"
psql (12.1)
SSL connection (protocol: TLSv1.3, cipher: TLS_AES_256_GCM_SHA384, bits:
256, compression: off)
Type "help" for help.

forumdb=>
```

Similarly, you can specify that you don't want an SSL connection at all. It is possible to force SSL mode to the `off` setting, `sslmode=disable`. This time, if `pg_hba.conf` has a `hostssl` mode, the connection will be rejected, while it will be served as a non-encrypted one if the `pg_hba.conf` file has a `host` line:

```
$ psql "postgresql://luca@localhost:5432/forumdb?sslmode=disable"
psql: error: could not connect to server: FATAL:  no pg_hba.conf entry for
host "127.0.0.1", user "luca", database "forumdb", SSL off
```

From the error, you can clearly see that there is no line that accepts a plain (`host` mode) connection in the `pg_hba.conf` file, or, on the other hand, that there are only `hostssl` lines.

Summary

In this chapter, we learned that PostgreSQL provides a very rich infrastructure for managing permissions associated with roles. Internally, PostgreSQL handles permissions on different database objects by means of ACLs, and every ACL contains information about the set of permissions, the users to whom permissions are granted, and the user who granted such permissions. In terms of tabular data, it is even possible to define column-based permissions and row-level permissions to exclude users from having access to a particular subset of data.

Permissions are granted by nested roles in a dynamically-inherited way or on-demand, leaving you the option to fine-tune how a role should exploit privileges.

With regard to security, we saw that PostgreSQL allows two different algorithms for password encryption, with SCRAM-SHA-256 being the most modern and robust. Lastly, when opportunely configured, the server can handle network connections via SSL, thereby encrypting all network traffic and data.

In the next chapter, you will learn all about transactions and how PostgreSQL manages them in a concurrent scenario, providing rock solid stability for your data.

References

- CREATE ROLE statement official documentation: https://www.postgresql.org/docs/12/sql-createrole.html
- ALTER ROLE statement official documentation: https://www.postgresql.org/docs/12/sql-alterrole.html

- DROP ROLE statement official documentation: https://www.postgresql.org/docs/12/sql-droprole.html
- GRANT statement official documentation: https://www.postgresql.org/docs/12/sql-grant.html
- REVOKE statement official documentation: https://www.postgresql.org/docs/12/sql-revoke.html
- PostgreSQL pg_roles catalog details: https://www.postgresql.org/docs/12/view-pg-roles.html
- PostgreSQL pg_authid catalog details: https://www.postgresql.org/docs/12/catalog-pg-authid.html
- PostgreSQL ACL documentation: https://www.postgresql.org/docs/12/ddl-priv.html
- PostgreSQL host-based access rule details: https://www.postgresql.org/docs/12/auth-pg-hba-conf.html
- PostgreSQL ACL utility functions: https://www.postgresql.org/docs/12/functions-info.html

11
Transactions, MVCC, WALs, and Checkpoints

This chapter introduces you to transactions, a fundamental part of every enterprise-level database system. PostgreSQL has very rich and standard-compliant transaction machinery that allows users to exactly define transaction properties, including nested transactions.

PostgreSQL relies heavily on transactions to keep data consistent across concurrent connections and parallel activities, and thanks to **Write-Ahead Logs** (**WALs**), PostgreSQL does its best to keep the data safe and reliable. Moreover, PostgreSQL implements **Multi-Version Concurrency Control** (**MVCC**), a way to maintain high concurrency between transactions.

The chapter can be split into two parts: the first one is more practical and provides concrete examples of what transactions are, how to use them, and how to understand MVCC. The second part is much more theoretical and explains how WALs work and how they allow PostgreSQL to recover even from a crash.

In this chapter, you will learn about the following topics:

- Introducing transactions
- Transaction isolation levels
- Explaining MVCC
- Savepoints
- Deadlocks
- How PostgreSQL handles persistency and consistency: WALs
- VACUUM

Technical requirements

In order to proceed, you need to know the following:

- How to issue SQL statements via `psql`
- How to connect to the cluster and a database
- How to check and modify the cluster configuration

The code for the chapter can be found in the following GitHub repository: `https://github.com/PacktPublishing/Learn-PostgreSQL`.

Introducing transactions

A transaction is an atomic unit of work that either succeeds or fails. Transactions are a key feature of any database system and are what allow a database to implement the ACID properties: atomicity, consistency, isolation, and durability. Altogether, the ACID properties mean that the database must be able to handle units of work on its whole (atomicity), store data in a permanent way (durability), without inter-mixed changes to the data (consistency), and in a way that concurrent actions are executed as if they were alone (isolation).

You can think of a transaction as a bunch of related statements that, in the end, will either all succeed or all fail. Transactions are everywhere in the database, and you have already used them even if you did not realize it: function calls, single statements, and so on are executed in a transaction block. In other words, every action you issue against the database is executed within a transaction, even if you did not ask for it explicitly. Thanks to this automatic wrapping of any statement into a transaction, the database engine can assure its data is always consistent and somehow protected from corruption, and we will see later in this chapter how PostgreSQL guarantees this.

Sometimes, however, you don't want the database to have control over your statements; rather, you want to be able to define the boundaries of transactions yourself, and of course, the database allows you to do it. For this reason, we call "implicit transactions" transactions the database starts for you without you needing to ask, and "explicit transactions" those that you ask the database to start.

Before we can examine both types of transactions and compare them, we need a little more background on transaction concepts.

First of all, any transaction is assigned a unique number, called the transaction identifier, or `xid` for short. The system automatically assigns an `xid` to newly created transactions – either implicit or explicit – and guarantees that no two transactions with the very same `xid` exist in the database.

The other main concept that we need to understand early in our transaction explanation is that PostgreSQL stores the `xid` that generates and or modifies a certain tuple within the tuple itself. The reason will be clear when we see how PostgreSQL handles transaction concurrency, so for the sake of this part, let's just assume that every tuple in every table is automatically labeled with the `xid` value of the transaction that created the tuple.

You can inspect what the current transaction is by means of the special function `txid_current()`. So, for example, if you ask your system a couple of simple statements such as the current time, you will see that every `SELECT` statement is executed as a different transaction:

```
forumdb=> SELECT current_time, txid_current();
    current_time    | txid_current
--------------------+--------------
 16:51:35.042584+01 |         4813
(1 row)

forumdb=> SELECT current_time, txid_current();
    current_time    | txid_current
--------------------+--------------
 16:52:23.028124+01 |         4814
(1 row)
```

As you can see from the preceding example, the system has assigned two different transaction identifiers, respectively `4813` and `4814`, to every statement, confirming that those statements have executed in different implicit transactions. You will probably get different numbers on your system.

If you inspect the special hidden column `xmin` in a table, you can get information about what transaction created the tuples; take the following example:

```
forumdb=> SELECT xmin, * FROM categories;
 xmin | pk |         title         |            description
------+----+-----------------------+-----------------------------------
  561 |  1 | DATABASE              | Database related discussions
  561 |  2 | UNIX                  | Unix and Linux discussions
  561 |  3 | PROGRAMMING LANGUAGES | All about programming languages
(3 rows)
```

As you can see, all the tuples in the preceding table have been created by the very same transaction, number `561`.

 PostgreSQL manages a few different hidden columns that you need to explicitly ask for when querying a table to be able to see them. In particular, every table has the `xmin`, `xmax`, `cmin`, and `cmax` hidden columns. Their use and aim will be explained later in this chapter.

Now that you know that every transaction is numbered and that such numbers are used to label tuples in every table, we can move forward and see the difference between implicit and explicit transactions.

Comparing implicit and explicit transactions

Implicit transactions are those that you don't ask for, but that the system applies to your statements. In other words, it is PostgreSQL that decides where the transaction starts and when it ends (transaction boundaries) and the rule is simple: every single statement is executed in its own separate transaction.

In order to better understand this concept, let's insert a few records into a table:

```
forumdb=> INSERT INTO tags( tag ) VALUES( 'linux' );
INSERT 0 1
forumdb=> INSERT INTO tags( tag ) VALUES( 'BSD' );
INSERT 0 1
forumdb=> INSERT INTO tags( tag ) VALUES( 'Java' );
INSERT 0 1
forumdb=> INSERT INTO tags( tag ) VALUES( 'Perl' );
INSERT 0 1
forumdb=> INSERT INTO tags( tag ) VALUES( 'Raku' );
INSERT 0 1
```

And let's query what the data in the table is:

```
forumdb=> SELECT xmin, * FROM tags;
 xmin | pk |  tag  | parent
------+----+-------+--------
 4824 |  9 | linux |
 4825 | 10 | BSD   |
 4826 | 11 | Java  |
 4827 | 12 | Perl  |
 4828 | 13 | Raku  |
(5 rows)
```

As you can see, the `xmin` field has a different (incremented) value for every single tuple inserted, which means a new transaction identifier (`xid`) has been assigned to the tuple or, more precisely, to the statement that executed `INSERT`. This means that *every single statement has executed in its own single-statement transaction.*

 The fact that you are seeing instances of `xid` incremented by a single unit is because on the machine used for the examples, there is no concurrency; that is, no other database activity is going on. However, you cannot make any predictions about what the next xid will be in a live system with different concurrent connections and running statements.

What if we had inserted all the preceding tags in one shot, being sure that if only one of them could not be stored for any reason, all of them would disappear? To this aim, we could use explicit transactions. An explicit transaction is a group of statements with a well-established transaction boundary: you issue a `BEGIN` statement to mark the start of the transaction, and either `COMMIT` or `ROLLBACK` to end the transaction. If you issue `COMMIT`, the transaction is marked as successful, therefore the modified data is stored permanently; on the other hand, if you issue `ROLLBACK`, the transaction is considered failed and all changes disappear.

Let's see this in practice – add another bunch of tags, but this time within a single explicit transaction:

```
forumdb=> BEGIN;
BEGIN
forumdb=> INSERT INTO tags( tag ) VALUES( 'PHP' );
INSERT 0 1
forumdb=> INSERT INTO tags( tag ) VALUES( 'C#' );
INSERT 0 1
forumdb=> COMMIT;
COMMIT
```

The only difference with respect to the previous bunch of `INSERT` statements is the explicit usage of `BEGIN` and `COMMIT`; since the transaction has committed, the data must be stored in the table:

```
forumdb=> SELECT xmin, * FROM tags;
 xmin | pk |  tag  | parent
------+----+-------+--------
 4824 |  9 | linux |
 4825 | 10 | BSD   |
 4826 | 11 | Java  |
 4827 | 12 | Perl  |
 4828 | 13 | Raku  |
 4829 | 14 | PHP   |
```

```
4829 | 15 | C#    |
(7 rows)
```

As you can see, not only is the data stored as we expected, but both the last rows have the very same transaction identifier; that is, 4829. This means that PostgreSQL has somehow merged the two different statements into a single one.

Let's see what happens if a transaction ends with a ROLLBACK statement – the final result will be that the changes must not be stored. As an example, modify the tag value of every tuple to full uppercase:

```
forumdb=> BEGIN;
BEGIN
forumdb=> UPDATE tags SET tag = upper( tag );
UPDATE 7
forumdb=> SELECT tag FROM tags;
  tag
-------
 LINUX
 BSD
 JAVA
 PERL
 RAKU
 PHP
 C#
(7 rows)

forumdb=> ROLLBACK;
ROLLBACK
forumdb=> SELECT tag FROM tags;
  tag
-------
 linux
 BSD
 Java
 Perl
 Raku
 PHP
 C#
(7 rows)
```

We first changed all the descriptions to uppercase, and the SELECT statement proves the database has done the job, but in the end, we changed our mind and issued a ROLLBACK function. At this point, PostgreSQL throws away our changes and keeps the pre-transaction state.

Therefore, we can summarize that every single statement is always executed as an implicit transaction, while if you need more control over what you need to atomically change, you need to open (BEGIN) and close (COMMIT or ROLLBACK) an explicit transaction.

Being in control of an explicit transaction does not mean that you will always have a choice about how to terminate it: sometimes PostgreSQL cannot allow you to COMMIT and consolidate a transaction because there are unrecoverable errors in it.
The most trivial example is when you do a syntax error:

```
forumdb=> BEGIN;
BEGIN
forumdb=> UPDATE tags SET tag = uppr( tag );
ERROR:  function uppr(text) does not exist
LINE 1: UPDATE tags SET tag = uppr( tag );
                                  ^
HINT:  No function matches the given name and argument types. You might
need to add explicit type casts.
forumdb=> COMMIT;
ROLLBACK
```

When PostgreSQL issues an error, it aborts the current transaction. Aborting a transaction means that, while the transaction is still open, it will not honor any following command nor COMMIT and will automatically issue a ROLLBACK command as soon as you close the transaction. Therefore, even if you try to work after a mistake, PostgreSQL will refuse to accept your statements:

```
forumdb=> BEGIN;
BEGIN
forumdb=> INSERT INTO tags( tag ) VALUES( 'C#' );
INSERT 0 1
forumdb=> INSERT INTO tags( tag ) VALUES( PHP );
ERROR:  column "php" does not exist
LINE 1: INSERT INTO tags( tag ) VALUES( PHP );
forumdb=> INSERT INTO tags( tag ) VALUES( 'Ocaml' );
ERROR:  current transaction is aborted, commands ignored until end of
transaction block
forumdb=> COMMIT;
ROLLBACK
```

Anyway, handling syntax errors or misspelled object names is not the only problem you can find when running a transaction, and after all, it is somehow quite simple to fix, but you can find that your transaction cannot continue because there is some data constraint that prevents the statement from completing successfully. Imagine we don't allow any tags with a description shorter than two characters:

```
forumdb=> ALTER TABLE tags
         ADD CONSTRAINT constraint_tag_length
         CHECK ( length( tag ) >= 2 );
ALTER TABLE
```

Consider a unit of work that performs two different INSERT statements as follows:

```
forumdb=> BEGIN;
BEGIN
forumdb=> INSERT INTO tags( tag ) VALUES( 'C' );
ERROR:  new row for relation "tags" violates check constraint
"constraint_tag_length"
DETAIL:  Failing row contains (17, C, null).
forumdb=> INSERT INTO tags( tag ) VALUES( 'C++' );
ERROR:  current transaction is aborted, commands ignored until end of
transaction block
forumdb=> COMMIT;
ROLLBACK
```

As you have seen, as soon as a DML statement fails, PostgreSQL aborts the transaction and refuses to handle any other statement. The only way you have to clear the situation is by ending the explicit transaction, and no matter the way you end it (either COMMIT or ROLLBACK), PostgreSQL will throw away your changes, rolling back the current transaction.

In the preceding examples, we have always shown COMMIT ending for a transaction, but it is clear that when you are in doubt about your data, changes you have made, or an unrecoverable error, you should issue ROLLBACK. We have shown COMMIT to make it clear that PostgreSQL will prevent erroneous work from successfully terminating.

So when are you supposed to use an explicit transaction? Every time you have a workload that must either succeed or fail, you have to wrap it in an explicit transaction. In particular, when losing a part of the work could compromise the remaining data, that is a good time to use a transaction. As an example, imagine an online shopping application: you surely do not want to charge your client before you have updated their cart and checked the availability of the products in storage. On the other hand, as a client, I would not want to get a message saying that my order has been confirmed just to discover that the payment has failed for any reason.

Therefore, since all the steps and actions have to be atomically performed (check the availability of the products, update the cart, take the payment, confirm the order), an explicit transaction is what we need to keep our data consistent.

Time within transactions

Transactions are *time-discrete*: the time does not change during a transaction. You can easily see this by opening a transaction and querying the current time multiple times:

```
forumdb=> BEGIN;
BEGIN
forumdb=> SELECT CURRENT_TIME;
    current_time
--------------------
 14:51:50.730287+01
(1 row)

forumdb=> SELECT pg_sleep_for( '5 seconds' );
 pg_sleep_for
--------------

(1 row)

forumdb=> SELECT CURRENT_TIME;
    current_time
--------------------
 14:51:50.730287+01
(1 row)

forumdb=> ROLLBACK;
ROLLBACK
```

If you really need a time-continuous source, you can use `clock_timestamp()`:

```
forumdb=> BEGIN;
BEGIN
forumdb=> SELECT CURRENT_TIME, clock_timestamp()::time;
    current_time     | clock_timestamp
--------------------+-----------------
 14:53:17.479177+01 | 14:53:22.152435
(1 row)

forumdb=> SELECT pg_sleep_for( '5 seconds' );
 pg_sleep_for
--------------
```

```
(1 row)

forumdb=> SELECT CURRENT_TIME, clock_timestamp()::time;
    current_time    | clock_timestamp
--------------------+------------------
 14:53:17.479177+01 | 14:53:33.022884


forumdb=> ROLLBACK;
ROLLBACK
```

How can we identify a transaction from another? Every transaction gets an identifier, as is explained in the following section.

More about transaction identifiers – the XID wraparound problem

PostgreSQL does not allow two transactions to share the same `xid` in any case. However, being an automatically incremented counter, `xid` will sooner or later do a wraparound, which means it will start counting over. This is known as the **xid wraparound problem** and PostgreSQL does a lot of work to prevent this from happening, as you will see later. But if the database is near the wraparound, PostgreSQL will start claiming it in the logs with messages like the following:

```
WARNING:  database "forumdb" must be vacuumed within 177009986 transactions
HINT:  To avoid a database shutdown, execute a database-wide VACUUM in
"forumdb".
```

If you carefully read the warning message, you will see that the system is talking about a shutdown: if the database undergoes a `xid` wraparound, data could be lost, so in order to prevent this, the system will automatically shut down if the `xid` wraparound is approaching.

There is, however, a way to avoid this automatic shutdown, by forcing a cleanup by means of running VACUUM. As you will see later in this chapter, one of the capabilities of VACUUM is to *freeze* old tuples so as to prevent the side effects of the `xid` wraparound, and therefore allowing the continuity of the database service. But what are the effects of the `xid` wraparound?

In order to understand such problems, we have to remember that every transaction is assigned a unique xid and that the next assignable xid is obtained by incrementing the last one assigned by a single unit.

This means that a transaction with a higher `xid` has started later than a transaction with a lower `xid`. In other words, a higher `xid` means the transaction is in the near future as opposed to a transaction with a lower xid. And since the `xid` is stored along with every tuple, a tuple with a higher `xmin` has been created later than a tuple with a lower `xmin`.

But when the `xid` overflows and therefore restarts its numbering from low numbers, transactions started later will appear with a lower `xid` than already running transactions, and therefore they will suddenly appear in the past. As a consequence, tuples with a lower transaction `xid` could also become in the past, instead of being in the future after the overflow, and therefore there will be a mismatch of the temporal workflow and tuple storage.

To avoid the `xid` wraparound, PostgreSQL implements a couple of tricks. First of all, the `xid` counter does not start from zero, but from the value 3. Values before 3 are reserved for internal use and no one transaction is allowed to store such a `xid`. Second, every tuple is enhanced with a status bit that indicates whether the tuple has been frozen or not: once a tuple has been frozen, its `xmin` must always be considered in the past, even if the value is greater than the current one.

Therefore, as the `xid` overflow is approaching, VACUUM performs a wide freeze execution, marking all the tuples in the past as frozen, so that even if the `xid` restarts its counting from lower numbers, the tuple already in the database will always appear in the past.

 In older PostgreSQL versions, VACUUM was literally removing the `xmin` value of the tuples to freeze substituting its value with the special value 2, which, being lower than the minimum usable value of 3, indicated that the tuple was in the past. However, when a forensic analysis is required, having the original `xmin` is valuable, and therefore PostgreSQL now uses a status bit to indicate whether the tuple has been frozen.

Virtual and real transaction identifiers

Being such an important resource, PostgreSQL is smart enough to avoid wasting transaction identifier numbers. In particular, when a transaction is initiated, the cluster uses a "virtual xid," something that works like an `xid` but is not obtained from the transaction identifier counter. In this way, every transaction does not consume an `xid` number from the very beginning. Once the transaction has done some work that involves data manipulation and changes, the virtual `xid` is transformed into a "real" `xid`, that is, one obtained from the `xid` counter.

Thanks to this extra work, PostgreSQL does not waste transaction identifiers on those transactions that do not strictly require strong identification. For example, there is no need to waste an `xid` on a transaction block like the following:

```
forumdb=> BEGIN;
BEGIN
forumdb=> ROLLBACK;
ROLLBACK
```

Since the preceding transaction does nothing at all, why should PostgreSQL involve all the `xid` machinery? There is no reason to use an `xid` that will not be attached to any tuple in the database and therefore will not interfere with any active snapshot.

There is, however, an important thing to note: the usage of the `txid_current()` function always materializes an `xid` even if the transaction has not got one yet. For that reason, PostgreSQL provides another introspection function named `txid_current_if_assigned()`, which returns `NULL` if the transaction is still in the "virtual `xid`" phase. It is important to note that PostgreSQL will not assign a real `xid` unless the transaction has manipulated some data, and this can easily be proven with a workflow like the following one:

```
forumdb=> BEGIN;
BEGIN
forumdb=> SELECT txid_current_if_assigned();
 txid_current_if_assigned
--------------------------

(1 row)

forumdb=> SELECT count(*) FROM tags;
 count
-------
     7
(1 row)

forumdb=> SELECT txid_current_if_assigned();
 txid_current_if_assigned
--------------------------

(1 row)

forumdb=> UPDATE tags SET tag = upper( tag );
UPDATE 7
forumdb=> SELECT txid_current_if_assigned();
 txid_current_if_assigned
--------------------------
```

```
                    4837
 (1 row)

forumdb=> SELECT txid_current();
 txid_current
 --------------
          4837
 (1 row)

forumdb=> ROLLBACK;
 ROLLBACK
```

At the beginning of the transaction, there is no `xid` assigned, and in fact `txid_current_if_assigned()` returns `NULL`. Even after a data read (that is, `SELECT`) the `xid` has not been assigned. However, as soon as the transaction performs some write activity (for example, an `UPDATE`), the `xid` is assigned and the results of both `txid_current_if_assigned()` and `txid_current()` are the same.

Multi-version concurrency control

What happens if two transactions, either implicit or explicit, try to perform conflicting changes over the same data? PostgreSQL must ensure the data is always consistent, and therefore it must have a way to "lock" (that is, block and protect) data subject to conflicting changes. Locks are a heavy mechanism that limits the concurrency of the system: the more locks you have, the more your transactions are going to wait to acquire the lock. To mitigate this problem, PostgreSQL implements MVCC, a well-known technique used in enterprise-level databases.

MVCC dictates that, instead of modifying an existing tuple within the database, the system has to replicate the tuple, apply the changes, and invalidate the original one. You can think of this as a copy-on-write mechanism used in operating filesystems such as ZFS.

To better understand what this means, let's assume the categories table has three tuples, and that we update one of them, to alter its description. What happens is that a new tuple, derived from the one we are going to apply UPDATE to, is inserted into the table, and the original one is invalidated:

Why is PostgreSQL and MVCC dealing with this extra work instead of doing an in-place update of the tuple? The reason is that this way, the database can cope with multiple versions of the same tuple, and every version is valid within a specific time window. This means that fewer locks are required to modify the data since the database is able to handle multiple versions of the same data at the same time and different transactions are going to see potentially different values.

For MVCC to work properly, PostgreSQL must handle the concept of snapshots: a snapshot indicates the time window in which a certain transaction is allowed to perceive data. A snapshot is, at its bare meaning, the range of transaction xids that define the boundaries of data available to a current transaction: every row in the database labeled with an xid within the range will be perceivable and usable by the current transaction. In other words, every transaction "sees" a dedicated subset of all the available data in the database.

The special function txid_current_snapshot() returns the minimum and maximum transaction identifiers that define the current transaction time boundaries. It becomes quite easy to demonstrate the concept with a couple of parallel sessions.
In the first session, let's run an explicit transaction, extract the identifier and the snapshot for future reference, and perform an operation:

```
-- session 1
forumdb=> BEGIN;
BEGIN
forumdb=> SELECT txid_current(), txid_current_snapshot();
 txid_current | txid_current_snapshot
--------------+-----------------------
         4928 | 4928:4928:
(1 row)

forumdb=> UPDATE tags SET tag = lower( tag );
UPDATE 5
```

As you can see in the preceding example, the transaction is number 4928 and its snapshot is bounded to itself, meaning that the transaction will see everything has been already consolidated in the database.

Now let's pause for a moment, and open another session to the same database – perform a single INSERT statement that is wrapped in an implicit transaction and get back the information about its xid:

```
forumdb=> INSERT INTO tags( tag ) VALUES( 'KDE' ) RETURNING txid_current();
 txid_current
--------------
     4929
(1 row)
```

The single-shot transaction has been assigned xid 4929, which is, of course, the very next xid available after the former explicit transaction (the system is running no other concurrent transactions to make it simpler to follow the numbering).
Go back to the first session and again inspect the information about the transaction snapshot:

```
-- session 1
forumdb=> SELECT txid_current(), txid_current_snapshot();
 txid_current | txid_current_snapshot
--------------+-----------------------
     4928     | 4928:4930:
(1 row)
```

This time, the transaction has grown its snapshot from itself to transaction 4930, which has not yet been started (txid_current_snapshot() reports its upper bound as non-inclusive). In other words, the current transaction now sees data consolidated even from a transaction that began after it, 4929. This can be even more explicit if the transaction queries the table:

```
-- session 1
forumdb=> SELECT xmin, tag FROM tags;
   xmin    |  tag
-----------+--------
  4928 | linux
  4928 | bsd
  4928 | java
  4928 | perl
  4928 | raku
  4929 | KDE
(6 rows)
```

As you can see, all the tuples but the last have been generated by the current transaction, and the last has been generated by xid 4929. But the preceding transaction is just a part of the story; while the first transaction is still incomplete, let's inspect the same table from another parallel session:

```
forumdb=> SELECT xmin, tag FROM tags;
   xmin   | tag
----------+-------
   4922 | linux
   4923 | BSD
   4924 | Java
   4925 | Perl
   4926 | Raku
   4929 | KDE
(6 rows)
```

All but the last tuple have different descriptions and, most notably, a different value for xmin from what transaction 4928 is seeing. What does it mean? It means that while the table has undergone an almost full rewrite of every tuple (an UPDATE on all but the last tuples), other concurrent transactions can still get access to the data in the table without having been blocked by a lock. This is the essence of MVCC: *every transaction perceives a different view of the storage, and the view is valid depending on the time window (snapshot) associated with the transaction.*

Sooner or later, the data on the storage has to be consolidated, and therefore when transaction 4928 completes the COMMIT of its work, the data in the table will become the truth that every transaction from there on will perceive:

```
-- session 1
forumdb=> COMMIT;
COMMIT

-- out from the transaction now
-- we all see consolidated data
forumdb=> SELECT xmin, tag FROM tags;
   xmin   | tag
----------+-------
   4928 | linux
   4928 | bsd
   4928 | java
   4928 | perl
   4928 | raku
   4929 | KDE
(6 rows)
```

MVCC does not always prevent the usage of locks: if two or more concurrent transactions start manipulating the same set of data, the system has to apply ordered changes, and therefore must force a lock on every concurrent transaction so that only one can proceed. It is quite simple to prove this with two parallel sessions similar to the preceding one:

```
-- session 1
forumdb=> BEGIN;
BEGIN
forumdb=> SELECT txid_current(), txid_current_snapshot();
 txid_current | txid_current_snapshot
--------------+-----------------------
         4930 | 4930:4930:
(1 row)

forumdb=> UPDATE tags SET tag = upper( tag );
UPDATE 6
```

In the meantime, in another session, execute the following statements:

```
-- session 2
forumdb=> BEGIN;
BEGIN
forumdb=> SELECT txid_current(), txid_current_snapshot();
 txid_current | txid_current_snapshot
--------------+-----------------------
         4931 | 4930:4930:
(1 row)

forumdb=> UPDATE tags SET tag = lower( tag );
-- BLOCKED!!!!
```

Transaction 4931 is locked because PostgreSQL cannot decide which data manipulation to apply. On one hand, transaction 4930 is applying uppercase to all the tags, but at the same time, transaction 4931 is applying lowercase to the very same data.

Since the two changes conflict, and the final result (that is, the result that will be consolidated in the database) depends on the exact order in which changes will be applied (and in particular on the last one applied), PostgreSQL cannot allow both transactions to proceed. Therefore, since 4930 applied the changes before 4931, the latter is suspended, waiting for transaction 4930 to complete either with success or failure. As soon as you end the first transaction, the second one will be unblocked (showing the message status for the UPDATE statement):

```
-- session 1
forumdb=> COMMIT;
COMMIT
```

```
-- session 2
UPDATE 6
-- unblocked, can proceed further ...
forumdb=>
```

Therefore, MVCC is not a silver bullet against lock usage but allows better concurrency in the overall usage of the database.

From the preceding description, it should be clear that MVCC comes at a cost: since the system has to maintain different tuple versions depending on the active transactions and their snapshots, the storage will literally grow over the effective size of consolidated data.

To prevent this problem, a specific tool named VACUUM, along with its background-running brother autovacuum, is in charge of scanning tables (and indexes) for tuple versions that can be thrown away, therefore reclaiming storage space. But when is a tuple version eligible for being destroyed by VACUUM? When there are no more transactions referencing the tuple xid (that is, xmin), that is when the tuple is no longer consolidated.

In the next section, you will explore transaction isolation levels, a way to control the expected transaction behavior when operating in a concurrent environment.

Transaction isolation levels

In a concurrent system, you could encounter three different problems:

- **Dirty reads**: A dirty read happens when the database is allowing a transaction to see work-in-progress data from other not-yet-finished transactions. In other words, data that has not been consolidated is visible to other transactions. No production-ready database allows that, and PostgreSQL is no exception: you are assured your transaction will only perceive data that has been consolidated and, in order to be consolidated, the transactions that created such data must be complete.
- **Nonrepeatable reads**: An unrepeatable read happens when the same query, within the same transaction, executed multiple times, perceives a different set of data. This essentially means that the data has changed between two sequential executions of the same query in the same transaction. PostgreSQL does not allow this kind of problem by means of snapshots: every transaction can perceive the snapshot of the data available depending on specific transaction boundaries.

- **Phantom reads**: A phantom read is somehow similar to an unrepeatable read, but what changes between the sequential execution of the same query is the size of the result set. This means that the data has not changed, but new data has been "appended" to the last execution result set.

The SQL standard provides four isolation levels that a transaction can adopt to prevent any of the preceding problems:

- **Read uncommitted**: The lowest level possible.
- **Read committed**: The default isolation level in PostgreSQL.
- **Repeatable read**: Useful for long jobs as the system does not see the effects of concurrent transactions; this offers us the possibility to work on a consistent snapshot during the entire execution of the transaction.
- **Serializable**: The strongest isolation level available.

Each level provides increasing isolation upon the previous level, so for example, Read committed enhances the behavior of Read uncommitted, Repeatable read enhances Read committed (and Read uncommitted), and Serializable enhances all of the previous levels.

PostgreSQL does not support all the preceding levels, as you will see in detail in the following subsections. You can always specify the isolation level you desire for the explicit transaction at the transaction's beginning; every isolation level has the very same name as reported in the preceding list. So, for example, the following begins a transaction in Read committed mode:

```
forumdb=> BEGIN TRANSACTION ISOLATION LEVEL REPEATABLE READ;
BEGIN
```

You can omit the optional keyword `TRANSACTION`, even if in our opinion this improves readability. It is also possible to explicitly set the transaction isolation level by means of a `SET TRANSACTION` statement. As an example, the following snippet produces the same effects as the preceding one:

```
forumdb=> BEGIN;
BEGIN
forumdb=> SET TRANSACTION ISOLATION LEVEL READ COMMITTED;
SET
```

It is important to note that the transaction isolation level cannot be changed once the transaction has started. In order to have an effect, SET TRANSACTION statement must be the very first statement executed in a transaction block. Every subsequent SET TRANSACTION statement that changes the already set isolation level will produce a failure and put the transaction in an aborting state, otherwise, if the subsequent SET TRANSACTION does not change the isolation level, they will have no effect and will produce no errors.

To better understand this case, the following is an example of an incorrect workflow where the isolation level is changed after the transaction has already executed a statement, even if it's not changing any data:

```
forumdb=> BEGIN;
BEGIN
forumdb=> SELECT count(*) FROM tags;
 count
-------
     7
(1 row)

-- a query has been executed, the SET TRANSACTION
-- is not anymore the very first command
forumdb=> SET TRANSACTION ISOLATION LEVEL SERIALIZABLE;
ERROR:  SET TRANSACTION ISOLATION LEVEL must be called before any query
```

In the following sections, we will discuss every isolation level in detail.

Read uncommitted

The Read uncommitted isolation level allows a transaction to be subjected to the dirty reads problem, which means it can perceive unconsolidated data from other incomplete transactions.

PostgreSQL does not support this isolation level, because, after all, it is not a true isolation level. In fact, Read uncommitted means that there is no isolation at all among transactions, and this is certainly a situation where interleaving data corruption happens.

You can set the isolation level explicitly, but PostgreSQL will ignore your will and set it silently to the most robust Read committed one.

Read Committed

The isolation level READ COMMITTED is the default one used by PostgreSQL: if you don't set a level, every transaction (implicit or explicit) will have this isolation level.

This level prevents dirty reads and allows the current transaction to see all the already consolidated data at the time every single statement in the transaction is executed. We have already seen this behavior in practice in the snapshot example.

Repeatable Read

The REPEATABLE READ isolation level imposes that every statement in the transaction will perceive only data already consolidated at the time the transaction started, or better, at the time the first statement of the transaction is started.

Serializable

The SERIALIZABLE isolation level imposes the REPEATABLE READ level and assures that two concurrent transactions will be able to successfully complete only if the end result would have been the same if the two transactions ran in sequential order.

In other words, if two (or more) transactions have the SERIALIZABLE isolation level and try to modify the same subset of data in a conflicting way, PostgreSQL will ensure that only one transaction can complete and will make the other fail.

Let's see this in action by creating an initial transaction and modifying a subset of data:

```
-- session 1
forumdb=> BEGIN TRANSACTION ISOLATION LEVEL SERIALIZABLE;
BEGIN
forumdb=> UPDATE tags SET tag = lower( tag );
UPDATE 7
```

To simulate concurrency, let's pause this transaction and open a new one in another session, applying other changes to the same set of data:

```
-- session 2
forumdb=> BEGIN TRANSACTION ISOLATION LEVEL SERIALIZABLE;
BEGIN
forumdb=> UPDATE tags SET tag = '[' || tag || ']';
-- blocked
```

Since the manipulated set of data is the same, the second transaction is locked as we saw in other examples before. Now assume the first transaction completes successfully:

```
-- session 1
forumdb=> COMMIT;
COMMIT
```

PostgreSQL realizes that also making the other transaction able to proceed would break the SERIALIZABLE promise because applying the transaction sequentially would produce different results depending on their order. Therefore, as soon as the first transaction commits, the second one is automatically aborted with a serializable error:

```
-- session 2
forumdb=> UPDATE tags SET tag = '[' || tag || ']';
ERROR:  could not serialize access due to concurrent update
```

What happens if the transaction manipulates data that apparently is not related? One transaction may fail again, in fact, let's modify one single tuple from one transaction:

```
-- session 1
forumdb=> BEGIN TRANSACTION ISOLATION LEVEL SERIALIZABLE;
BEGIN
forumdb=> UPDATE tags SET tag = '{' || tag || '}' WHERE tag = 'java';
UPDATE 1
```

In the meantime, modify exactly one other transaction from another session:

```
-- session 2
forumdb=> BEGIN TRANSACTION ISOLATION LEVEL SERIALIZABLE;
BEGIN
forumdb=> UPDATE tags SET tag = '[' || tag || ']' WHERE tag = 'perl';
UPDATE 1
```

This time, there is no locking of the second transaction because the touched tuples are completely different. However, as soon as the first transaction executes a COMMIT, the second transaction is no longer able to COMMIT by itself:

```
-- session 2 (assume session 1 has issued COMMIT)
forumdb=> COMMIT;
ERROR:  could not serialize access due to read/write dependencies among
transactions
DETAIL:  Reason code: Canceled on identification as a pivot, during commit
attempt.
HINT:  The transaction might succeed if retried.
```

This is quite a common problem when using serializable transactions: the application or the user must be ready to execute their transaction over and over because PostgreSQL could make it fail due to the serializability of the workflows.

Explaining MVCC

xmin is only a part of the story of managing MVCC. PostgreSQL labels every tuple in the database with four different fields named xmin (already described), xmax, cmin, and cmax. Similar to what you have learned about xmin, in order to make those fields appear in a query result, you need to explicitly reference them; for instance:

```
forumdb=> SELECT xmin, xmax, cmin, cmax, * FROM tags ORDER BY tag;
 xmin | xmax | cmin | cmax | pk | tag  | parent
------+------+------+------+----+------+--------
 4854 |    0 |    0 |    0 | 24 | c++  |
 4853 |    0 |    0 |    0 | 23 | java |
 4852 |    0 |    0 |    0 | 22 | perl |
 4855 |    0 |    0 |    0 | 25 | unix |
(4 rows)
```

The meaning of xmin has been already described in a previous section: it indicates the transaction identifier of the transaction that created the tuple. The xmax field, on the other hand, indicates the xid of the transaction that invalidated the tuple, for example, because it has deleted the data. The cmin and cmax fields indicate respectively the command identifiers that created and invalidated the tuple within the same transaction (PostgreSQL numbers every statement within a transaction starting from zero).

Why is it important to keep track of the statement identifier (cmin, cmax)? Since the lowest isolation level that PostgreSQL applies is Read Committed, every single statement (that is, command) in a transaction must see the snapshot of the data consolidated when the command is started.

You can see the usage of cmin and cmax within the same transaction in the following example. First of all, we begin an explicit transaction, then we insert a couple of tuples with two different INSERT statements; this means that the created tuples will have a different cmin:

```
forumdb=> BEGIN;
BEGIN
```

```
forumdb=> SELECT xmin, xmax, cmin, cmax, tag, txid_current()
          FROM tags ORDER BY tag;
```

```
 xmin | xmax | cmin | cmax | tag  | txid_current
------+------+------+------+------+--------------
 4854 |   0  |   0  |   0  | c++  |      4856
 4853 |   0  |   0  |   0  | java |      4856
 4852 |   0  |   0  |   0  | perl |      4856
 4855 |   0  |   0  |   0  | unix |      4856
(4 rows)

-- first writing command (number 0)
forumdb=> INSERT INTO tags( tag ) values( 'raku' );
INSERT 0 1

-- second writing command (number 1)
forumdb=> INSERT INTO tags( tag ) values( 'lua' );
INSERT 0 1

-- fourth command within transaction (number 3)
forumdb=> SELECT xmin, xmax, cmin, cmax, tag, txid_current()
          FROM tags ORDER BY tag;

 xmin | xmax | cmin | cmax | tag  | txid_current
------+------+------+------+------+--------------
 4854 |   0  |   0  |   0  | c++  |      4856
 4853 |   0  |   0  |   0  | java |      4856
 4856 |   0  |   1  |   1  | lua  |      4856
 4852 |   0  |   0  |   0  | perl |      4856
 4856 |   0  |   0  |   0  | raku |      4856
 4855 |   0  |   0  |   0  | unix |      4856
(6 rows)
```

So far, within the same transaction, the two new tuples inserted have an xmin that is the same as txid_current(). Obviously, those tuples have been created by the same transaction. However, please note that the second tuple, being in the second writing command, has a cmin that holds 1 (command counting starts from zero).

Therefore, PostgreSQL knows every tuple when it has been created by means of a transaction and command within that transaction.

Let's move on with our transaction: declare a cursor that holds a query against the tags table and delete all tuples but two. The transaction session continues as follows:

```
forumdb=> DECLARE tag_cursor CURSOR FOR SELECT xmin, xmax, cmin, cmax, tag,
txid_current() FROM tags ORDER BY tag;
DECLARE CURSOR

forumdb=> DELETE FROM tags WHERE tag NOT IN ( 'perl', 'raku' );
DELETE 4
```

```
forumdb=> SELECT xmin, xmax, cmin, cmax, tag, txid_current()
          FROM tags ORDER BY tag;
 xmin | xmax | cmin | cmax | tag  | txid_current
------+------+------+------+------+--------------
 4852 |    0 |    0 |    0 | perl |         4856
 4856 |    0 |    0 |    0 | raku |         4856
(2 rows)
```

As you can see, the table now holds only two tuples – this is the expected behavior after all.

But the cursor has started before the DELETE statement, and therefore it must perceive the data as it was before the DELETE statement. In fact, if we ask the cursor what data it can obtain, we see that it returns all the tuples as they were before the DELETE statement:

```
forumdb=> FETCH ALL FROM tag_cursor;
 xmin | xmax | cmin | cmax | tag  | txid_current
------+------+------+------+------+--------------
 4854 | 4856 |    2 |    2 | c++  |         4856
 4853 | 4856 |    2 |    2 | java |         4856
 4856 | 4856 |    0 |    0 | lua  |         4856
 4852 |    0 |    0 |    0 | perl |         4856
 4856 |    0 |    0 |    0 | raku |         4856
 4855 | 4856 |    2 |    2 | unix |         4856
(6 rows)
```

There is an important thing to note: every deleted tuple has a value in xmax that holds the current transaction identifier (4856), meaning that this very transaction has deleted the tuples. However, the transaction has not committed yet, therefore the tuples are still there but are marked to be tied to the snapshot that ends in 4856. Moreover, the deleted tuples have a cmax that holds the value 2, which means that the tuples have been deleted from the third writing command in the transaction.

Since the cursor has been defined before the statement, it is able to "see" the tuples as they were, even if PostgreSQL knows exactly from which point in time they have disappeared.

Readers may have noted that cmin and cmax hold the same value, and that is due to the fact that the fields are overlapping the very same storage.

In the following section, you are going to see how to disassemble a transaction into smaller pieces by means of savepoints.

Savepoints

A savepoint is a way to split a transaction into smaller blocks that can be rolled back independently of each other. Thanks to savepoints, you can divide a big transaction (one transaction with multiple statements) into smaller chunks, allowing a subset of the bigger transaction to fail without having the overall transaction fail. PostgreSQL does not handle transaction nesting, so you cannot issue a nested set of BEGIN, nor COMMIT/ROLLBACK statements. Savepoints allow PostgreSQL to mimic the nesting of transaction blocks.

Savepoints are marked with a mnemonic name, which you can use to commit or rollback. The name must be unique within the transaction, and if you reuse the same over and over, the previous savepoints with the same name will be discarded. Let's see an example:

```
forumdb=> BEGIN;
BEGIN
forumdb=> INSERT INTO tags( tag ) VALUES ( 'Eclipse IDE' );
INSERT 0 1
forumdb=> SAVEPOINT other_tags;
SAVEPOINT
forumdb=> INSERT INTO tags( tag ) VALUES ( 'Netbeans IDE' );
INSERT 0 1
forumdb=> INSERT INTO tags( tag ) VALUES ( 'Comma IDE' );
INSERT 0 1
forumdb=> ROLLBACK TO SAVEPOINT other_tags;
ROLLBACK
forumdb=> INSERT INTO tags( tag ) VALUES ( 'IntelliJIdea IDE' );
INSERT 0 1
forumdb=> COMMIT;
COMMIT


forumdb=> SELECT tag FROM tags WHERE tag like '%IDE';
        tag
--------------------
 Eclipse IDE
 IntelliJIdea IDE
(2 rows)
```

In the preceding transaction, the first statement does not belong to any savepoint and therefore follows the life of the transaction itself. After the other_tags savepoint is created, all the following statements follow the lifecycle of the savepoint itself, therefore once ROLLBACK TO SAVEPOINT is issued, the statements within the savepoint are discarded. After that, other statements belong to the outer transaction, and therefore follow the lifecycle of the transaction itself. In the end, the result is that everything that has been executed outside the savepoint is stored in the table.

Once you have defined a savepoint, you can also change your mind and release it, so that statements within the savepoint follow the same lifecycle of the main transaction. Here's an example:

```
forumdb=> BEGIN;
BEGIN
forumdb=> SAVEPOINT editors;
SAVEPOINT
forumdb=> INSERT INTO tags( tag ) VALUES ( 'Emacs Editor' );
INSERT 0 1
forumdb=> INSERT INTO tags( tag ) VALUES ( 'Vi Editor' );
INSERT 0 1
forumdb=> RELEASE SAVEPOINT editors;
RELEASE
forumdb=> INSERT INTO tags( tag ) VALUES ( 'Atom Editor' );
INSERT 0 1
forumdb=> COMMIT;
COMMIT

forumdb=> SELECT tag FROM tags WHERE tag LIKE '%Editor';
      tag
--------------
 Emacs Editor
 Vi Editor
 Atom Editor
(3 rows)
```

When RELEASE SAVEPOINT is issued, it is like the savepoint has disappeared and therefore the two INSERT statements follow the main transaction lifecycle. In other words, it is like the savepoint has never been defined.

In a transaction, you can have multiple savepoints but once you roll back a savepoint, you roll back all the savepoints that follow it:

```
forumdb=> BEGIN;
BEGIN
forumdb=> SAVEPOINT perl;
SAVEPOINT
forumdb=> INSERT INTO tags( tag ) VALUES ( 'Rakudo Compiler' );
INSERT 0 1
forumdb=> SAVEPOINT gcc;
SAVEPOINT
forumdb=> INSERT INTO tags( tag ) VALUES ( 'Gnu C Compiler' );
INSERT 0 1
forumdb=> ROLLBACK TO SAVEPOINT perl;
ROLLBACK
forumdb=> COMMIT;
```

```
COMMIT

forumdb=> SELECT tag FROM tags WHERE tag LIKE '%Compiler';
 tag
-----
(0 rows)
```

As you can see, even if the transaction has issued a COMMIT, everything that has been done after the perl savepoint, to which the transaction has rolled back, has been rolled back to.

In other words, rolling back to a savepoint means you roll back everything after said savepoint.

Transactions can lead to a situation where the cluster is unable to proceed. These situations are named *deadlocks* and are described in the next section.

Deadlocks

A deadlock is an event that happens when different transactions depend on each other in a circular way. Deadlocks are, to some extent, normal events in a concurrent database environment and nothing an administrator should worry about, unless they become extremely frequent, meaning there is some dependency error in the applications and the transactions.

When a deadlock happens, there is no choice but to terminate the locked transactions. PostgreSQL has a very powerful deadlock detection engine that does exactly this job: it finds stalled transactions and, in the case of a deadlock, terminates them (producing ROLLBACK).

In order to produce a deadlock, imagine two concurrent transactions applying changes to the very same tuples in a conflicting way. For example, the first transaction could do something like the following:

```
-- session 1
forumdb=> BEGIN;
BEGIN
forumdb=> SELECT txid_current();
 txid_current
--------------
         4875
(1 row)

forumdb=> UPDATE tags SET tag = 'Perl 5' WHERE tag = 'perl';
UPDATE 1
```

And in the meantime, the other transaction performs the following:

```
-- session 2
forumdb=> BEGIN;
BEGIN
forumdb=> SELECT txid_current();
 txid_current
--------------
         4876
(1 row)


forumdb=> UPDATE tags SET tag = 'Java and Groovy' WHERE tag = 'java';
UPDATE 1
```

So far, both transactions have updated a single tuple without conflicting with each other. Now imagine that the first transaction tries to modify the tuple that the other transaction has already changed; as we have already seen in previous examples, the transaction will remain locked, waiting to acquire the lock on the tuple:

```
-- session 1
forumdb=> UPDATE tags SET tag = 'The Java Language' WHERE tag = 'java';
-- locked
```

If the second transactions tries, on the other hand, to modify a tuple already touched by the first transaction, it will be locked waiting for the lock acquisition:

```
-- session 2
forumdb=> UPDATE tags SET tag = 'Perl and Raku' WHERE tag = 'perl';
ERROR:   deadlock detected
DETAIL:   Process 78918 waits for ShareLock on transaction 4875; blocked by
process 80105.
Process 80105 waits for ShareLock on transaction 4876; blocked by process
78918.
HINT:   See server log for query details.
CONTEXT:   while updating tuple (0,1) in relation "tags"
```

This time, however, PostgreSQL realizes the two transactions cannot solve the problem because they are waiting on a circular dependency, and therefore decides to kill the second transaction in order to give the first one a chance to complete. As you can see from the error message, PostgreSQL knows that transaction 4875 is waiting for a lock hold by transaction 4876 and vice versa, so there is no solution to proceed but killing one of the two.

Being natural events in a concurrent transactional system, deadlocks are something you have to deal with, and your applications must be prepared to replay a transaction in case they are forced to ROLLBACK by deadlock detection.

Deadlock detection is a complex and resource-expensive process, therefore PostgreSQL does it on a scheduled basis. In particular, the `deadlock_timeout` configuration parameter expresses how often PostgreSQL should search for dependency among stalled transactions. By default, this value is set at 1 second, and is expressed in milliseconds:

```
forumdb=> SELECT name, setting
          FROM pg_settings
          WHERE name like '%deadlock%';
       name        | setting
-------------------+---------
 deadlock_timeout  | 1000
(1 row)
```

Decreasing this value is often a bad idea: while your applications and transactions will fail sooner, your cluster will be forced to consume extra resources in dependency analysis.

In the following section, you will discover how PostgreSQL ensures that data is made persistent on storage, even in the case of a crashing cluster.

How PostgreSQL handles persistency and consistency: WALs

In the previous sections, you have seen how to interact with explicit transactions, and most notably how PostgreSQL executes every single statement within a transaction.

PostgreSQL goes to a lot of effort internally to ensure that consolidated data on storage reflects the status of the committed transactions. In other words, data can be considered consolidated only if the transaction that produced (or modified) it has been committed. But this also means that, once a transaction has been committed, its data is "safe" on storage, no matter what happens in the future.

PostgreSQL manages transactions and data consolidations by means of **Write-Ahead Logs** (**WALs**). This section introduces you to the concept of WALs and their use within PostgreSQL.

Write-Ahead Logs (WALs)

Before we dig into the details, it is required to briefly explain how PostgreSQL internally handles data. Tuples are stored in mass storage – usually, a disk – under the $PGDATA/base directory, in files named only by numbers. When a transaction requests access to a particular set of tuples, PostgreSQL loads the data from the $PGDATA/base directory and places the requested data in one or more shared buffers. The shared buffers are an in-memory copy of the on-disk data, and all the transactions access the shared data because they provide much more performance and do not require every single transaction to seek the data out of the storage.

The next figure shows the loading of a few data pages into the shared buffers memory location:

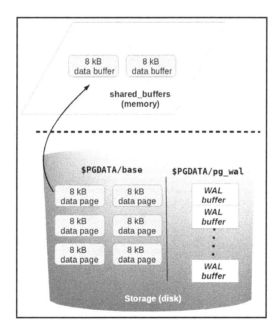

When a transaction modifies some data, it does so by modifying the in-copy memory, which means it modifies the shared buffers area.

At this point, the in-memory copy of the data does not correspond to the stored version, and it is here that PostgreSQL has to guarantee consistency and persistency without losing performance.

What happens is that the data is kept in memory but is marked as dirty, meaning that it is a copy not yet synchronized with the on-disk original source. Once the changes to a dirty buffer have been committed, PostgreSQL consolidates the changes in the WALs and keeps the dirty buffer in memory to be served as the most recent available copy for other transactions.

Sooner or later, PostgreSQL will push the dirty buffer to the storage, replacing the original copy with the modified version, but a transaction usually does not know and does not care about when this is going to happen.

The following diagram explains the preceding workflow: the red buffer has been modified by a transaction and therefore does not match what is on disk anymore; however, when the transaction issues a COMMIT, the changes are forced and flushed to the WALs:

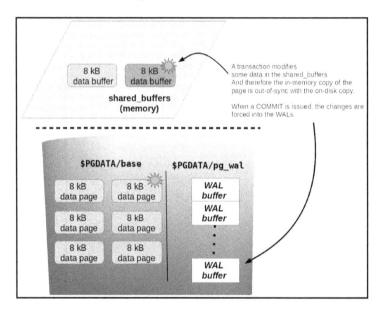

Why is the WAL space supposed to be faster than overwriting the original data block in the $PGDATA/base directory? The trick is that in order to find the exact position on the disk storage where the block has to be overwritten, PostgreSQL should have to perform what is called a random-seek, which is a costly I/O operation. On the other hand, the WALs are sequentially written as a journal, and therefore there is no need to perform a random-seek. Therefore, writing the WALs prevents the I/O performance degradation and allows PostgreSQL to overwrite the data block at a future time, when for instance, the cluster is not overloaded and has I/O bandwidth available.

Every time a transaction performs a COMMIT, its actions and modified data are permanently stored in a piece of the WAL, in particular a specific part of the current WAL segment (more on this later). Therefore, PostgreSQL can reproduce the transaction and its effects in order to perform the very same data changes.

This, however, does not suffice in making PostgreSQL reliable: PostgreSQL makes a big effort to ensure the data actually hits the disk storage. In particular, during the writing of the WALs, PostgreSQL isolates itself from the outside world, disabling operating system signals, so that it cannot be interrupted. Moreover, PostgreSQL issues fsync(2), a particular operating system call that forces the filesystem cache to flush data on disk.

PostgreSQL does all of this in order to ensure that the data physically hits the disk layer, but it must be clear that if the filesystem, or the disk controller (that is, the hardware), lies, the data could not be physically on the disk. This is important, but PostgreSQL cannot do anything about that and has to trust what the operating system (and thus the hardware) reports back as feedback.

In any case, COMMIT will return success to the invoking transaction if and only if PostgreSQL has been able to write the changes on the disk. Therefore, at the transaction level, if a COMMIT succeeds (that is, there is no error), the data has been written in the WALs, and therefore can be assumed to be "safe" on the storage layer.

WALs are split into so-called segments. A segment is a file made of exactly 16 MB of changes in the data. While it is possible to modify the size of segments during initdb, we strongly discourage this and will assume every segment is 16 MB.
This means that PostgreSQL writes, sequentially, a single file at a time (that is, a WAL segment) and when this has reached the size of 16 MB, it is closed and a new 16 MB file is created. The WAL segements (or WALs for short) are stored in the pg_wal directory under $PGDATA. Every segment has a name made up of hexadecimal digits, and 24 characters long. The first 8 characters indicate the so called "time-line" of the cluster (something related to replication), the second 8 digits indicate an increasing sequence number named the Log Sequence Number (LSN for short), and the last 8 digits provide the offset within the LSN. Here's an example:

```
$ sudo -u postgres ls -1 $PGDATA/pg_wal
000000070000024700000A8
000000070000024700000A9
000000070000024700000AA
000000070000024700000AB
000000070000024700000AC
000000070000024700000AD
...
```

In the previous content of the `pg_wal`, you can see that every WAL segment has the same timeline, number `7`, and the LSN is `247`. Every file, then, has a different offset with the first one being `A8`, the second `A9`, and so on. As you can imagine, WAL segment names are not made for humans, but PostgreSQL knows exactly how and in which file it has to search for information.

Sooner or later, depending on the memory resources and usage of the cluster, the data in memory will be written back to its original disk positions, meaning that the WALs are serving only as temporary safe storage on disk. The reason for that is not only tied to a performance bottleneck, as already explained, but also to allow data restoration in the event of a crash.

WALs as a rescue method in the event of a crash

When you cleanly stop a running cluster, for example, by means of `pg_ctl`, PostgreSQL ensures that all dirty data in memory is flushed to the storage in the correct order, and then halts itself.

But what happens if the cluster is uncleanly stopped, for example by means of a power failure?

This event is named a **crash**, and once PostgreSQL starts over, it performs a so-called **crash-recovery**. In particular, PostgreSQL understands it has stopped in an unclean way, and therefore the data on the storage could not be the last version that existed when the cluster terminated its activity. But PostgreSQL knows that all committed data is at least present in the WALs, and therefore starts reading the WALs in what is called **WAL-replay**, and adjusts the data on the storage according to what is in the WALs. Until the crash recovery has completed, the cluster is not usable and does not accept connections; once the crash recovery has finished, the cluster knows that the data on the storage has been made coherent and therefore normal operations can start again.

This process allows the cluster to somehow self-heal after an external event has caused the lifecycle to abort. This makes it clear that the main aim of the WALs is not to avoid performance degradations, but rather to ensure the cluster is able to recover after a crash. And in order to be able to do that, it must have data written permanently to the storage, but thanks to the sequential way in which WALs are written, data is made persistent with less I/O penalties.

Checkpoints

Sooner or later, the cluster must make every change that has already been written in the WALs also available in the data files, that is, it has to write tuples in a random-seek way. These writes happen at very specific times named checkpoints. A checkpoint is a point in time at which the database makes an extra effort to ensure that everything already present in the WALs is also written in the correct position in the data storage.

The following diagram helps with understanding what happens during a CHECKPOINT:

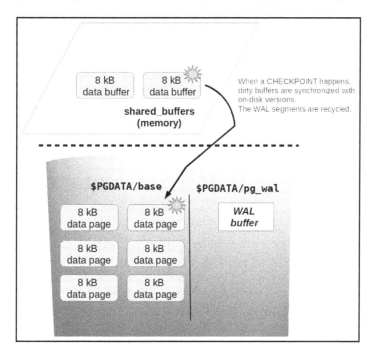

But why should the database make this extra synchronization effort?

If the synchronization does not happen, the WALs will keep growing and thus consume storage space. Moreover, if the database crashes for any reason, the WAL-replay must walk across a very long set of WALs.

Thanks to checkpoints, instead, the cluster knows that in the event of a crash, it has to synchronize data between the storage and the WALs only after the last checkpoint is successfully performed. In other words, the storage space and time required to replay the WALs are reduced from the crash instant to the last checkpoint.

But there is another advantage: since after a checkpoint PostgreSQL knows that the data in the WALs has been synchronized with the data in the storage, it can throw away already synchronized WALs. In fact, even in the event of a crash, PostgreSQL will not need any WAL part that precedes the last checkpoint at all. Therefore, PostgreSQL performs WAL recycling: after a checkpoint, a WAL segment is reused as an empty segment for the upcoming checkpoint.

Thanks to this machinery, the space required to store WAL segments will pretty much remain the same during the cluster lifecycle because, at every checkpoint, segments will be reused. Most notably, in the event of a crash, the number of WAL segments to replay will be the total number of those produced since the last checkpoint.

> PostgreSQL 13 is able to provide you with some information about how many WAL segments a specific query is going to consume, that is, how much data is inserted into the WALs due to the execution of a query. The special command EXPLAIN (detailed in Chapter 13, *Indexes and Performance Optimization*) can provide you with the WAL information.

Checkpoint configuration parameters

The database administrator can fine-tune the checkpoints, meaning they can decide when and how often a checkpoint can happen. Since checkpoints are consolidating points, the more often they happen, the less time will be required to recover from a crash. On the other hand, the more seldomly they are executed, the more the database will not suffer from I/O bottlenecks. In fact, when a checkpoint is reached, the database must force every dirty buffer from memory to disk, and this usually means that an I/O spike is introduced; during such a spike, other concurrent database activities, such as getting new data from the storage, will be penalized because the I/O bandwidth is temporarily exhausted from the checkpoint activity.

For the preceding reasons, it is very important to carefully tune checkpoints and in particular, their tuning must reflect the cluster workload.

Checkpoints can be tuned by means of three main configuration parameters that interact with each other and that are explained in the following subsections.

checkpoint_timeout and max_wal_size

Checkpoint frequency can be tuned by two orthogonal parameters: max_wal_size and checkpoint_timeout.

The `max_wal_size` parameter dictates how much space the `pg_wal` directory can occupy. Since at every checkpoint the WAL segments are recycled, the `pg_wal` directory tends to occupy the very same size eventually. Tuning the `max_wal_size` parameter specifies after how many data changes the checkpoint must be completed, and therefore this parameter is a "quantity" specification.

`checkpoint_timeout` expresses after how much time the checkpoint must be forced.

The two parameters are orthogonal, meaning that the first that happens triggers the checkpoint execution: your database produces data changes over the `max_wal_size` parameter or when the `checkpoint_timeout` time has elapsed.

As an example, let's take a system with the following settings:

```
forumdb=> SELECT name, setting, unit FROM pg_settings
          WHERE name IN ( 'checkpoint_timeout', 'max_wal_size' );
        name         | setting | unit
---------------------+---------+------
 checkpoint_timeout  | 300     | s
 max_wal_size        | 1024    | MB
(2 rows)
```

After 300 seconds (5 minutes) a checkpoint is triggered unless, in the meantime, 1024 MB of data has been changed. Therefore if your database is not doing much activity, a checkpoint is triggered by `checkpoint_timeout`, while in the case that the database is heavily accessed, a checkpoint is triggered for every 1 GB of data produced.

Checkpoint throttling

In order to avoid an I/O spike at the execution of a checkpoint, PostgreSQL introduced a third parameter named `checkpoint_completion_target`, which can handle values between 0 and 1. This parameter indicates the amount of time the checkpoint can delay the writing of dirty buffers for. In particular, the time provided to complete a checkpoint is computed as `checkpoint_timeout` x `checkpoint_completion_target`.

For example, if `checkpoint_completion_target` is set to `0.2` and `checkpoint_timemout` is 300 seconds, the system will have 60 seconds to write all the data. The system calibrates the required I/O bandwidth to fulfill the dirty buffers' writing.

Therefore, if you set `checkpoint_completion_target` to 0, you are going to see spikes in the checkpoint executions, with the consequence of high usage of I/O bandwidth, while setting the parameter to 1 means you are going to see continuous I/O activity with low I/O bandwidth.

Manually issuing a checkpoint

It is always possible for the cluster administrator to manually start a checkpoint process: the PostgreSQL statement `CHECKPOINT` starts all the activities that would normally happen at `checkpoint_timeout` or `max_wal_size`.

The checkpoint being such an invasive operation, why should someone want to perform it manually? One reason could be to ensure that all the data on the disk has been synchronized, for example, before starting a streaming replication or a file-level backup.

In the following section, you will learn about the VACUUM process, the technique that allows PostgreSQL to reclaim unused space, removing no longer visible tuples.

VACUUM

In the previous sections, you have learned how PostgreSQL exploits MVCC to store different versions of the same data (tuples) that different transactions can perceive depending on their active snapshot. However, keeping different versions of the same tuples requires extra space with regard to the last active version, and this space could fill your storage sooner or later. To prevent that, and reclaim storage space, PostgreSQL provides an internal tool named vacuum, the aim of which is to analyze stored tuple versions and remove the ones that are no longer perceivable.

 Remember: a tuple is not perceivable when there are no more active transactions that can reference the version, which means having the tuple version within their snapshot.

Vacuum can be an I/O-intensive operation since it must reclaim no more used disk space, and therefore can be an invasive operation. For that reason, you are not supposed to run vacuum very frequently and PostgreSQL also provides a background job, named autovacuum, which can run a vacuum for you depending on the current database activity.

The following subsections will show you both manual and automatic vacuum.

Manual VACUUM

Manual vacuum can be run against a single table, a subset of table columns, or a whole database, and the synopsis is as follows:

```
VACUUM [ FULL ] [ FREEZE ] [ VERBOSE ] [ ANALYZE ] [ table_and_columns [,
...] ]
```

There are three main versions of VACUUM that perform progressively more aggressive refactoring:

- *Plain* VACUUM (the default) does a micro-space-reclaim, which means it throws away dead tuple versions but does not defragment the table, and therefore the final effect is no space being reclaimed.
- VACUUM FULL performs a whole table rewrite, throwing away dead tuples and removing defragmentation, thus also reclaiming disk space.
- VACUUM FREEZE marks already consolidated tuples as frozen, preventing the xid wraparound problem.

VACUUM cannot be executed within a transaction, nor a function or procedure. The extra options VERBOSE and ANALYZE provide a verbose output and perform a statistic update of the table contents (this is useful for performance gain) respectively.

In order to see the effects of VACUUM, let's build a simple example. First of all, ensure that autovacuum is set to off. If it's not, edit the $PGDATA/postgresql.conf configuration file and set the parameter to off, then restart the cluster. After that, inspect the size of the tags table:

```
forumdb=> SHOW autovacuum;
 autovacuum
------------
 off
(1 row)

forumdb=> SELECT relname, reltuples, relpages, pg_size_pretty(
pg_relation_size( 'tags' ) )
FROM pg_class WHERE relname = 'tags' AND relkind = 'r';
 relname | reltuples | relpages | pg_size_pretty
---------+-----------+----------+----------------
 tags    |         6 |        1 | 8192 bytes
(1 row)
```

As you can see, the table has only six tuples and occupies a single data page on disk, of the size 8 KB. Now let's populate the table with about 1 million random tuples:

```
forumdb=> INSERT INTO tags( tag )
SELECT 'FAKE-TAG-#' || x
FROM generate_series( 1, 1000000 ) x;
INSERT 0 1000000
```

Since we have stopped autovacuum, PostgreSQL does not know the real size of the table, and therefore we need to perform a manual ANALYZE to inform the cluster about the new data in the table:

```
forumdb=> ANALYZE tags;
ANALYZE
forumdb=> SELECT relname, reltuples, relpages, pg_size_pretty(
pg_relation_size( 'tags' ) )
FROM pg_class WHERE relname = 'tags' AND relkind = 'r';
 relname |  reltuples  | relpages | pg_size_pretty
---------+-------------+----------+----------------
 tags    | 1.00001e+06 |    6370  | 50 MB
```

It is now time to invalidate all the tuples we have inserted, for example, by overwriting them with an UPDATE (which, due to MVCC, will duplicate the tuples):

```
forumdb=> UPDATE tags SET tag = lower( tag ) WHERE tag LIKE 'FAKE%';
UPDATE 1000000
```

The table now still has around 1 million valid tuples, but the size has almost doubled because every tuple now exists in two versions, one of which is dead:

```
forumdb=> ANALYZE tags;
ANALYZE
forumdb=> SELECT relname, reltuples, relpages, pg_size_pretty(
pg_relation_size( 'tags' ) )
FROM pg_class WHERE relname = 'tags' AND relkind = 'r';
 relname |  reltuples  | relpages | pg_size_pretty
---------+-------------+----------+----------------
 tags    | 1.00001e+06 |   12739  | 100 MB
(1 row)
```

We have now built something that can be used as a test lab for VACUUM. If we execute plain VACUUM, every single data page will be freed of dead tuples but pages will not be reconstructed, so the number of data pages will remain the same, and the final table size on storage will be the same too:

```
forumdb=> VACUUM VERBOSE tags;
...
```

```
INFO:  "tags": found 1000000 removable, 1000006 nonremovable row versions
in 12739 out of 12739 pages

VACUUM

forumdb=> ANALYZE tags;
ANALYZE

forumdb=> SELECT relname, reltuples, relpages, pg_size_pretty(
pg_relation_size( 'tags' ) )
FROM pg_class WHERE relname = 'tags' AND relkind = 'r';
 relname |   reltuples  | relpages | pg_size_pretty
---------+-------------+----------+----------------
 tags    | 1.00001e+06 |   12739 | 100 MB
(1 row)
```

VACUUM informs us that 1 million tuples can be safely removed, while 1 million (plus the original 6 tuples) cannot be removed because they represent the last active version. However, after this execution, the table size has not changed: all data pages are essentially fragmented.

So what is the aim of plain VACUUM? This kind of VACUUM provides new free space on every single page, so the table can essentially sustain 1 million new tuples without changing its own size. We can prove this by performing the same tuple invalidation we have already done:

```
forumdb=> UPDATE tags SET tag = upper( tag ) WHERE tag LIKE 'fake%';
UPDATE 1000000
forumdb=> ANALYZE tags;
ANALYZE
forumdb=> SELECT relname, reltuples, relpages, pg_size_pretty(
pg_relation_size( 'tags' ) )
FROM pg_class WHERE relname = 'tags' AND relkind = 'r';
 relname |   reltuples  | relpages | pg_size_pretty
---------+-------------+----------+----------------
 tags    | 1.00001e+06 |   12739 | 100 MB
(1 row)
```

As you can see, nothing has changed in the number of tuples, pages, and table size. Essentially, it went like this: we introduced 1 million new tuples in the beginning, then we updated all of them, making the 1 million become 2 million, then we used VACUUM on the table, lowering the number again to 1 million but leaving the free space already allocated so that the table was occupying space for 2 million but only half of that storage was full. After that, we created 1 million new tuple versions but the system did not need to allocate more space because there was enough free, even if scattered across the whole table.

On the other hand, VACUUM FULL not only frees the space within the table, but also reclaims all such space, compacting the table to its mimimum size. If we execute VACUUM FULL right now, at least 50 MB of data space will be reclaimed because 1 million tuples will be thrown away:

```
forumdb=> VACUUM FULL VERBOSE tags;
INFO:  vacuuming "public.tags"
INFO:  "tags": found 1000000 removable, 1000006 nonremovable row versions
in 12739 pages
DETAIL:  0 dead row versions cannot be removed yet.
CPU: user: 0.18 s, system: 0.61 s, elapsed: 1.03 s.
VACUUM
forumdb=> ANALYZE tags;
ANALYZE
forumdb=> SELECT relname, reltuples, relpages, pg_size_pretty(
pg_relation_size( 'tags' ) )
FROM pg_class WHERE relname = 'tags' AND relkind = 'r';
 relname |  reltuples  | relpages | pg_size_pretty
---------+-------------+----------+----------------
 tags    | 1.00001e+06 |     6370 | 50 MB
(1 row)
```

The output of VACUUM FULL is pretty much the same as plain VACUUM: it shows that 1 million tuples can be thrown away. The end result, however, is that the whole table has gained the space occupied by said tuples. It is important to remember, however, that, while tempting, VACUUM FULL forces a complete table rewrite and therefore pushes a lot of work down to the I/O system, thus incurring potential performance penalties.

It is possible to summarize the main effects of VACUUM in diagrams. Imagine a situation like the one depicted in the following diagram, where a table is occupying two data pages, respectively with four and three valid tuples:

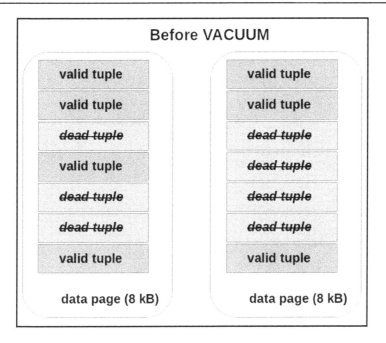

If plain VACUUM executes, the total number of pages will remain the same but every page will free the space occupied by dead tuples and will compact valid tuples together, as shown in the following diagram:

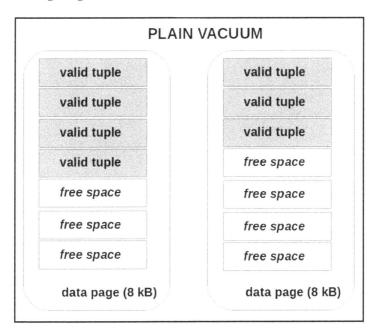

If VACUUM FULL executes, the table's data pages are fully rewritten to compact all valid tuples together. In this situation, the second page of the table results is empty, and therefore is discarded, and therefore there is a gain in the space consumption on the storage device. The situation becomes the one depicted in the following diagram:

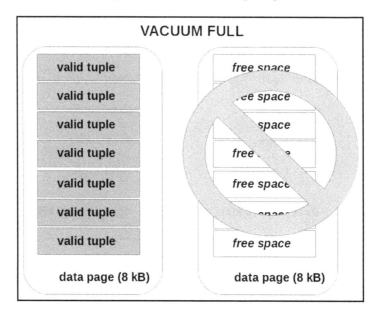

In the event that you are approaching an xid wraparound, VACUUM FREEZE solves the problem by marking the tuples as "always in the past."

For other usages of VACUUM, please see the official documentation.

Automatic VACUUM

Since PostgreSQL 8.4, there is a background job named *autovacuum*, which is responsible for running VACUUM on behalf of the system administrator.

The idea is that, VACUUM being an I/O-intensive operation, a background job can perform small micro-vacuums without interfering with the normal database activity.

Usually, you don't have to worry about autovacuum, since it is enabled by default and has general settings that can be useful in many scenarios. However, there are different settings that can help you to fine-tune autovacuum. A system with a good autovacuum configuration usually does not need manual VACUUM, and often the traits of a manual VACUUM are that autovacuum must be configured to run more frequently.

The main settings for autovacuum can be inspected from the `$PGDATA/postgresql.conf` configuration file or, as usual, the `pg_settings` catalog. The most important configuration parameters are the following:

- `autovacuum` enables or disables the autovacuum background machinery. There is no reason, beyond doing experiments as we did in the previous section, to keep autovacuum disabled.
- `autovacuum_vacuum_threshold` indicates how many new tuple versions will be allowed before autovacuum can be activated on a table. The idea is that we don't want autovacuum to trigger if only a small amount of tuples have changed in a table, because that will produce an I/O penalty without an effective gain in space.
- `autovacuum_vacuum_scale_factor` indicates the amount, as a percentage, of tuples that have to be changed before autovacuum performs a concrete vacuum on a table. The idea is that the more the table grows, the more autovacuum will wait for dead tuples before it performs its activities.
- `autovacuum_cost_limit` is a value that measures the maximum threshold over which the background process must suspend itself to resume later on.
- `autovacuum_cost_delay` indicates how many milliseconds (in multiples of ten) autovacuum will be suspended to not interfere with other database activities. The suspension is performed only when the cost delay is reached.

Essentially, the activity of autovacuum goes like this: if the number of changed tuples is greater than `autovacuum_vacuum_threshold + ( table-tuples * autovacuum_vacuum_scale_factor )`, the autovacuum process activates. It then performs a vacuum on the table measuring the amount of work. If the amount of work reaches what `autovacuum_cost_limit` is set to, the process suspends itself for `autovacuum_cost_delay` milliseconds, and then resumes and proceeds further. Any time autovacuum reaches the threshold, it suspends itself, producing the effect of an incremental vacuum.

But how does autovacuum compute the cost of the activity it is doing? There are a set of tunable values that express how much it costs to fetch a new data page, to scan a dirty page, and so on:

```
forumdb=> SELECT name, setting   FROM pg_settings
    WHERE name like 'vacuum_cost%';
          name           | setting
-------------------------+---------
 vacuum_cost_delay       | 10
 vacuum_cost_limit       | 10000
 vacuum_cost_page_dirty  | 20
```

```
vacuum_cost_page_hit    | 1
vacuum_cost_page_miss   | 10
```

Such values are used for both manual VACUUM and autovacuum, with the exception that autovacuum has its own autovacuum_vacuum_cost_limit, which is usually is set to 200.

Manual VACUUM is never subjected to the cost machinery and therefore performs until it finishes its job.

Similar parameters exist for the ANALYZE part because the autovacuum background process performs VACUUM ANALYZE and therefore you have autovacuum_analyze_threshold and autovacuum_analyze_scale_factor, which are in charge of defining the window of activity for the ANALYZE part (that is related to updating the statics on the content of the table).

You can have more than one background process doing the autovacuum activity. In particular, the autovacuum_max_workers parameter defines how many background processes PostgreSQL can start in parallel to perform autovacuum activities. On a single database, there will be only one worker active in a specific instant, therefore it does not make sense to raise this value over the number of actively used databases in the system.

Summary

PostgreSQL exploits MVCC to enable high concurrent access to the underlying data, and this means that every transaction perceives a snapshot of the data while the system keeps different versions of the same tuples. Sooner or later, invalid tuples will be removed and the storage space will be reclaimed. On one hand, MVCC provides better concurrency, but on the other hand, it requires extra effort to reclaim the storage space once transactions no longer reference dead tuples. PostgreSQL provides VACUUM with this aim and also has a background process named autovacuum to periodically and non-invasively reclaim storage space and keep the system clean and healthy.

In order to improve I/O and reliability, PostgreSQL stores data in a journal written sequentially, the WAL. The WAL is split into segments, and at particular time intervals, named checkpoints, all the dirty data in memory is forced to a specified position in the storage and WAL segments are recycled.

In this chapter, you have learned about WAL and MVCC internals, as well as transaction boundaries and savepoints. You have also seen how to impose a specific transaction isolation level that, depending on your needs, can protect your data against concurrent updates of the same tuples.

In the next chapter, you will discover how PostgreSQL can be extended beyond its normal functionalities by means of pluggable modules named extensions.

References

- PostgreSQL Transaction Isolation Levels, official documentation: `https://www.postgresql.org/docs/12/sql-set-transaction.html`
- PostgreSQL Transaction Isolation Level SERIALIZABLE, official documentation: `https://www.postgresql.org/docs/current/transaction-iso.html#XACT-SERIALIZABLE`
- PostgreSQL Savepoints, official documentation: `https://www.postgresql.org/docs/12/sql-savepoint.html`
- PostgreSQL VACUUM, official documentation: `https://www.postgresql.org/docs/12/sql-vacuum.html`

12
Extending the Database - the Extension Ecosystem

Extensions are a powerful way of packaging together related database objects, such as tables, functions, and routines, making the management of the objects as a single unit easier. Extensions allow you and other developers to literally extend the already rich PostgreSQL set of features by providing a clear, concise, and accurate way of installing, upgrading, and removing features and objects. In this chapter, you will see what extensions are and how they can be installed, upgraded, or removed by means of automated tools or manually. Moreover, you will learn how to build your own extension from scratch so that you will be immediately productive in packaging your own scripts and tools to distribute across other databases and PostgreSQL instances.

The chapter consists of the following topics:

- Introducing extensions
- Managing extensions
- Exploring the PGXN client
- Installing extensions
- Creating your own extension

Introducing extensions

SQL is a declarative language that allows you to create and manipulate objects, as well as data. You can group SQL statements into scripts so that you can run the scripts in a more predictable and reproducible way. However, such scripts are seen by PostgreSQL as a sequence of unrelated commands, that is, you are responsible for correlating such commands into appropriate scripts. Things get even worse when you have to deal with foreign languages or binary libraries; the cluster knows nothing about your aim and how every single object is related to each other. Luckily, extensions help in getting order out of chaos.

An extension is a packaged set of files that can be installed into the cluster in order to provide more functionalities, therefore to "extend" the current cluster set of features.

An extension can be something general, like a new data type, a new index type, or a service to send emails directly from within PostgreSQL, or it can be something really specific to a particular use case, like a set of tables and data to provide ad hoc configuration. An extension does not have any opinion on how you are going to use it, and therefore you are free to install and forget it or use it in every database of your cluster.

The main aim of the extension mechanism is to provide a common interface for administering new features. Thanks to extensions, you have a common set of statements to deploy, install, upgrade, and remove an extension as a whole thing within the cluster. It does not matter whether your extension is made up of a single function or a whole set of linked objects, the extension mechanism will handle all the objects at once, making the administration easier.

PostgreSQL has built a whole ecosystem around the concept of extensions, and therefore not only does it provide statements to manage extensions, but also a platform to build new extensions and convert existing scripts into extensions. Then, extensions can be made publicly available by means of a global repository known as the **PostgreSQL eXtensions Network** (**PGXN**).

You can think of PostgreSQL extensions as being like libraries in programming languages, such as modules for Perl or, similarly, gems in Ruby, JARs in Java, and so on. Similarly, the PGXN infrastructure can be thought of as the CPAN for Perl (or PEAR to PHP, and so on).

The extension ecosystem

The beauty of extensions is that they provide a uniform way to bundle modules that can be deployed (installed) and used in PostgreSQL. Developers are free to contribute to expanding the number of modules available for PostgreSQL, and this has rapidly grown to what is now a full ecosystem.

Similar to programming languages, like Perl, Python, and others, PostgreSQL can now be customized by means of add-ons and modules that share a common infrastructure and architecture and can be managed by the same statements without any regard for the features they provide.

Extensions are mainly collected in the **PostgreSQL eXtension Network** (**PGXN**), a repository that can be queried to get information about an extension or to download an extension (and a particular version of it), and can be updated with new modules.

 The PGXN is as similar as CPAN to Perl, CTAN to LaTeX, PEAR to PHP, and so on.

While you can find PostgreSQL extensions all around the internet, chances are you will interface with PGXN almost every time you need a new extension. PGXN is not a simple website or a code repository, but it is a full specification about four main parts: a *search engine*, an *extension manager*, an *API* (application programming interface), and a *client*.

The **search engine** allows users to search for the PGXN content for a specific extension. The **manager** is responsible for accepting new extensions (or new extension versions) and letting users obtain them (that is, distributing the extensions). The **API** defines how applications can interact with the manager and the search engine, and therefore how a client can be built. There are two main clients available—the PGXN website and the `pgxnclient` command-line application. While we will discuss the `pgxnclient` application in the following subsections, you will see an example of usage of the PGXN website later in the chapter.

Extensions are built on top of the **PostgreSQL eXtension System** (**PGXS**), which is a basic set of rules an extension must adhere to in order to expose a common manageable interface. In particular, PostgreSQL provides a uniform Makefile that every extension should use to provide a set of common functionalities to install, upgrade, and remove an extension. You can inspect the PGXS base Makefile, finding its location with `pg_config`:

```
$ pg_config --pgxs
/usr/local/lib/postgresql/pgxs/src/makefiles/pgxs.mk
```

`pgxs.mk` is the base makefile that provides common functionalities to every extension, and its usage will become more clear when we show how to create an extension from scratch.

Extension components

An extension is made up of two main components—a **control file** and a **script file**:

- The control file provides information about the extension and how to manage it, for instance, where and how to install it, how to upgrade it, and so on. The control file is somehow the metadata of the extension.
- The script file is a SQL file that contains statements to create database objects that are part of the extension. To some extent, this is the content of the extension. The script file can, in turn, load other files that complete the extension, like a shared library and the like.

When you ask PostgreSQL to install an extension, the system inspects the control file to get information about the extension, ensures the extension has not been already installed, and then proceeds to execute the script file within a transaction. As the end result, you have the extension available on your database.

Every extension has a version so that you can decide precisely which version to install. If you do not specify any target version, PostgreSQL will suppose you want to interact with the latest version available.

Extensions are installed in the share-directory of the cluster, usually found by executing the `pg_config` command with the `--sharedir` option. Here's an example:

```
$ pg_config --sharedir
/postgres/12/share/postgresql
```

All the files that make up the extension will be placed in the shared directory, and the cluster expects the files to be available there to the user that runs the cluster (usually the operating system user `postgres`). Once the files are available to the cluster, the extension must be selectively installed in every database that needs it; remember that PostgreSQL provides very strong isolation between databases, and therefore an extension loaded into a database is not automatically available in another database. However, please remember that template databases (see `Chapter 1`, *Introduction to PostgreSQL*) can be used as a skeleton for newly created databases, and therefore once you install an extension in a template database, you will find such an extension already available in all the other created databases.

The control file

An extension control file must have a name that is related to the extension and the `.control` suffix. For example, a valid name could be `learnpg12.control`.

The control file is a text file where you can specify `directives`, which are instructions and metadata to let PostgreSQL handle the extension installation. Every directive has a name and a value. The most common directives are as follows:

- `directory` specifies the path to the extension script path.
- `default_version` specifies the version of the extension to install when the user does not specify any.
- `comment` is a description of the extension and its aim.
- `requires` is an optional list of other extensions needed to install and use this, and therefore represents a dependency list.
- `schema` is an SQL schema into which extension objects will be installed.
- `relocatable` indicates whether the extension can be moved into a user-selected schema.
- `superuser` indicates whether the extension can also be installed by non-superuser accounts (defaults to `yes`, meaning that only superusers can install the extension).

There must be at least one control file per extension, and such a file is known as the *main control file*. However, an extension can have additional control files (named *secondary control files*).

Every secondary control file must target a specific version and must have the same name as the main control file with the version number prefixed with double dashes; for instance, if the main control file is `learnpg12.control`, the secondary files could be `learnpg12--1.1.control`, `learnpg12--1.2.control`, and so on.

The script file

The script file contains plain SQL used to create extension objects. An extension object could be a table, a trigger, a function, or a binding for an external language.

Every script file must be named after the extension name and with a suffix of `.sql`; the version of the extension is specified with a number preceded by a double dash. As an example, the file `learnpg12--1.0.sql` creates objects for version `1.0` of the extension.

There must be at least one script file per extension, but it is possible to specify more than one: in such cases, every additional file must include the version to upgrade to and the final target version. For example, the file `learnpg12--1.0-1.1.sql` provides an upgrade from version `1.0` to version `1.1`.

As already specified, the script file is executed in a transaction and therefore cannot interact with the transaction boundaries (that is, it can issue neither a `COMMIT` nor a `ROLLBACK`). Similarly, executing in a transaction, a script file is prevented from executing anything that cannot be executed in a transaction block (for example, utility commands such as `VACUUM`).

Managing extensions

Every extension is managed at a database level, meaning that every database that needs an extension must manage such an extension life cycle. In other words, there is not a per-cluster way of managing an extension and applying it to every database within the cluster.

Extensions are mainly managed by means of three SQL statements: `CREATE EXTENSION`, `DROP EXTENSION`, and `ALTER EXTENSION`, to respectively install an extension in a database, remove the extension from the database, and modify extension attributes or upgrade them.

Every extension is specified by a mnemonic and a version; if a version is not specified, PostgreSQL assumes you want to deal with the latest available version or the one that is already installed.

In the following subsections, each of the three management statements will be explained.

Creating an extension

The `CREATE EXTENSION` statement allows you to install an existing extension into the current database.

The synopsis of the statement is as follows:

```
CREATE EXTENSION [ IF NOT EXISTS ] extension_name
    [ WITH ] [ SCHEMA schema_name ]
             [ VERSION version ]
             [ FROM old_version ]
             [ CASCADE ]
```

The extension name is the mnemonic for the extension, and as you can see, you can specify the version number of the extension to install. In the case that the extension depends on any other extension, the `CASCADE` option allows the system to automatically execute a recursive `CREATE EXTENSION` for the dependency. You can decide the schema into which the extension objects must be placed, and of course, that makes sense only for such extensions that can be relocated.

The `FROM` keyword is used for creating an extension with a non-extension module, and the latter is an obsolete way of packaging PostgreSQL objects and therefore will not be explained here.

Lastly, as you can imagine, `IF NOT EXITS` allows the command to gracefully fail in the case that the extension has been already installed. More precisely, doing nothing in the case that the extension has been already installed in the database.

In order to better see how `CREATE EXTENSION` works, assume we want to install the PL/Perl procedural language in the `forumdb` database; since the PL/Perl extension is available as the PostgreSQL `contrib` module, you should have the extension already available within the cluster. Therefore, in order to install it, you have to do the following:

```
forumdb=# CREATE EXTENSION plperl;
CREATE EXTENSION
```

Please note that the PL/Perl extension (mnemonic `plperl`) requires installation by means of the database administrator. If you try to install the same extension again, the command fails unless you use the IF NOT EXISTS clause:

```
forumdb=# CREATE EXTENSION plperl;
ERROR:  extension "plperl" already exists

forumdb=# CREATE EXTENSION IF NOT EXISTS plperl;
NOTICE:  extension "plperl" already exists, skipping
CREATE EXTENSION
```

As another easy example, we can install the `pg_stat_statements` extension at a specific version:

```
forumdb=# CREATE EXTENSION pg_stat_statements VERSION '1.4';
CREATE EXTENSION
```

Viewing installed extensions

In the `psql` terminal, it is possible to get a list of installed extensions by means of the `\dx` special command:

```
forumdb=# \dx
                                List of installed extensions
        Name         | Version |   Schema    |
Description
---------------------+---------+-------------+-----------------------------
----------------------------
 pg_stat_statements | 1.4      | public      | track execution statistics of
all SQL statements executed
 plperl             | 1.0      | pg_catalog  | PL/Perl procedural language
 plpgsql            | 1.0      | pg_catalog  | PL/pgSQL procedural language
(3 rows)
```

The very same information can be found out from the special catalog `pg_extension`, which can be joined with `pg_namespace` to extract human-readable information about the schema the extension is living in:

```
forumdb=# SELECT x.extname, x.extversion, n.nspname
          FROM pg_extension x JOIN pg_namespace n
          ON n.oid = x.extnamespace;

      extname        | extversion | nspname
---------------------+------------+------------
```

```
plpgsql              | 1.0         | pg_catalog
plperl               | 1.0         | pg_catalog
pg_stat_statements | 1.4         | public
(3 rows)
```

Finding out available extension versions

It is possible to inspect the cluster to get information about available extension versions, which means versions you can actually install in a database. The special catalog `pg_available_extension_versions` allows you to get all the available versions for any available extension. As an example, the `pg_stat_statements` extension has the following values available in the cluster:

```
forumdb=# SELECT name, version
          FROM pg_available_extension_versions
          WHERE name = 'pg_stat_statements';

        name          | version
--------------------+---------
 pg_stat_statements | 1.4
 pg_stat_statements | 1.5
 pg_stat_statements | 1.6
(3 rows)
```

It is useful to know that the `pg_stat_statements` extension can be installed in a version between `1.4` and `1.6`.

 You should always install the latest version of an extension, that is, the one with the highest version number, unless you are forced to install a specific version for backward compatibility.

Altering an existing extension

The `ALTER EXTENSION` statement is very rich and complex and allows you to fully modify an existing extension. The statement allows four main changes to an existing extension:

- Upgrading the extension to a new version
- Setting the schema of a relocatable extension
- Adding a database object to the extension
- Removing a database object from the extension

In order to upgrade an already installed extension, you must specify the UPDATE clause, specifying the target version number. As a simple example, imagine we want to upgrade the pg_stat_statements extension from version 1.4 to version 1.6; assuming the upgrade files are already installed in the cluster, you can simply do the following:

```
forumdb=# ALTER EXTENSION pg_stat_statements UPDATE TO '1.6';
ALTER EXTENSION

forumdb=# \dx pg_stat_statements
                              List of installed extensions
       Name         | Version | Schema |                      Description
--------------------+---------+--------+----------------------------------
------------------------
 pg_stat_statements | 1.6     | public | track execution statistics of all
SQL statements executed
(1 row)
```

Moving a relocatable extension from one schema to another is done by specifying the SET SCHEMA clause, for example:

```
forumdb=# ALTER EXTENSION pg_stat_statements SET SCHEMA my_schema;
ALTER EXTENSION

forumdb=# \dx pg_stat_statements
                                List of installed extensions
       Name         | Version |  Schema   |
Description
--------------------+---------+-----------+----------------------------------
--------------------------
 pg_stat_statements | 1.6     | my_schema | track execution statistics of
all SQL statements executed
(1 row)
```

That will move all the extension objects into the schema my_schema, which was created earlier.

If you want to remove an existing database object from one extension, for instance, a table, you can use the DROP clause followed by the type of the object and, of course, its name. As an example, if we remove the view pg_stat_statements from the extension with the same name, we can specify the object type (VIEW) after the DROP clause as follows:

```
forumdb=# ALTER EXTENSION pg_stat_statements
          DROP VIEW my_schema.pg_stat_statements;
ALTER EXTENSION
```

What happens is that the view is still there, but unrelated to the extension, and therefore the view and the extension now have a different life cycle. In other words, changing the extension does not imply any more changes to the view itself because the latter is now a user-defined object, not an extension-defined one.

Of course, it is possible to add a new object to an extension with the ADD clause, which works as the opposite of the DROP one and requires the type and name of the object. For instance, to add the pg_stat_statements view back to the extension, it is possible to do the following:

```
forumdb=# ALTER EXTENSION pg_stat_statements
          ADD VIEW my_schema.pg_stat_statements;
ALTER EXTENSION
```

You can also add your own objects to the extension, so for example, adding a new table to the extension means that the extension will undergo the extension life cycle:

```
forumdb=# ALTER EXTENSION pg_stat_statements
          ADD TABLE my_schema.foo;
ALTER EXTENSION
```

The foo table is now part of the extension, and as such, it cannot be manipulated anymore with statements that do not take care of the extension. For instance, if you try to delete the table, PostgreSQL will prevent you from damaging the extension:

```
forumdb=# DROP TABLE my_schema.foo;
ERROR:  cannot drop table my_schema.foo because extension
pg_stat_statements requires it
HINT:  You can drop extension pg_stat_statements instead.
```

You can now remove the table from the extension, or drop the whole extension.

As you have seen, having extensions packages objects together in a way that prevents a single object being alone in a different life cycle.

Removing an existing extension

DROP EXTENSION deletes an extension from the current database. The synopsis of the statement is the following:

```
DROP EXTENSION [ IF EXISTS ] name [, ...] [ CASCADE | RESTRICT ]
```

The command supports the IF EXISTS clause as many other statements do. Moreover, it is possible to specify more than one name of the extension to be removed from the database.

The CASCADE option also removes database objects that depend on the objects of the extension, while its counterpart RESTRICT makes the command fail if there are other objects that still depend on this extension. As an example, the following statement removes two extensions in a single pass, also removing all the objects that depend on those extensions:

```
forumdb=# DROP EXTENSION plperl, plpgsql CASCADE;
NOTICE:  drop cascades to function get_max(integer,integer)
DROP EXTENSION
```

As you can see, since the user-defined function get_max() was dependent on one of the two extensions, the CASCADE option made the process to drop the function too.

To summarize, you have learned how to manually manage an extension, from installing it to upgrading it or removing it; in the next section, you will learn how to perform the same steps in a more automated way.

Exploring the PGXN client

The PGXN client is an external application, written in Python, that works as a command-line interface to PGXN. The application, named pgxnclient, works by means of commands, which are actions such as install, download, uninstall, and so on, allowing a database administrator to instrument PGXN and work with extensions.

To some extent, pgxnclient works the same as the command cpan (or cpanm) for Perl, zef for Raku, pip for Python, and so on.
Being an external application means that pgxnclient is not distributed with PostgreSQL, and therefore you need to install it on your machines before you can use it. Installing pgxnclient is not mandatory in order to use PostgreSQL extensions, but it can make your life a lot easier.

In the following subsections, you will see how to install pgxnclient on main Unix and Unix-like operating systems, but before that, it is important to let you know that, once installed, you will find two executables on your system: pgxn and pgxnclient. You can think of those executables as aliases of one another, even if this is not really true (one wraps the other); however, you can use either one you please, obtaining the very same result. In this chapter, we will use pgxn as the main executable.

Installing pgxnclient on Debian GNU/Linux and derivates

pgxnclient is packaged for Debian GNU/Linux and derivates, and that means you can simply ask apt to install it:

```
$ sudo apt install pgxnclient
Reading package lists... Done
Building dependency tree
Reading state information... Done
The following packages were automatically installed and are no longer
required:
  libecpg-compat3 libecpg-dev libecpg6 libgd-perl libgdbm5 libicu60
libperl5.26 libpgtypes3 libpython3.6 libpython3.6-minimal libpython3.6-
stdlib
  libreadline7 libtinfo5 perl-modules-5.26 pgadmin4-doc
Use 'sudo apt autoremove' to remove them.
The following NEW packages will be installed
  pgxnclient
0 to upgrade, 1 to newly install, 0 to remove and 2 not to upgrade.
Need to get 31,4 kB of archives.
After this operation, 162 kB of additional disk space will be used.
Get:1 http://apt.postgresql.org/pub/repos/apt disco-pgdg/main amd64
pgxnclient all 1.3-1.pgdg19.04+1 [31,4 kB]
Fetched 31,4 kB in 1s (54,5 kB/s)
Selecting previously unselected package pgxnclient.
(Reading database ... 392532 files and directories currently installed.)
Preparing to unpack .../pgxnclient_1.3-1.pgdg19.04+1_all.deb ...
Unpacking pgxnclient (1.3-1.pgdg19.04+1) ...
Setting up pgxnclient (1.3-1.pgdg19.04+1) ...
Processing triggers for man-db (2.8.5-2) ...
```

Once the program has been installed, you can simply test it with the `--version` option, which will print the version number you installed:

```
$ pgxn --version
pgxnclient 1.3
```

Installing pgxnclient on Fedora Linux

pgxnclient is packaged for Fedora as well, so you can install it with the operating system package manager:

```
$ sudo dnf install -y pgxnclient
...

Total download size: 105 k
Installed size: 406 k
Downloading Packages: pgxnclient-1.3-1.f30.x86_64.rpm
...
Installed:
  pgxnclient-1.3-1.f30.x86_64
Complete!
```

Once the process has completed, you can query the application to verify it is actually working:

```
$ pgxn --version
pgxnclient 1.3
```

Installing pgxnclient on FreeBSD

pgxnclient is packaged for FreeBSD, so you can install it via the pkg tool or via the ports. The fastest way is by means of pkg, and all you have to do is ask to install the program:

```
$ sudo pkg install --yes pgxnclient
Updating FreeBSD repository catalogue...
FreeBSD repository is up to date.
All repositories are up to date.
New packages to be INSTALLED:
        pgxnclient: 1.2.1_3

Number of packages to be installed: 1

108 KiB to be downloaded.
[1/1] Fetching pgxnclient-1.2.1_3.txz: 100%  108 KiB 110.4kB/s    00:01
Checking integrity... done (0 conflicting)
[1/1] Installing pgxnclient-1.2.1_3...
[1/1] Extracting pgxnclient-1.2.1_3: 100%
```

Once the process has finished, you can query the tool for its version to see if it is working:

```
$ pgxn --version
pgxnclient 1.2.1
```

Installing pgxnclient from sources

You can always install `pgxnclient` from sources, even if this is suggested only if you are on an operating system that does not provide a packaged version, or if the version is out of date with regard to your needs. You can download a compressed version of the latest release from the official project GitHub repository, for example:

```
$ wget https://github.com/pgxn/pgxnclient/archive/v1.3.zip
Connecting to github.com (github.com)|140.82.118.4|:443... connected.
HTTP request sent, awaiting response... 302 Found
Location: https://codeload.github.com/pgxn/pgxnclient/zip/v1.3 [following]
--2020-02-28 19:53:23--
https://codeload.github.com/pgxn/pgxnclient/zip/v1.3
Resolving codeload.github.com (codeload.github.com)... 140.82.114.9
Connecting to codeload.github.com
(codeload.github.com)|140.82.114.9|:443... connected.
HTTP request sent, awaiting response... 200 OK
Length: unspecified [application/zip]
Saving to: 'v1.3.zip'

2020-02-28 19:53:24 (313 KB/s) - 'v1.3.zip' saved [101925]
```

Once you have the compressed archive, you need to decompress it and enter the directory that will be created with it – named after the version of PXGN you have downloaded – in our case, `pgxnclient-3.1`. Once you are in the directory, executing the Python script `setup.py` will allow you to install the application:

```
$ unzip v1.3.zip
...
$ cd pgxnclient-1.3
$ sudo python setup.py install
...
Installed /usr/lib/python2.7/site-packages/pgxnclient-1.3-py2.7.egg
Processing dependencies for pgxnclient==1.3
Searching for six==1.12.0
Best match: six 1.12.0
Processing six-1.12.0-py2.7.egg
six 1.12.0 is already the active version in easy-install.pth

Using /usr/lib/python2.7/site-packages/six-1.12.0-py2.7.egg
Finished processing dependencies for pgxnclient==1.3
```

Once you have completed the installation, you can query the application to verify that it is working:

```
$ pgxn --version
pgxnclient 1.3
```

The pgxnclient command-line interface

The PGXN client application provides a command-line interface similar to other command-based applications, such as `cpan` and `git`. You can get a list of the main commands by asking for help:

```
$ pgxn help
usage: pgxn [--version] [--help] COMMAND ...

Interact with the PostgreSQL Extension Network (PGXN).

optional arguments:
  --version  print the version number and exit
  --help     show this help message and exit

available commands:
  COMMAND    the command to execute. The complete list is available using
             `pgxn help --all`. Builtin commands are:
    check    run a distribution's test
    download
             download a distribution from the network
    help     display help and other program information
    info     print information about a distribution
    install  download, build and install a distribution
    load     load a distribution's extensions into a database
    mirror   return information about the available mirrors
    search   search in the available extensions
    uninstall
             remove a distribution from the system
    unload   unload a distribution's extensions from a database
```

Usually, you will use the following subset of commands:

- `search` to search for distributions by means of keywords.
- `info` to have a closer look at an extension.
- `download` to download (but not install) an extension.
- `install` to download and install an extension in the cluster.
- `load` to execute CREATE EXTENSION against a specific database.
- `unload` to execute DROP EXTENSION against a specific database.
- `uninstall` to remove an extension from a cluster.

The smallest set of commands you will probably use are `search`, `install`, and `uninstall`.

For every command, you can get more detailed help if you specify the command as an argument to the `help` command itself. For example, to get more information about the `search` command, you can do the following:

```
$ pgxn help search
usage: pgxn search [--help] [--mirror URL] [--verbose] [--yes]
                   [--docs | --dist | --ext]
                   TERM [TERM ...]

search in the available extensions

positional arguments:
  TERM           a string to search

optional arguments:
  --help         show this help message and exit
  --docs         search in documentation [default]
  --dist         search in distributions
  --ext          search in extensions

global options:
  --mirror URL   the mirror to interact with [default:
https://api.pgxn.org/]
  --verbose      print more information
  --yes          assume affirmative answer to all questions
```

In the following sections, you will see how to use PXGN effectively to install an extension.

Installing extensions

Usually, the workflow for getting an extension up and running involves a few steps. First, you need to find out which extension to use, which version, and the compatibility with your cluster. Once you have found out the extension you need, you have to install it in the cluster.

Installing it in the cluster really means *deploying* it in the cluster, that is, moving all the extension-related files and libraries into the shared directory of the cluster so that PostgreSQL can seek the code required to run the extension.

Lastly, you need to create the extension in every single database that needs it. Creating an extension is like enabling the usage of the extension within a specific database.

In order to demonstrate the usage of an extension, we will install *Orafce*, the Oracle compatibility functions extension. Describing the whole extension is not the aim of this section, so let's just say that this extension provides a set of functions, data types, and other stuff that makes PostgreSQL look like an Oracle database so that migrating an Oracle-based application becomes easier.

The following subsections describe every single step required to get the extension up and running.

Installing the extension via pgxnclient

Usually, the first step in installing an extension is getting details about it– that means searching for an extension. In this particular case, we already know what extension we are looking for, but let's search it via `pgxn`:

```
$ pgxn search --ext orafce
orafce 3.9.0
      Oracle's compatibility functions and packages
```

The `search` command explores the ecosystem to find every extension related to our search criteria – in this particular case, the extension name (`--ext`). Thanks to `pgxn`, we now know that we need to install `orafce` version `3.9.0`, the latest stable version available at the time of writing.

Once you have decided which extension you need, you can run the `install` command of `pgxn` to let the installation proceed. The installation workflow includes downloading, compiling (if needed) the source tree, packaging it, and placing it in the shared directory of the PostgreSQL cluster.

You can inspect the ongoing process in very rich detail thanks to the `--verbose` option, and if you are using `pgxn` with a different user from the one that runs the cluster, you can use the `--sudo` option to inform `pgxn` to switch the user when needed:

```
$ pgxn install orafce --verbose --sudo
DEBUG: running pg_config --libdir
DEBUG: running command: ['/usr/local/bin/pg_config', '--libdir']
DEBUG: testing if /usr/local/lib is writable
DEBUG: opening url: http://api.pgxn.org/index.json
DEBUG: opening url: http://api.pgxn.org/dist/orafce.json
INFO: best version: orafce 3.9.0
...
/usr/bin/install -c -m 755  orafce.so '/usr/local/lib/postgresql/orafce.so'
/usr/bin/install -c -m 644 .//orafce.control
```

```
'/usr/local/share/postgresql/extension/'
/usr/bin/install -c -m 644 .//orafce--3.9.sql .//orafce--3.2--3.3.sql
.//orafce--3.3--3.4.sql .//orafce--3.4--3.5.sql .//orafce--3.5--3.6.sql
.//orafce--3.6--3.7.sql .//orafce--3.7--3.8.sql .//orafce--3.8--3.9.sql
'/usr/local/share/postgresql/extension/'
/usr/bin/install -c -m 644 .//README.asciidoc .//COPYRIGHT.orafce
.//INSTALL.orafce '/usr/local/share/doc/postgresql/extension/'
```

Installing the extension manually

The starting point is the PGXN website, where you can search for a specific extension by name or by keywords. Once you browse the PGXN site, you have a textbox where you can insert the keyword for the search, and since we already know the extension name, we can choose **Extensions** from the pull-down menu. The web interface is shown in the following screenshot:

The result of our search will be displayed, as shown in the following screenshot, so we can enter the extensions page with all the information and the documentation for the installation process:

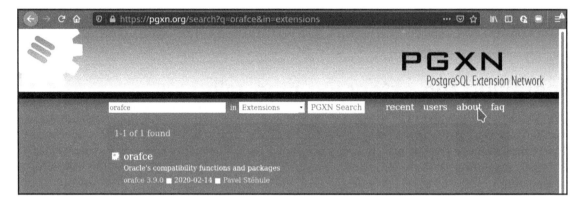

Once we have found the extension we are looking for, we can download it by clicking on the download icon from the page like the one shown in the following screenshot. The result is that we will download a compressed `zip` file with all the stuff related to the extension:

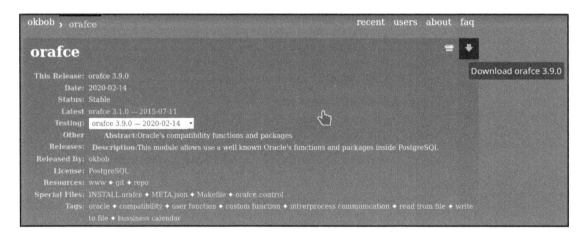

In order to proceed further, you first have to decompress the archive you downloaded:

```
$ unzip orafce-3.9.0.zip
Archive:  orafce-3.9.0.zip
   creating: orafce-3.9.0/
...
```

Now you can enter the directory created for this extension and compile it:

```
$ cd orafce-3.9.0
$ make
cc -Wall -Wmissing-prototypes -Wpointer-arith -Wdeclaration-after-statement
-Werror=vla -Wendif-labels -Wmissing-format-attribute -Wformat-security -
fno-strict-aliasing -fwrapv -Wno-unused-command-line-argument -O2 -pipe  -
fstack-protector-strong -fno-strict-aliasing  -fPIC -DPIC -I. -I./ -
I/usr/local/include/postgresql/server -
I/usr/local/include/postgresql/internal  -I/usr/local/include -
I/usr/local/include -I/usr/local/include  -c -o parse_keyword.o
parse_keyword.c
cc -Wall -Wmissing-prototypes -Wpointer-arith -Wdeclaration-after-statement
-Werror=vla -Wendif-labels -Wmissing-format-attribute -Wformat-security -
fno-strict-aliasing -fwrapv -Wno-unused-command-line-argument -O2 -pipe  -
fstack-protector-strong -fno-strict-aliasing  -fPIC -DPIC -I. -I./ -
I/usr/local/include/postgresql/server -
I/usr/local/include/postgresql/internal  -I/usr/local/include -
I/usr/local/include -I/usr/local/include  -c -o convert.o convert.c
...
```

Once the compilation process has finished, you can install the extension (but this will require you to either run it as the operating system user that manages the cluster or via sudo):

```
$ sudo make install
/bin/mkdir -p '/usr/local/lib/postgresql'
/bin/mkdir -p '/usr/local/share/postgresql/extension'
/bin/mkdir -p '/usr/local/share/postgresql/extension'
/bin/mkdir -p '/usr/local/share/doc/postgresql/extension'
/usr/bin/install -c -m 755  orafce.so '/usr/local/lib/postgresql/orafce.so'
/usr/bin/install -c -m 644 .//orafce.control
'/usr/local/share/postgresql/extension/'
/usr/bin/install -c -m 644 .//orafce--3.9.sql .//orafce--3.2--3.3.sql
.//orafce--3.3--3.4.sql .//orafce--3.4--3.5.sql .//orafce--3.5--3.6.sql
.//orafce--3.6--3.7.sql .//orafce--3.7--3.8.sql .//orafce--3.8--3.9.sql
'/usr/local/share/postgresql/extension/'
/usr/bin/install -c -m 644 .//README.asciidoc .//COPYRIGHT.orafce
.//INSTALL.orafce '/usr/local/share/doc/postgresql/extension/'
```

And you are done: the extension has been deployed to the cluster and you can use it in your databases via the CREATE EXTENSION statement.

Using the installed extension

Once the extension has been installed – that means *deployed* to the PostgreSQL cluster either manually or via `pgxn` – you can create the extension in every single database you need it for.

The Orafce extension must be created by a superuser, so you need to connect to the database as an administrator in order to execute the `CREATE EXTENSION` statement:

```
$ psql  -U postgres forumdb
psql (12.1, server 12.2)
Type "help" for help.

forumdb=# CREATE EXTENSION orafce;
CREATE EXTENSION
```

If you now inspect the extensions installed in the database, you will see the freshly created Orafce at version 3.9 – the same as we found when searching the extension with `pgxn` or on the website:

```
forumdb=# \dx
                                                    List of installed
extensions
   Name   | Version |   Schema   |
Description
---------+---------+------------+------------------------------------
---------------------------------------------------
  orafce  | 3.9     | public    | Functions and operators that emulate a
subset of functions and packages from the Oracle RDBMS
  plpgsql | 1.0     | pg_catalog | PL/pgSQL procedural language
(2 rows)
```

Once the extension has been installed in the database, every user can use it. As a simple test, you can query the `DUAL` table that Oracle has and that Orafce created for your legacy queries to continue to run:

```
$ psql -U luca forumdb
psql (12.1, server 12.2)
Type "help" for help.

forumdb=> SELECT 1 FROM dual;
 ?column?
----------
        1
(1 row)
```

Removing an installed extension

It could happen that you don't need an extension anymore, and therefore you want to remove it from your cluster.

If a database does not need the extension and its related stuff anymore, you can issue a DROP EXTENSION statement and the extension will disappear from your database. Of course, if the extension has been installed as a database superuser, you need to issue the statement as a superuser too. With regard to the Orafce example, as a superuser, you can do the following:

```
$ psql -U postgres forumdb
psql (12.1, server 12.2)
Type "help" for help.

forumdb=# DROP EXTENSION orafce;
DROP EXTENSION
```

As you can imagine, inspecting the extension list does not show the Orafce entry anymore, and all the features, including the DUAL table, have disappeared:

```
forumdb=# \dx
ù                    List of installed extensions
  Name    | Version |   Schema   |          Description
----------+---------+------------+------------------------------
 plpgsql  | 1.0     | pg_catalog | PL/pgSQL procedural language
(1 row)

forumdb=# SELECT 1 FROM DUAL;
ERROR:  relation "dual" does not exist
LINE 1: SELECT 1 FROM DUAL;
```

Having removed an extension from a single database does not remove it from other databases where you have executed an explicit CREATE EXTENSION function. It doesn't remove the extension files and libraries from the cluster share directory either.
The exact way of removing (undeploying) the extension from your cluster depends on the way you first installed it in the cluster.

Removing an extension via pgxncliet

The `uninstall` command of `pgxn` performs the exact opposite actions to the install one: it removes all files related to an extension. The command-line options are the same, and this leads us to execute a command as simple as the following one:

```
$ pgxn uninstall orafce --sudo --verbose
DEBUG: running pg_config --libdir
DEBUG: running command: ['/usr/local/bin/pg_config', '--libdir']
DEBUG: testing if /usr/local/lib is writable
DEBUG: opening url: http://api.pgxn.org/index.json
DEBUG: opening url: http://api.pgxn.org/dist/orafce.json
INFO: best version: orafce 3.9.0
...
rm -f '/usr/local/lib/postgresql/orafce.so'
rm -f '/usr/local/share/postgresql/extension'/orafce.control
rm -f '/usr/local/share/postgresql/extension'/orafce--3.9.sql
'/usr/local/share/postgresql/extension'/orafce--3.2--3.3.sql
'/usr/local/share/postgresql/extension'/orafce--3.3--3.4.sql
'/usr/local/share/postgresql/extension'/orafce--3.4--3.5.sql
'/usr/local/share/postgresql/extension'/orafce--3.5--3.6.sql
'/usr/local/share/postgresql/extension'/orafce--3.6--3.7.sql
'/usr/local/share/postgresql/extension'/orafce--3.7--3.8.sql
'/usr/local/share/postgresql/extension'/orafce--3.8--3.9.sql
rm -f '/usr/local/share/doc/postgresql/extension'/README.asciidoc
'/usr/local/share/doc/postgresql/extension'/COPYRIGHT.orafce
'/usr/local/share/doc/postgresql/extension'/INSTALL.orafce
```

As you can see from the bottom lines, all the files have been removed from the cluster shared directory. The extension is therefore gone forever, and if you need to install it again, you will need to restart from the very first step.

Removing a manually installed extension

You need to use `make` again, this time with the `uninstall` command, from the directory where you extracted the downloaded compressed archive:

```
$ cd orafce-3.9.0
$ sudo make uninstall
rm -f '/usr/local/lib/postgresql/orafce.so'
rm -f '/usr/local/share/postgresql/extension'/orafce.control
rm -f '/usr/local/share/postgresql/extension'/orafce--3.9.sql
'/usr/local/share/postgresql/extension'/orafce--3.2--3.3.sql
'/usr/local/share/postgresql/extension'/orafce--3.3--3.4.sql
```

```
'/usr/local/share/postgresql/extension'/orafce--3.4--3.5.sql
'/usr/local/share/postgresql/extension'/orafce--3.5--3.6.sql
'/usr/local/share/postgresql/extension'/orafce--3.6--3.7.sql
'/usr/local/share/postgresql/extension'/orafce--3.7--3.8.sql
'/usr/local/share/postgresql/extension'/orafce--3.8--3.9.sql
rm -f '/usr/local/share/doc/postgresql/extension'/README.asciidoc
'/usr/local/share/doc/postgresql/extension'/COPYRIGHT.orafce
'/usr/local/share/doc/postgresql/extension'/INSTALL.orafce
```

To summarize, you have seen how to deal with an extension by means of the PGXN client or manually by obtaining it through the PGXN infrastructure. In the following section, you will learn how to build your own extension.

Creating your own extension

In this section, we will build an extension from scratch, so that you will better understand how they are made up. The idea is to let you know how to convert even your own SQL scripts into an extension, with all the advantages that an extension can provide in terms of manageability.

Defining an example extension

In order to demonstrate how to build your own extension, we are going to create a simple extension that applies to the forum database, providing some more features. In particular, we are going to define an extension named tagext that will provide a utility function that, given a particular tag within the tag table, will return the full path to that tag with all ancestors. For example, the tag Linux is a child of the tag Operating Systems and therefore the path to the tag Linux is Operating System > Linux.

In particular, we want our extension to provide us with a function named tag_path that, given a tag, provides the tag path as in the following example:

```
forumdb=> SELECT tag_path( 'Kubuntu' );
                    tag_path
----------------------------------------------
 Operating Systems > Linux > Ubuntu > Kubuntu
(1 row)
```

In the following sections, you will see how to reach the preceding result by implementing the example extension.

Creating extension files

Let's start with the control file first, where we insert some basic information about our extension:

```
comment = 'Tag Programming Example Extension'
default_version = '1.0'
superuser       = false
relocatable     = true
```

The preceding control file, named `tagext.control`, contains a comment that describes the extension to other administrators, specifies the `default_version`, that is the version to be installed if none is specified by the user, and dictates that this extension can be installed by any user (`superuser = false`) and moved to any schema the user wishes to (`relocatable = true`).

Then comes the Makefile, that is, the file that will build and install the extension:

```
EXTENSION = tagext
DATA = tagext--1.0.sql

PG_CONFIG = pg_config
PGXS := $(shell $(PG_CONFIG) --pgxs)
include $(PGXS)
```

The Makefile is very simple and can be used as a skeleton for other extension Makefiles. In particular, we define the name of the extension we are going to manage via this Makefile, as well as the file to use for producing the extension content: this is specified in the `DATA` variable and therefore we are instrumenting the system to use the `tagext--1.0.sql` file to be used to create the objects this extension provides.

The trailing lines define the use of the `PGXS` build infrastructure and in particular, are used to include the PGXS base Makefile, which is computed from the output of the `pg_config` command.

With all the infrastructure in place, it is now possible to define the content of the extension, therefore the `tagext--1.0.sql` file contains the definition of a function (see `Chapter 7, Server Side Programming`) that, given a specific tag, returns the text representation of the tag path with all the ancestors:

```
CREATE OR REPLACE FUNCTION tag_path( tag_to_search text )
RETURNS TEXT
AS $CODE$
DECLARE
```

```
    tag_path text;
    current_parent_pk int;
BEGIN

    tag_path = tag_to_search;

    SELECT parent
    INTO   current_parent_pk
    FROM   tags
    WHERE  tag = tag_to_search;

    -- here we must loop
    WHILE current_parent_pk IS NOT NULL LOOP
        SELECT parent, tag || ' > ' || tag_path
        INTO   current_parent_pk, tag_path
        FROM   tags
        WHERE  pk = current_parent_pk;
    END LOOP;

    RETURN tag_path;
END
$CODE$
LANGUAGE plpgsql;
```

The function works by taking a tag as an argument, then querying the `tags` table to get the parent primary key, and then looping on every parent tag. At every loop, the `tag_path` text string is enriched by the parent tag name, so that the end result is to have a string like `parent 1 > parent 2 > child`.

Once all the files are ready, we will have a situation like the following, with the Makefile, the control file, and the extension content file:

```
$ ls -1
Makefile
tagext--1.0.sql
tagext.control
```

Installing the extension

Having all the pieces in place, it is possible to use the `Makefile` to install (deploy) the extension in the cluster:

```
$ sudo make install
/bin/mkdir -p '/usr/local/share/postgresql/extension'
/bin/mkdir -p '/usr/local/share/postgresql/extension'
```

```
/usr/bin/install -c -m 644 .//tagext.control
'/usr/local/share/postgresql/extension/'
/usr/bin/install -c -m 644 .//tagext--1.0.sql
'/usr/local/share/postgresql/extension/'
```

And it is now possible to install the extension in the `forumdb` database by means of CREATE EXTENSION and then try to execute the function the extension defines:

```
forumdb=> CREATE EXTENSION tagext;
CREATE EXTENSION

forumdb=> \dx tagext
                List of installed extensions
  Name   | Version | Schema |            Description
---------+---------+--------+----------------------------------
 tagext  | 1.0     | public | Tag Programming Example Extension
(1 row)


forumdb=> SELECT tag_path( 'Kubuntu' );
                  tag_path
---------------------------------------------
 Operating Systems > Linux > Ubuntu > Kubuntu
(1 row)
```

The function works and can build up a tag tree or path for the specified tag, with all its ancestors, as well as PostgreSQL reporting that the extension is at version 1.0.

Creating an extension upgrade

Imagine we want to enrich our extension function so that the user is able to specify the tag separator in the path output. We can produce a new version of the function, drop the old one, and allow the user to upgrade the extension with the new content.

Let's start by creating an upgrade of the content of the extension, that is, the function the extension provides. First of all, create a file named `tagext--1.0--1.1.sql` and place the following content in it:

```
DROP FUNCTION IF EXISTS tag_path( text );

CREATE OR REPLACE FUNCTION tag_path( tag_to_search text,
                                     delimiter text DEFAULT ' > ' )
RETURNS TEXT
AS $CODE$
DECLARE
```

```
  tag_path text;
  current_parent_pk int;
BEGIN

  tag_path = tag_to_search;

  SELECT parent
  INTO   current_parent_pk
  FROM   tags
  WHERE  tag = tag_to_search;

  -- here we must loop
  WHILE current_parent_pk IS NOT NULL LOOP
      SELECT parent, tag || delimiter || tag_path
      INTO   current_parent_pk, tag_path
      FROM   tags
      WHERE  pk = current_parent_pk;
  END LOOP;

  RETURN tag_path;
END
$CODE$
LANGUAGE plpgsql;
```

The file first performs a drop of the older version of the function (if that exists and has been installed by the previous version of this extension). After that, a new function with an additional optional parameter is created. The function does exactly the same job as the previous one, but this time it exploits the variable delimiter to separate multiple tags.

Since we added a new file to the extension, we need to inform the Makefile about the file and therefore we have to add the new file to the DATA variable so that the Makefile content looks like the following:

```
EXTENSION = tagext
DATA = tagext--1.0.sql tagext--1.0--1,1.sql

PG_CONFIG = pg_config
PGXS := $(shell $(PG_CONFIG) --pgxs)
include $(PGXS)
```

Performing an extension upgrade

With the new `Makefile` and the `tagext--1.0--1.1.sql` files, the situation on the disk looks like the following:

```
$ ls -1
Makefile
tagext--1.0--1.1.sql
tagext--1.0.sql
tagext.control
```

It is therefore now possible to install (deploy) the extension to the cluster, again running an `install` command:

```
$ sudo make install
/bin/mkdir -p '/usr/local/share/postgresql/extension'
/bin/mkdir -p '/usr/local/share/postgresql/extension'
/usr/bin/install -c -m 644 .//tagext.control
'/usr/local/share/postgresql/extension/'
/usr/bin/install -c -m 644 .//tagext--1.0.sql .//tagext--1.0--1.1.sql
'/usr/local/share/postgresql/extension/'
```

And within the database, it is possible to upgrade the extension with `ALTER EXTENSION`:

```
forumdb=> ALTER EXTENSION tagext UPDATE TO '1.1';
ALTER EXTENSION

forumdb=> \dx tagext
                List of installed extensions
  Name   | Version | Schema |            Description
---------+---------+--------+---------------------------------
 tagext  | 1.1     | public | Tag Programming Example Extension
(1 row)
```

As you can see, the extension version is now at `1.1`, so it is possible to invoke the `tag_path` function with or without the new argument:

```
forumdb=> SELECT tag_path( 'Kubuntu' );
                      tag_path
---------------------------------------------
 Operating Systems > Linux > Ubuntu > Kubuntu
(1 row)

forumdb=> SELECT tag_path( 'Kubuntu', ' --> ' );
                       tag_path
-----------------------------------------------
 Operating Systems --> Linux --> Ubuntu --> Kubuntu
(1 row)
```

You now have the knowledge to manage the whole life cycle of your own extensions.

Summary

This chapter has introduced you to the extension ecosystem, a very rich and powerful system to package related objects and manage them as a single unit. Extensions provide a way to add new features to your cluster and your databases and most notably provide a clear and concise way of building updates and repeatable installation, therefore easing the distribution of the features to other clusters and databases.

Thanks to the PGXS building infrastructure, creating an extension from scratch is comprehensive and quite easy, while thanks to tools such as `pgxnclient`, managing a lot of extensions can be automated.

In the next chapter, you will learn how to take care of the status and performance of your cluster.

References

- PostgreSQL official documentation about extensions: `https://www.postgresql.org/docs/12/extend-extensions.html`
- PostgreSQL official documentation about the extension build system (PGXS): `https://www.postgresql.org/docs/12/extend-pgxs.html`
- The `pgxnclient` official repository: `https://pypi.org/project/pgxnclient/`
- The `pgxnclient` official documentation: `https://pgxn.github.io/pgxnclient/`
- *PostgreSQL 11 Server Side Programming – Quick Start Guide*, Packt Publishing

13
Indexes and Performance Optimization

Performance tuning is one of the most complex tasks in the daily job of a database administrator. SQL is a declarative language, and therefore it does not define how to access the underlying data – that responsibility is left to the database engine. PostgreSQL, therefore, must select, for every statement, the best available access to the data.

A particular component, the planner, is responsible for deciding on the best among all the available paths to the underlying data and another component, the optimizer, is responsible for executing the statement with such a particular access plan.

The aim of this chapter is to teach you how PostgreSQL executes a query, how the planner computes the best execution plan, and how you can help in improving the performance by means of indexes.

You will learn about the following topics in this chapter:

- Execution of a statement
- Indexes
- The EXPLAIN statement
- An example of query tuning
- ANALYZE and how to update statistics
- Auto-explain

Technical requirements

You need to know the following:

- How to execute queries against the database
- How to execute **data description language** (**DDL**) statements

The code for this chapter can be found in the following GitHub repository: `https://github.com/PacktPublishing/Learn-PostgreSQL`.

Execution of a statement

SQL is a declarative language: you ask the database to execute something on the data it contains, but you do not specify how the database is supposed to complete the SQL statement. For instance, when you ask to get back some data, you execute a `SELECT` statement, but you only specify the clauses that specify which subset of data you need, not how the database is supposed to pull the data from its persistent storage. You have to trust the database – in particular, PostgreSQL – to be able to do its job and get you the fastest path to the data, always, under any circumstance of workload. The good news is that PostgreSQL is really good at doing this and is able to understand (and to some extent, interpret) your SQL statements and its current workload to provide you with access to the data in the fastest way.

However, finding the fastest path to the data often requires an equilibrium between searching for the absolute fastest path and the time spent in reasoning about this path; in other words, PostgreSQL sometimes chooses a compromise to get you data in a fast-enough way, even if that is not the absolute fastest one.

Sometimes, on the other hand, PostgreSQL cannot understand very well how to find the fastest path to the data, and therefore, it needs some help from the database administrator. Usually, a slow statement hides a miswritten statement, which means a statement written with wrong or in-contrast clauses. Other times, a slow query is due to PostgreSQL reasoning about the wrong size of the dataset it has to handle. In all these cases, the database administrator has to provide some tuning in the database or the statements to help PostgreSQL make the best decisions.

In order to be able to help your cluster optimize your statements, you need to understand how PostgreSQL handles a SQL statement first. In the following section, you will learn all the fundamentals about how a SQL statement is converted into a set of actions that PostgreSQL executes to manage data.

Execution stages

A SQL statement – a query, for short – is handled in four main stages:

1. The first stage is parsing; a dedicated component, the parser, handles the textual form of the statement (the SQL text) and verifies whether it is correct or not. If the statement has any syntax errors, the execution stops at this early stage; otherwise, the parser disassembles the statement into its main part, for example, the list of involved tables and columns, the clauses to filter data, sorting, and so on.

2. Once the parser has completed successfully, the statement goes to the second stage: the rewriting phase. The rewriter is responsible for applying any syntactic rules to rewrite the original SQL statement into what will be effectively executed. In particular, the rewriter is responsible for applying rules (refer to `Chapter 8`, *Triggers and Rules*). When the rewriter has completed its task, producing the effective statement that the database is going to handle, this statement passes to the next stage: optimization.

3. At the optimization phase, the query is handled by the optimizer, which is responsible for finding the very fastest path to the data that the statement needs. Finding the fastest path to the data is not a simple task: the optimizer must decide how, from among all the available access methods, such as indexes, to get to the data. As you can imagine, reasoning and iterating among all the available access methods consumes time and resources, so the task of the optimizer is not only to find out the fastest access method but also to find it out in a short time.

4. Lastly, when the optimizer has decided how to access the data, the query goes to the last phase: execution. The execution phase is handled by the executor component, which is responsible for effectively going to the storage and retrieving (or inserting) the data using the access method decided by the executor.

Predicate pushdowns are a nice example of optimization, and pretty easy to understand, even for a beginner.

To summarize, a single SQL statement goes through four stages, all shown in the following diagram: a parsing phase that checks the syntax of the statement, a rewriting phase that transforms the query into something more specific, the optimization phase that decides how to access the data requested by the query, and lastly, the execution phase, which gets physical access to the data. This can be visualized in the following diagram:

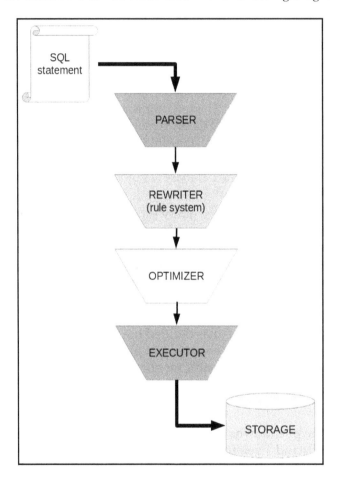

The database administrator can only hook in the optimization phase, trying to help PostgreSQL better understand the statement and optimize it correctly whenever PostgreSQL is not doing an optimal job. The following section takes a closer look at the optimizer, in order to prepare you for the ways you can tune your queries and your database to handle queries in a smarter and faster way.

The optimizer

The optimizer is the component responsible for deciding what to use to access the data as quickly as possible. How can the optimizer choose among the different ways to access the data? The optimizer decides to depend on the concept of cost: every way to access the data is assigned a cost and the way that has the lowest cost wins and is chosen as the best access method.

It is for this reason that the PostgreSQL optimizer is called a cost-based optimizer.

PostgreSQL is configured to assign a specific cost to every operation it performs: seeking data from the storage, performing some CPU-based operation (for example, sorting in memory), and so on. The optimizer iterates over all the possible ways of accessing data and mangling it to return to the user the desired result, computing the total amount of cost for every way – that is the sum of the costs of every operation PostgreSQL will perform. After this, the plan with the lowest cost is passed to the executor as a sequence of actions to perform, and thus data is managed.

This is only half of the story, though. There are cases where the job of the optimizer is really simple: if there is only an access method, it is trivial to decide how to access data. However, there are statements that involve so many objects and tables that iterating over all the possibilities would require a lot of time, so much time that the result will be overtaken by the time spent in computing the optimal way to access the data. For this reason, if the statement involves more than 12 table joins, the optimizer does not iterate all the possibilities but rather executes a genetic algorithm to find a compromise way to access the data. The compromise is between the time spent in computing the path to the data and finding a not-too-bad access path.

Since PostgreSQL 9.6, the executor can also perform data access using parallel jobs. This means, for instance, that retrieving a very large set of data can be performed by dividing the amount of work between different parallel workers (for example, threads).

In all the cases, the optimizer divides the set of actions to pass to the executor in nodes: a node is an action to execute in order to provide the final or an intermediate result. For example, say you execute a generic query asking for data in a specific order, as follows:

```
SELECT * FROM categories ORDER BY description;
```

The optimizer will pass two actions to the executor, and thus the nodes: one to retrieve all the data and one to sort the data.

In the following subsections, we will present the main nodes that the optimizer considers and passes to the executor. We will start from the sequential nodes – those nodes that will be executed with a single job – and then we will see how PostgreSQL builds parallelism on top of them.

Nodes that the optimizer uses

In this section, we will present the main nodes you can encounter in the optimizer plan. There are different nodes for every operation that can be performed, and for every different access method that PostgreSQL accepts.

It is important to note that nodes are stackable: the output of a node can be used as the input to another node. This allows the construction of very complex execution plans made by different nodes, which can produce a fine-grain access method to the data.

Sequential nodes

Sequential nodes are those nodes that will be executed sequentially, one after the other, in order to achieve the final result. The main nodes are listed here and will be explained in the following subsections:

- **Sequential Scan**
- **Index Scan**, **Index Only Scan**, and **Bitmap Index Scan**
- **Nested Loop**, **Hash Join**, and **Merge Join**
- The **Gather** and **Merge** parallel nodes

Sequential Scan

Sequential Scan (**Seq Scan**) is the only node that is always available to the optimizer and the executor, in particular when there is no other valuable alternative. In a sequential scan, the executor will go to the beginning of the dataset on the disk – for example, the beginning of the file corresponding to a table – and will read all the data one block after the other in sequential order.

This node is, for example, always used when you ask for the contents of a table without any particular filtering clause, such as in the following example:

```
SELECT * FROM categories;
```

The **Sequential Scan** node is also used when the filtering clause is not very limiting in the query so that the end result will be to get almost the whole table contents. In such a case, the database can perform a sequential read-all operation faster, throwing away those tuples that are filtered out by the query clauses.

Index nodes

An index scan has access to the data that involves an index in order to find quickly the requested dataset. In PostgreSQL, all indexes are secondary, meaning that they live alongside the table; therefore, you will have in storage a data file for the table and one for every index you build on the table. This means that an index scan always requires two distinct accesses to the storage: one to read the disk and extract the information of where in the table the requested tuples are, and another to access the disk to seek the tuples pointed out by the index.

From this, it should be clear that PostgreSQL avoids using indexes when they are not useful, which is when the previously mentioned double storage access accounts for more disadvantages than advantages.

However, when PostgreSQL believes that accessing the data through an index could be valuable, it will produce an index node that can specialize in three different types.

Index Scan is, as the name suggests, the "classical" index access method: PostgreSQL reads the chosen index, and from that, it goes seeking the tuples, reading again from the storage.

Index Only Scan is a particular degeneration of **Index Scan**: if the requested data only involves columns that belong to the index, PostgreSQL is smart enough to avoid the second trip to storage since it can extract all the required information directly from the index.

The last type of index-based node you can encounter is **Bitmap Index Scan**: PostgreSQL builds a memory bitmap of where tuples that satisfy the statement clauses are, and then this bitmap is used to get from the storage to the tuples. **Bitmap Index Scan** is usually associated with **Bitmap Heap Scan**, as you will see in the examples in the following sections.

Join nodes

When PostgreSQL performs a join between two (or more) tables, it uses one out of three possible nodes. In this section, we will describe these join nodes, considering a join between two tables: an outer table (to the left of a join) and an inner one (the table on the right side of a join).

The most simple node to understand is **Nested Loop**: both tables are scanned in a sequential or indexed-based method and every tuple is checked to see whether there is a match. Essentially, the algorithm can be described by the following piece of pseudo-Java code:

```
for ( Tuple o : outerTable )
    for ( Tuple i : innerTable )
      if ( o.matches( i ) )
          appendTupleToResultDataSet ( o, i );
```

As you can see from the preceding pseudo-code, **Nested Loop** is named after the nesting of the loops it performs in order to evaluate every tuple between the inner and the outer tables.

Anyway, a nested loop is not forced to perform a sequential scan on both tables, and, in fact, depending on the context, every table could be walked in a sequential or indexed-based access method. However, the core of the nested loop does not change: there will always be a nested double loop to search for matches among the tuples.

The following diagram shows the behavior of a **Nested Loop** join:

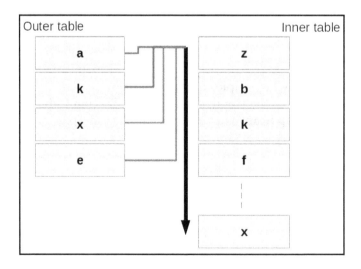

PostgreSQL chooses **Nested Loop** only if the inner table is small enough so that looping every time against it does not introduce any particular penalties.

Another way to perform a join is by using a **Hash Join** node: the inner table is mapped into a hash, which is a set of buckets containing the tuples of the table; the outer table is then walked and for every tuple extracted from the outer table, the hash is searched to see whether there is a match. The following piece of pseudo-Java code illustrates the mechanics of **Hash Join**:

```
Hash innerHash = buildHash( innerTable );
for ( Tuple o : outerTable )
    if ( innerHash.containsKey( buildHash( o ) ) )
        appendTupleToResultDataSet( o, i );
```

As you can see from the preceding example, the first step involves having the dinner table and then walking across the outer table to see whether any of its tuples can match the values in the hash map of the inner table.

The following diagram shows the **Hash Join** algorithm:

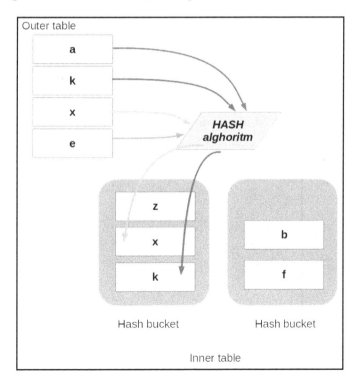

The last type of join you can encounter in PostgreSQL is **Merge Join**. As the name suggests, **Merge Join** involves a step of sorting: both the tables are first sorted by the join key(s), and then they are walked sequentially. For every tuple of the outer table, all the tuples that do match in the inner table are extracted. Since both tables are sorted, a non-matching tuple indicates that it is time to move on to the next join key.

The following pseudo-Java code illustrates the algorithm of a merge join:

```
outerTable = sort( outerTable );
innerTable = sort( innerTable );
int innerIdx = 0;

for ( Tuple o : outerTable )
    for ( ; innerIdx < innerTable.length(); innerIdx++ ){
        Tuple i = innerTable[ innerIdx ];
      if ( o == i )
         appendTupleToResultSet( o, i );
      else
         break;
    }
```

As you can see, once the tables have been sorted, a tuple is extracted from the outer table and is compared with all the tuples within the inner table. As soon as the tuples do not match anymore, another tuple from the outer table is extracted and the inner table restarts its loop from the previous position. In other words, both tables are walked exactly once and only one time.

The following diagram depicts **Merge Join**, where both the tables are displayed after the sorting step:

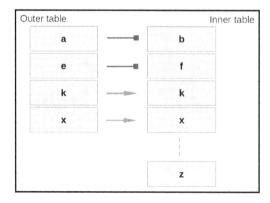

We will now move on to parallel nodes.

Parallel nodes

Parallel nodes are those nodes that PostgreSQL can execute that distribute the amount of work among parallel processes, therefore getting to the final result faster. It is important to note that parallel execution is not always the right choice: there is a set time to distribute the job among parallel processes, as well as the time and resources needed to return the results of every single process. For this reason, PostgreSQL enables parallel execution of certain nodes only if the estimated parallel version will provide a benefit over sequential execution.

As a simple example, consider a case where you have a very tiny table made by only a few tuples, such as four. If you require all the table content, the resources and time spent in launching and synchronizing parallel processes will be much greater than going directly to the table and getting back the result dataset sequentially. The rule of thumb is: if the requested dataset is small enough, PostgreSQL will never choose parallel execution.

It is important to understand the fact that just because the planner produces a parallel plan, which is an execution plan made of parallel nodes, it does not mean that the executor will follow this parallelism. There could be conditions, in particular at runtime, that prevent PostgreSQL for executing a parallel plan, even if that would be the optimal choice (for instance, PostgreSQL does not have enough room to spawn the required amount of parallel processes).

In the following subsection, you will learn what the main parallel nodes available are.

Gather nodes

A parallel execution plan always involves two types of **Gather** nodes: a plain **Gather** node and a **Gather Merge** node.

Gather nodes are responsible for collecting back results from parallel execution nodes, assembling them together to produce the final result. The difference is that a **Gather Merge** node requires the parallel processes to provide it sorted output so that the assembling of the set of results is done following the ordering of the data.

A plain **Gather** node does not require the sorting of the batch results, so it simply assembles all the pieces together to provide the final result.

Parallel scans

All the main nodes that you can find in a sequential access method can be made parallel.

In a **Parallel Seq** scan, all the data pages of a table are split across all the available parallel processes, so every process walks through a smaller set of the table in a sequential way.

In a **Parallel Index** scan and a **Parallel Index Only** scan, blocks of the index are assigned to every parallel process so that every process walks through a subset of the index and, therefore, of the table.

Last, in a **Parallel Bitmap Heap** scan, a bitmap of the table is first built, and then the map is split across different parallel processes.

Parallel joins

When PostgreSQL decides to go for a parallel join method, it tries to keep the inner table accessed in a non-parallel way (assuming such a table is small enough) and performs parallel access to the outer table using one of the nodes presented in the last section.

However, in the case of **Hash Join**, the inner table is computed as a hash by every parallel process, which therefore requires every parallel process working on the outer table to compute the same results for the inner table. For that reason, there is also **Parallel Hash Join**, which allows a hash map of the inner table to be computed in parallel by every process working on the outer table.

Parallel aggregations

When the final result set is made by the aggregation of different parallel subqueries, there must be a parallel aggregation, which is the aggregation of every single parallel part.

This aggregation happens in different steps: first, there is a **Partial Aggregate** node, done by every parallel process that produces a partial result set. After that, a **Gather** node (or **Gather Merge**) collects all the partial results and passes the whole set to the **Finalize Aggregate** node, which sequentially assembles the final result.

When does the optimizer choose a parallel plan?

As already stated, PostgreSQL does not even consider a parallel plan as a choice if the expected size of the result set is too small. In particular, if the table to seek data for has a dimension lower than the `min_parallel_table_scan_size` parameter(defaults to 8 MB), or the index to walk through is smaller than `min_parallel_index_scan_size` (defaults to 512 kB), PostgreSQL will not take into account a parallel plan at all.

You can force PostgreSQL to perform a parallel plan, even if the preceding values are not satisfied, with an extra configuration parameter – `force_parallel_mode` – which is set to off by default.

In any case, when PostgreSQL considers the parallel plan to be an option, it does not default to using it: it rather evaluates carefully the costs of a sequential plan and the costs of the parallel plan to see whether it is still worth the extra setup effort.

There are, however, other restrictions to the application of a parallel plan: PostgreSQL must ensure not to spawn too many parallel processes, so if there are already too many parallel processes working on the system, the parallel execution will not be considered as an option. Moreover, any statement that produces a data write – that is, anything different from a SELECT statement – will not be a valid candidate for a parallel plan, as well as any statement that can be suspended and resumed, such as the usage of a cursor.

Lastly, any query that involves the invocation of a function marked as PARALLEL UNSAFE will not produce a parallel plan candidate.

Utility nodes

Besides the already-introduced nodes that are used to access the data in a single table – or in multiple ones, in the case of joins – there are also some utility nodes that are used in a plan to achieve the final result.

When your statement involves an ordering of the result that is a clause such as ORDER BY, the planner inserts a **Sort** node. If the query has an output limitation, such as a LIMIT clause, a **Limit** node is inserted in the plan to reduce the final result set.

In that case, instead of a UNION ALL statement, the node used is an **Append** one (remember that UNION ALL allows duplicated tuples, while UNION does not).

If your statement involves the aggregation of different queries, like UNION, a **Distinct** node is inserted. The very same node has another feature: it can serve a DISTINCT tuple selection.

When a statement uses a GROUP BY clause, the planner inserts a **GroupAggregate** node responsible for the tuple squashing. Similarly, when the statement involves a window function (refer to Chapter 7, *Server Side Programming*), the planner introduces a **WindowAgg** node for managing the tuple aggregation required by the window function.

In the case of a **Common Table Expression** (CTE), the planner introduces a **CTEScan** node responsible for the join between the CTE subquery and the real table. If a join requires the materialization of a dataset – that is, if there is the need to simulate a table from a set of query results – the planner introduces a **Materialize** node.

Node costs

Every node is associated with a cost, which is the estimation of how expensive, in terms of computational resources, the execution of the node will be. Of course, every node has a variable cost that depends on the type and quantity of the input, as well as the node type.

PostgreSQL provides a list of costs, expressed in arbitrary units, for the main type of operations that a node can perform. Computing the cost of a node is, therefore, the computation of the cost of the single operations that the node performs multiplied by the number of times these operations are repeated, and this depends on the size of the data that the node has to evaluate.

The costs can be adjusted in the cluster configuration – that is, in the `postgresql.conf` main file or in the `pg_settings` catalog. In particular, it is possible to query the cluster about the main costs involved in a node execution:

```
forumdb=> SELECT name, setting
            FROM pg_settings
            WHERE name LIKE 'cpu%\_cost'
                OR name LIKE '%page\_cost'
            ORDER BY setting DESC;

         name          | setting
-----------------------+---------
 random_page_cost      | 4
 seq_page_cost         | 1
 cpu_tuple_cost        | 0.01
 cpu_index_tuple_cost  | 0.005
 cpu_operator_cost     | 0.0025
```

The preceding are the default costs for a fresh installation of PostgreSQL, and you should not change any of the preceding values unless you are really sure about what you are doing. Remember that the costs are what make the planner choose from different plans, so setting the costs to wrong values will lead the optimizer to adopt the wrong execution plans.

As you can see from the preceding list of costs, the base for all the optimizer computation is the cost of a single data page access in sequential mode: this value is set to the unit of cost. CPU costs, which are costs related to the analysis of a tuple already in memory, are much smaller than a unit, while the access to the storage in a random way is much more expensive than sequential access.

Costs can change depending on the computation power of your system; in particular, having enterprise-level SSD storage disks can make you decrease `random_page_cost` to `1.5`, which is almost the same as a sequential page cost.

Changing the optimizer cost is very difficult and is discouraged in pretty much all scenarios, and it is not within the scope of this book. Rather, you are going to understand how the planner estimates the costs of accessing the data.

Later in this chapter, you will see how the preceding costs are applied to compute the cost of a query plan.

In the following section, you will learn about indexes, the way PostgreSQL can access your data using a faster path.

Indexes

An index is a data structure that allows faster access to the underlying table so that specific tuples can be found quickly. Here, "quickly" means faster than scanning the whole underlying table and analyzing every single tuple.

PostgreSQL supports different types of indexes, and not all types are optimal for every scenario and workload. In the following sections, you will discover the main types of indexes that PostgreSQL provides, but in any case, you can extend PostgreSQL with your own indexes or indexes provided by extensions.

An index in PostgreSQL can be built on a single column or multiple columns at once; PostgreSQL supports indexes with up to 32 columns.

An index can cover all the data in the underlying table, or can index specific values only – in that case, the index is known as "partial." For example, you can decide to index only those values of certain columns that you are going to use the most.

An index can also be unique, meaning that it is used to ensure the uniqueness of the values it indexes, such as, for example, the primary keys of a table. Moreover, an index can be built on top of a user-defined function, which means the index is going to index the return values of those functions.

 In order to be used in an index, a user-defined function must be declared as `IMMUTABLE`, which means its output must be the same for the very same input.

PostgreSQL is able to mix and match indexes together; therefore, multiple different indexes can be used to satisfy the query plan. Thanks to this important feature of PostgreSQL, you don't have to define all the possible column permutation indexes, since PostgreSQL will try to mix unrelated indexes together.

In the following subsections, you will learn all the available indexes types in a PostgreSQL 12 cluster, as well as how to create or drop an index.

Index types

The default index PostgreSQL uses is **Balanced Tree** (**B-Tree**), a particular implementation of a balanced tree that keep its depth constant even with large increases in the size of the underlying table. A B-Tree index can be used for most operators and column types, even string comparison in LIKE-based queries, but is effective only if the pattern starts with a fixed string. The B-Tree index also supports the UNIQUE condition and is therefore used to build the primary key indexes.

One drawback of the B-Tree index is that it does copy the whole column(s)' values into the tree structure, and therefore if you use B-Tree to index large values (for example, long strings), the index will rapidly grow in size and space.

Another type of index that PostgreSQL provides is the hash index: this index is built on the result of a hash function for the value of the column(s). It is important to note that the hash index can be used only for equality operators, not for range nor disequality operators. In fact, being an index built on a hash function, the index cannot compare two hash values to understand their ordering; only the equality (which produces the very same hash value) can be evaluated.

Block Range Index (BRIN) is a particular type of index that is based on the range of values in data blocks on storage. The idea is that every block has a minimal and maximal value, and the index then stores a couple of values for every data block on the storage. When a particular value is searched from a query, the index knows in which data block the values can be found, but all the tuples in the block must be evaluated.

Therefore, this type of index is not as accurate as a B-Tree and is called lossy, but it is much smaller in size with respect to all the other types of indexes since it only stores a couple of values for every data block.

GIN is a type of index that instead of pointing to a single tuple points to multiple values, and to some extent to an array of values. Usually, this kind of index is used in full text search scenarios, where you are indexing a written text where there are multiple duplicated keys (for example, the same word or term) that point to different places (for example, different phrases and lines).

Then comes Generalized Index Search Tree (GIST), which is a platform on top of which new index types can be built. The idea is to provide a pluggable infrastructure where you can define operators and features that can index a data structure. An example is SP-GIST, a spatial index used in geographical applications.

Creating an index

Indexes can be created by means of the CREATE INDEX statement, which has the following synopsis:

```
CREATE [ UNIQUE ] INDEX [ CONCURRENTLY ] [ [ IF NOT EXISTS ] name ] ON [
ONLY ] table_name [ USING method ]
    ( { column_name | ( expression ) } [ COLLATE collation ] [ opclass ] [
ASC | DESC ] [ NULLS { FIRST | LAST } ] [, ...] )
    [ INCLUDE ( column_name [, ...] ) ]
    [ WITH ( storage_parameter = value [, ... ] ) ]
    [ TABLESPACE tablespace_name ]
    [ WHERE predicate ]
```

Indexes are identified by a mnemonic name, similar to the tables that they are related to. It is interesting to note that the index name is always unqualified, which means it does not includes the schema where the index is going to live: an index is always found within the very same schema as the underlying table. However, it is possible to store an index in another tablespace than that of the underlying table, and this can be useful to store important indexes in faster storage. The statement supports the IF NOT EXISTS clause to abort the creation in a gentle way if an index with the same name already exists.

The UNIQUE clause specifies that the index is going to verify the uniqueness of its columns. The WHERE clause allows the creation of a partial index, which is an index that contains information only about those tuples that satisfy the WHERE condition(s).

The INCLUDE clause allows you to specify some extra columns of the underlying table that are going to be stored in the index, even if not indexed. The idea is that if the index is useful for an index-only scan, you can still get extra information without the trip to the underlying table. Of course, having a covering index (which is the name of an INCLUDE clause index) means that the index is going to grow in size and, at the same time, every tuple update could require extra index update effort.

The USING clause allows the specification of the type of index to be built, and if none is specified, the default B-Tree is used.

The ONCURRENTLY clause allows the creation of an index in a concurrent way: when an index is in its building phase, the underlying table is locked against changes so that the index can finish its job of indexing the tuple values. In a concurrent index creation, the table allows changes even during index creation, but once the index has been built, another pass on the underlying table is required to "adjust" what has changed in the meantime.

In order to make it more practical, let's see how to build a simple index on the posts table. Let's say we want to index the category a post belongs to:

```
forumdb=> CREATE INDEX idx_post_category
          ON posts( category );
CREATE INDEX
```

The preceding code will create an index named idx_post_category on the table posts, using the single-column category and the default index type (B-Tree).

The following does something similar, creating a multi-column index:

```
forumdb=> CREATE INDEX idx_author_created_on
          ON posts( author, created_on );
CREATE INDEX
```

It is important to note that, when creating multi-column indexes, you should always place the most selective columns first. PostgreSQL will consider a multi-column index from the first column onward, so if the first columns are the most selective, the index access method will be the cheapest. In the preceding example, assuming we want to search for a combination of authors and dates, we could expect many authors to publish on a specific day, so the date (the created_on column) is not going to be very selective, at least not as selective as the specific author; it is for that reason that we pushed the created_on column to the right in the column list.

If we would like to create a hash index, we could do something such as the following:

```
forumdb=> CREATE INDEX idx_post_created_on
          ON posts USING hash ( created_on );
CREATE INDEX
```

Of course, such an index will be useful only for equality comparison, so a query such as the following will never use the preceding index:

```
SELECT * FROM posts WHERE created_on < CURRENT_DATE;
```

But a query like the following could use the hash index:

```
SELECT * FROM posts WHERE created_on = CURRENT_DATE;
```

This is because we are asking for equality comparison.

Inspecting indexes

Indexes are "attached" to their underlying tables, and so `psql` shows the defined indexes whenever you ask it to describe a table with the `\d` special command:

```
forumdb=> \d posts

                             Table "public.posts"
      Column       |            Type           | Collation | Nullable |
Default
-------------------+---------------------------+-----------+----------+--------
---------------------
 pk                | integer                   |           | not null |
generated always as identity
 title             | text                      |           |          |
 content           | text                      |           |          |
 author            | integer                   |           | not null |
 category          | integer                   |           | not null |
 reply_to          | integer                   |           |          |
 created_on        | timestamp with time zone  |           |          |
CURRENT_TIMESTAMP
 last_edited_on    | timestamp with time zone  |           |          |
CURRENT_TIMESTAMP
 editable          | boolean                   |           |          | true
Indexes:
    "posts_pkey" PRIMARY KEY, btree (pk)
    "idx_author_created_on" btree (author, created_on)
    "idx_post_category" btree (category)
    "idx_post_created_on" hash (created_on)
```

As you can see from the preceding snippet of code, the command shows all the available indexes with their method (for example, `btree`) and a list of the columns the index is built on top of.

The `pg_index` special catalog contains information about the indexes and their main attributes, and so it can be queried to get the very information (and more) that is provided by `psql`. In particular, since an index is registered into `pg_class` with the special `relkind` value of `i`, we can join `pg_class` and `pg_index` to get detailed information in a statement, as follows:

```
forumdb=> SELECT relname, relpages, reltuples,
          i.indisunique, i.indisclustered, i.indisvalid,
          pg_catalog.pg_get_indexdef(i.indexrelid, 0, true)
          FROM pg_class c JOIN pg_index i on c.oid = i.indrelid
          WHERE c.relname = 'posts';

-[ RECORD 1 ]---+-----------------------------------------------------------
------------------
relname         | posts
relpages        | 43
reltuples       | 2000
indisunique     | t
indisclustered  | f
indisvalid      | t
pg_get_indexdef | CREATE UNIQUE INDEX posts_pkey ON posts USING btree (pk)
-[ RECORD 2 ]---+-----------------------------------------------------------
------------------
relname         | posts
relpages        | 43
reltuples       | 2000
indisunique     | f
indisclustered  | f
indisvalid      | t
pg_get_indexdef | CREATE INDEX idx_post_category ON posts USING btree
(category)
...
```

The `indisunique` column is set to true if the index has been created with the UNIQUE clause, as it happens for the primary key index. `indisvalid` is a Boolean value that indicates whether the index is usable or not (as you will see later on, you can decide to disable an index for any reason). Since you can cluster a table against an index – that is, you can sort the table depending on a specific index – `indisclustered` indicates whether the table is clustered against the specific index.

The `pg_get_indexdef()` special function provides a textual representation of the CREATE INDEX statement used to produce every index and can be very useful to decode and learn how to build complex indexes.

Therefore, either using the `psql \d` command or querying `pg_index`, you can get details about existing indexes and their status.

Dropping an index

In order to discard an index, you need to use the `DROP INDEX` statement, which has the following synopsis:

```
DROP INDEX [ CONCURRENTLY ] [ IF EXISTS ] name [, ...] [ CASCADE | RESTRICT ]
```

The statement accepts the name of the index, and can drop more than one index at the same time if you specify multiple names on the same statement.

The `CONCURRENTLY` clause prevents the command from acquiring an exclusive lock on the underlying table, preventing other queries from accessing the table until the index has been dropped.

The `CASCADE` option drops the index and all other objects that depend on the index, while the `RESTRICT` option is its counterpart and prevents the index from being dropped if any object still insists on the index. The `RESTRICT` clause is the default.

Lastly, the `IF EXISTS` option allows the command to gracefully abort if the index has already been dropped.

Invalidating an index

It is possible to invalidate an index, which is a way to tell PostgreSQL to not consider that index at all without dropping the index and building it again. This can be useful in situations where you are studying your cluster's behavior and want to force the optimizer to choose another path to access the data that does not include a specific index, or it can be necessary because there is a problem with an index.

In order to invalidate an index, you have to directly manipulate the `pg_index` system catalog to set the `indisvalid` attribute to false. For example, in order to suspend the usage of the `idx_author_created_on` index, you have to do an update against `pg_index`, as follows:

```
forumdb=# UPDATE pg_index SET indisvalid = false
          WHERE indexrelid = ( SELECT oid FROM pg_class
                               WHERE relkind = 'i'
```

```
                                    AND relname = 'idx_author_created_on' );
UPDATE 1

forumdb=# \d posts
...
Indexes:
    "posts_pkey" PRIMARY KEY, btree (pk)
    "idx_author_created_on" btree (author, created_on) INVALID
    "idx_post_category" btree (category)
    "idx_post_created_on" hash (created_on)
...
```

 You need to invalidate an index as an administrator user, even if you are the user that created the index. This is due to the fact that you need to manipulate the system catalog, which is something that is restricted to only administrator users.

As you can see, the index is then marked as `INVALID`, to indicate that PostgreSQL will not ever try to consider it for its execution plans. You can, of course, reset the index to its original status, making the same update as the preceding and setting the `indisvalid` column to a true value.

Rebuilding an index

Since an index is detached from the data stored in the table, it is possible that the information within the index gets corrupted or somehow out of date. This is not a normal condition, and it does not happen in day-to-day usage of the database. However, knowing how you can rebuild an index is important knowledge because it helps you prevent anomalies and allows you to revalidate indexes that have been kept out of date (because they were not valid).

You can always rebuild an index starting from the data in the underlying table by use of the `REINDEX` command, which has the following synopsis:

```
REINDEX [ ( VERBOSE ) ] { INDEX | TABLE | SCHEMA | DATABASE | SYSTEM } [
CONCURRENTLY ] name
```

You can decide to rebuild a single index by means of the `INDEX` argument followed by the name of the index, or you can rebuild all the indexes of a table by means of the `TABLE` argument followed, as you can imagine, by the table name.

Going further, you can rebuild all the indexes of all the tables within a specific schema by means of the SCHEMA argument (followed by the name of the schema) or the whole set of indexes of a database using the DATABASE argument and the name of the database you want to reindex. Lastly, you can also rebuild indexes on system catalog tables by means of the SYSTEM argument.

You can execute REINDEX within a transaction block but only for a single index or table, which means only for the INDEX and TABLE options. All the other forms of the REINDEX command cannot be executed in a transaction block.

The CONCURRENTLY option prevents the index from acquiring exclusive locks on the underlying table in a way similar to that of building a new index.

The EXPLAIN statement

EXPLAIN is the statement that allows you to see how PostgreSQL is going to execute a specific query. You have to pass the statement you want to analyze to EXPLAIN, and the execution plan will be shown.

There are a few important things to know before using EXPLAIN:

- It will only show the best plan, which is the one with the lowest cost among all the evaluated plans.
- It will not execute the statement you are asking the plan for, therefore the EXPLAIN execution is fast and pretty much constant each time.
- It will present you with all the execution nodes that the executor will use to provide you with the dataset.

Let's see an example of EXPLAIN in action to better understand. Imagine we need to understand the execution plan of the SELECT * FROM categories statement. In this case, you need to prefix the statement with the EXPLAIN command, as follows:

```
forumdb=> EXPLAIN SELECT * FROM categories;
                          QUERY PLAN
---------------------------------------------------------
 Seq Scan on categories  (cost=0.00..1.01 rows=1 width=68)
(1 row)
```

As you can see, the output of EXPLAIN reports the query plan. There is a single execution node, of the Seq Scan type, followed by the table against which the node is executed (on categories). In the output of the EXPLAIN command, you will find all the types of execution nodes already discussed in the previous sections.

For every node, EXPLAIN will report some more information between parentheses: the cost, the number of rows, and the width. The cost is the amount of effort required to execute the node, and is always expressed as a "startup cost" and a "final cost." The startup cost is how much work PostgreSQL has to do before it begins executing the node; in the preceding example, the cost is 0, meaning the execution of the node can begin immediately. The final cost is how much effort PostgreSQL has to do to provide the last bit of the dataset – that is, to complete the execution of the node.

The rows field indicates how many tuples the node is expected to provide in the final dataset, and is a pure estimation. Being an estimation, the value will never be correct and you have to keep in mind that it will never be zero: when PostgreSQL estimates a very low number of tuples, it always provides 1 as the number of rows.

Lastly, the width field indicates how many bits every tuple will occupy, as an average. Essentially, this information is used to estimate the network traffic that the query will produce: in the preceding example, it is possible to estimate 68 bytes per tuple.

Now consider another example, just to get used to the EXPLAIN output: the query changes a little to produce a few more nodes, as follows:

```
forumdb=> EXPLAIN
          SELECT title
          FROM categories ORDER BY description DESC;
                          QUERY PLAN
-------------------------------------------------------------
 Sort  (cost=1.02..1.02 rows=1 width=64)
   Sort Key: description DESC
   ->  Seq Scan on categories  (cost=0.00..1.01 rows=1 width=64)
(3 rows)
```

Here, we have two different nodes: the first at the top is the Sort node (due to the ORDER BY clause), and the second node is Seq Scan, as in the previous example. Please note that there are three output rows, so how do we determine which rows are nodes and which are not? The first row in the plan is always a node, and the other node rows are indented to the right and have an arrow as a prefix (->). The other lines in the plan provide information about the node they are under; therefore, in the preceding example, the Sort Key row is additional information to the Sort node.

Another approach to distinguish node rows from additional information is to consider that every node line has the cost, rows, and width attribute in parentheses.

Once you have discovered the nodes of the query, you have to find out the very first node, which is usually the most indented one, and also the one with the lowest startup cost: in the preceding example, Seq Scan is the first node executed. This node does the very same thing explained in the previous example: it forces the executor to go to the table on the physical storage and retrieve, in sequential order, all the table content. One thing, however, is different in the preceding example: the average width has decreased, and this is due to the fact that the query does not require all the columns of every tuple, only the title one.

Once the sequential scan node has completed, its output is used as input for the Sort node, which performs the desired ORDER BY operation. As you can easily read, the sort key of the original statement is printed to provide you with enough information to understand what the executor will sort data on. The Sort node has a startup cost that is greater (better or pretty much the same) as the previous node's final cost: the sequential scan has a final cost of 1.01 and the sort starts with a cost of 1.02. This emphasizes again how nodes are executed in a pipeline, and also tells you that the sort cannot start before the other node has completed. The sort node has a final cost equal to the startup cost, meaning that this node is straightforward for PostgreSQL to be executed.

You can try to execute EXPLAIN on different statements to see how a plan changes, and which nodes can be generated, in order to be used to recognize the nodes and the resource information.

In the following subsections, you will see different options to explain a statement.

EXPLAIN output formats

By default, EXPLAIN provides a text-based output, but it can also provide much more structured output in XML, JSON, and YAML. These other formats are not only useful when you have to cope with external tools and applications but can also be useful because they provide more information for tuning a query plan.

You can specify the format you want with the FORMAT option followed by the name of the format, which can be TEXT, XML, JSON, or YAML. Take the following example:

```
forumdb=> EXPLAIN ( FORMAT JSON ) SELECT * FROM categories;
                QUERY PLAN
-----------------------------------------
  [                                     +
```

```
  {                                         +
    "Plan": {                               +
      "Node Type": "Seq Scan",       +
      "Parallel Aware": false,         +
      "Relation Name": "categories",+
      "Alias": "categories",            +
      "Startup Cost": 0.00,             +
      "Total Cost": 1.01,               +
      "Plan Rows": 1,                    +
      "Plan Width": 68                   +
    }                                        +
  }                                          +
]
(1 row)
```

As you can see, the JSON format provides not only a different structure to the query plan but also a different and more rich set of information. For example, from the preceding example, we can see that the query has been executed in a nonparallel mode (`Parallel Aware = false`).

If you need to parse the `EXPLAIN` output with an application or tool, you should stick to one of the structured formats, not the default text one.

EXPLAIN ANALYZE

The `ANALYZE` mode of `EXPLAIN` enhances the command by effectively running the query to explain. Therefore, the command does a double task: it prints out the best plan to execute the query and it runs the query, also reporting back some statistical information.

To better understand the concept, consider the output of `EXPLAIN ANALYZE` compared to the output of a plain `EXPLAIN` command:

```
forumdb=> EXPLAIN SELECT * FROM categories;
                          QUERY PLAN
---------------------------------------------------------
 Seq Scan on categories  (cost=0.00..1.01 rows=1 width=68)
(1 row)

forumdb=> EXPLAIN ANALYZE SELECT * FROM categories;
            QUERY PLAN
-----------------------------------------------------------------------
--------------------------
 Seq Scan on categories  (cost=0.00..1.01 rows=1 width=68) (actual
time=0.023..0.025 rows=1 loops=1)
 Planning Time: 0.102 ms
```

```
 Execution Time: 0.062 ms
 (3 rows)
```

The output of the EXPLAIN ANALYZE command is enhanced by the "actual" part of every node: the executor reports back how the execution of the node went exactly. Therefore, while EXPLAIN can only estimate the costs of a node, the EXPLAIN ANALYZE provides feedback on the execution time (expressed in milliseconds), the effective number of rows, and how many times a node has been executed (loops).

The node time is expressed, similarly to the cost, in a startup time and a final time, which is the time taken for the node to complete its execution. Therefore, in the preceding example, PostgreSQL took 0.023 milliseconds to "warm-up" and completed the query execution in 0.025 milliseconds, so the node required 0.002 milliseconds to complete its job.

At the very end of the command output, EXPLAIN ANALYZE provides overall time information, which includes the planning time, which is the time the optimizer has spent producing the best candidate access plan, and the execution time, which is the total time spent running the query (excluding the parsing and planning time).

Therefore, the preceding example took 0.062 milliseconds to "fetch data" and 0.102 milliseconds in deciding how to fetch the data, so the query took 0.164 milliseconds.

> When the data to access is really small, the planning time is longer than the execution time.

The execution time also includes time spent running BEFORE triggers, while AFTER triggers are not counted because their function is executed once the plan has completed.

The planning time, similarly, accounts only for the time spent in producing the best access plan, not the time required to process rules and writing of the statement, as well as parsing.

> EXPLAIN ANALYZE executes always the query you want to analyze, therefore in order to avoid side-effects, you should wrap EXPLAIN ANALYZE in a transaction and rollback the work once the analysis has complete.

EXPLAIN ANALYZE can also be invoked by passing ANALYZE as an option to EXPLAIN, as follows:

```
forumdb=> EXPLAIN ( ANALYZE ) SELECT * FROM categories ORDER BY title DESC;
```

The option form of EXPLAIN ANALYZE is handy when you want to adds other options to EXPLAIN, as shown in the following subsection.

EXPLAIN options

EXPLAIN provides a rich set of options, most of which can only be used in the ANALYZE form. All of the EXPLAIN options presented in this section are Boolean, which means they can be turned on and off but nothing more.

The VERBOSE option allows every node to report more detailed information, such as the list of the output columns, even when not specified. For example, even if the query does not explicitly ask for the list of columns, note how thanks to VERBOSE, you can find out which columns a node will provide to the output dataset:

```
forumdb=> EXPLAIN (VERBOSE on) SELECT * FROM categories;
                            QUERY PLAN
------------------------------------------------------------------
 Seq Scan on public.categories  (cost=0.00..1.01 rows=1 width=68)
   Output: pk, title, description
(2 rows)
```

The COSTS option, which is turned on by default, shows the costs part of a node. As an example, turning it off removes the startup and final costs, as well as the average width and the number of rows:

```
forumdb=> EXPLAIN (COSTS off) SELECT * FROM categories;
       QUERY PLAN
-----------------------
 Seq Scan on categories
(1 row)

forumdb=> EXPLAIN (COSTS on) SELECT * FROM categories;
                   QUERY PLAN
----------------------------------------------------
 Seq Scan on categories  (cost=0.00..1.01 rows=1 width=68)
(1 row
```

The TIMING option, which is on by default, shows the effective execution time when EXPLAIN is invoked with ANALYZE. In other words, setting TIMING to off means that the output of EXPLAIN will not show the time of the query execution. For example, note in the following EXPLAIN statement how the actual time is missing from the output:

```
forumdb=> EXPLAIN (ANALYZE on, TIMING off) SELECT * FROM categories;
                            QUERY PLAN
```

```
------------------------------------------------------------------------
--------
 Seq Scan on categories  (cost=0.00..1.01 rows=1 width=68) (actual rows=1
loops=1)
 Planning Time: 0.091 ms
 Execution Time: 0.050 ms
(3 rows)
```

The SUMMARY option reports the total time spent in planning for the execution and the time spent for the query execution so that you can get an idea of how much effort the planner has used to find out the best execution plan. As an example, note how the planning and execution times are reported at the end of the output:

```
forumdb=> EXPLAIN (ANALYZE, SUMMARY on) SELECT * FROM categories;
                                      QUERY PLAN
------------------------------------------------------------------------
-------------------------
 Seq Scan on categories  (cost=0.00..1.01 rows=1 width=68) (actual
time=0.021..0.022 rows=1 loops=1)
 Planning Time: 0.097 ms
 Execution Time: 0.059 ms
(3 rows)
```

The BUFFERS option, which defaults to off, provides information about the data buffers the query used to complete. For example, note how there is buffer-related information on the execution node in the following query:

```
forumdb=> EXPLAIN (ANALYZE, BUFFERS on) SELECT * FROM categories;

                                      QUERY PLAN
------------------------------------------------------------------------
-------------------------
 Seq Scan on categories  (cost=0.00..1.01 rows=1 width=68) (actual
time=0.014..0.015 rows=1 loops=1)
   Buffers: shared hit=1
 Planning Time: 0.076 ms
 Execution Time: 0.042 ms
(4 rows)
```

The buffer information is not trivial to analyze and can be split into two parts: a prefix and a suffix.

The prefix can be any of the following:

- `shared`, meaning a PostgreSQL shared buffer, which is the database in-memory cache
- `temp`, meaning temporary memory (used for sorting, hashing, and so on)
- `local`, meaning temporary database objects space (for instance, temporary tables)

The suffix can be any of the following:

- `hit`, providing the number of memory successes
- `read`, providing the number of buffers read from the storage (and therefore not in the cache)
- `dirtied`, the number of buffers modified by the query
- `written`, the number of buffers removed from the PostgreSQL cache and written to disk
- `lossy`, the number of buffers that PostgreSQL has checked in memory in a second pass

Combining the prefix and the suffix provides information on the buffers. For example, in the previous query, the buffer line contained `shared hit=1`, which reads as "1 buffer has been successfully found in the database cache, no more operations on buffers are required."

The `WAL` option, available since PostgreSQL 13, provides information about the `WAL` usage of a statement. Clearly, this option also requires the `ANALYZE` one to actually execute the query; the system is not able to estimate the `WAL` traffic. As an example, consider the following query, which adds a bunch of fake usernames to the `users` table:

```
forumdb=> EXPLAIN (WAL on, ANALYZE on, FORMAT yaml)
  insert into users( username, gecos, email)
  select 'username'||v, v, v||'@c.b.com' from generate_series(1, 100000) v;
                  QUERY PLAN
------------------------------------------
  - Plan:
  ...
      WAL Records: 303959                    +
      WAL FPI: 0                             +
      WAL Bytes: 25730509                    +
  ...
```

As you can see, the output reports information about the number of `WAL` records that have been generated, the number of `WAL` **Full Page Images** (**FPIs**), and the number of bytes written into the `WAL` logs.

An example of query tuning

In the previous section, you have learned how to use EXPLAIN to understand how PostgreSQL is going to execute a query; it is now time to use EXPLAIN in action to tune some slow queries and improve performance.

This section will show you some basic concepts of the day-to-day usage of EXPLAIN as a powerful tool to determine where and how to instrument PostgreSQL in doing faster data access. Of course, query tuning is a very complex subject and often requires repeated trial-based optimization, so the aim of this section is not to provide you with true knowledge about query tuning but rather a basic understanding of how to improve your own database and queries.

Sometimes, tuning a query involves simply rewriting it a way that is more comfortable – or better, more comprehensible –to PostgreSQL, but most often, that means using an appropriate index to speed up access to the underlying data.

Let's start with a simple example: we want to extract all the posts ordered by creation day, so the query is as follows:

```
SELECT * FROM posts ORDER BY created_on;
```

We can pass it to EXPLAIN to get an idea about how PostgreSQL is going to execute it:

```
forumdb=> EXPLAIN SELECT * FROM posts ORDER BY created_on;
                              QUERY PLAN
-----------------------------------------------------------------
 Sort  (cost=1009871.99..1022372.04 rows=5000000 width=55)
   Sort Key: created_on
   ->  Seq Scan on posts  (cost=0.00..111729.20 rows=5000000 width=55)
(3 rows)
```

 In our demo database, there are 1,000 users (authors), each with 5,000 posts. The posts table requires around 480 megabytes of disk space.

As you can see, the first node to be executed is the sequential scan (the initial cost is 0), which is going to produce 5,000,000 tuples as output. Why a sequential scan? First of all, there is no filtering clause – we want to retrieve all the data stored in the table – and second, there is no access method on the table (there are no indexes).

Since we asked to sort the output, the following node to execute is a sorting node, which produces the very same number of tuples as a result.

How much does it take to complete the preceding query? EXPLAIN ANALYZE can help us answer that question:

```
forumdb=> EXPLAIN ANALYZE SELECT * FROM posts ORDER BY created_on;
                                                    QUERY PLAN
----------------------------------------------------------------------------
------------------------------------------------------
 Sort  (cost=1009871.99..1022372.04 rows=5000020 width=55) (actual
time=31321.217..32806.986 rows=5000000 loops=1)
   Sort Key: created_on
   Sort Method: external merge  Disk: 367024kB
   ->  Seq Scan on posts  (cost=0.00..111729.20 rows=5000020 width=55)
(actual time=0.078..2546.672 rows=5000000 loops=1)
 Planning Time: 0.148 ms
 Execution Time: 33138.966 ms
```

The pure execution time is more than 33 seconds. Is it possible to reduce the total amount of time by building a specific index on the created_on field:

```
forumdb=> CREATE INDEX idx_posts_date ON posts( created_on );
CREATE INDEX
forumdb=> EXPLAIN ANALYZE SELECT * FROM posts ORDER BY created_on;
                                                    QUERY PLAN
----------------------------------------------------------------------------
-----------------------------------------------------
 Index Scan using idx_posts_date on posts  (cost=0.43..376844.33
rows=5000000 width=55) (actual time=0.768..8694.662 rows=5000000 loops=1)
 Planning Time: 6.357 ms
 Execution Time: 9128.668 ms
(3 rows)
```

The query is now running at almost one-third of the time required without the index, and, in fact, the query plan has changed from a sequential scan to an index scan with the freshly created new index.

Of course, the newly created index has a penalty in terms of storage space: as you can imagine, the increase in speed comes with an extra space cost that can be checked as follows:

```
forumdb=> SELECT pg_size_pretty( pg_relation_size( 'posts' ) ) AS
table_size,
  pg_size_pretty( pg_relation_size( 'idx_posts_date' ) ) AS index_size;
 table_size | index_size
------------+-----------
```

```
 482 MB     | 107 MB
(1 row)
```

The index requires around 22% of the disk space occupied by the whole table data.

Now, whether this extra disk space is too much or not depends on your resources and your final aim: in the preceding case, assuming you are executing the query quite often, the increased speed is justified by the extra space.

Let's now concentrate on a more typical query: finding out all the posts of a specific user in a specific period of time. The resulting query will be something like the following one, assuming a 2-day period:

```
SELECT p.title, u.username
FROM posts p
JOIN users u ON u.pk = p.author
WHERE u.username = 'fluca1978'
AND   daterange( CURRENT_DATE - 2, CURRENT_DATE ) @> p.created_on::date
```

How does PostgreSQL execute the preceding query? Again, EXPLAIN can help us understand what the database thinks the best query plan is:

```
forumdb=> EXPLAIN
    SELECT p.title, u.username
    FROM posts p
    JOIN users u ON u.pk = p.author
    WHERE u.username = 'fluca1978'
    AND   daterange( CURRENT_DATE - 2, CURRENT_DATE ) @>
p.created_on::date;

                                   QUERY PLAN
-----------------------------------------------------------------------------
---------------
 Nested Loop  (cost=0.28..187049.79 rows=25 width=22)
   Join Filter: (p.author = u.pk)
   ->  Index Scan using users_username_key on users u  (cost=0.28..8.29
rows=1 width=14)
         Index Cond: (username = 'fluca1978'::text)
   ->  Seq Scan on posts p  (cost=0.00..186729.00 rows=25000 width=16)
         Filter: (daterange((CURRENT_DATE - 2), CURRENT_DATE) @>
(created_on)::date)
```

The very first node to be executed is a sequential scan on the posts table, where the date filter is applied. This node will produce around 25,000 tuples that are going to be merged in a nested loop with the resulting tuples from the index scan node on the users' table. Since the username is a unique attribute of the users' table, PostgreSQL knows it can find a single username using an index, and so it does. This makes the result of the users' table really small, a single tuple, and therefore, it is possible to use a nested loop to join the two result sets. PostgreSQL is expecting 25 rows as the final result.

How can we check whether this is correct? EXPLAIN ANALYZE can help us compare what PostgreSQL estimates with the reality:

```
forumdb=> EXPLAIN ANALYZE
    SELECT p.title, u.username
    FROM posts p
    JOIN users u ON u.pk = p.author
    WHERE u.username = 'fluca1978'
    AND   daterange( CURRENT_DATE - 2, CURRENT_DATE ) @>
p.created_on::date;


                                                    QUERY PLAN
--------------------------------------------------------------------------
--------------------------------------------------------------
 Nested Loop  (cost=0.28..187049.79 rows=25 width=22) (actual
time=152.156..5008.775 rows=10 loops=1)
   Join Filter: (p.author = u.pk)
   Rows Removed by Join Filter: 9990
   ->  Index Scan using users_username_key on users u  (cost=0.28..8.29
rows=1 width=14) (actual time=1.174..1.203 rows=1 loops=1)
         Index Cond: (username = 'fluca1978'::text)
   ->  Seq Scan on posts p  (cost=0.00..186729.00 rows=25000 width=16)
(actual time=150.959..4755.512 rows=10000 loops=1)
         Filter: (daterange((CURRENT_DATE - 2), CURRENT_DATE) @>
(created_on)::date)
         Rows Removed by Filter: 4990000
 Planning Time: 0.582 ms
 Execution Time: 5012.352 ms
```

The final result is near enough to what PostgreSQL estimated: instead of 25 final tuples, there were only 10. This small error propagates from the sequential scan on the posts table, where the planner estimated 25,000 rows instead of 10,000. Also, it is interesting to note that the query execution requires 5 seconds.

What happens if we add an index to the author column of the table posts?

```
forumdb=> CREATE INDEX idx_posts_author ON posts( author );
CREATE INDEX
forumdb=> EXPLAIN ANALYZE
    SELECT p.title, u.username
    FROM posts p
    JOIN users u ON u.pk = p.author
    WHERE u.username = 'fluca1978'
    AND   daterange( CURRENT_DATE - 2, CURRENT_DATE ) @>
p.created_on::date;


                                                          QUERY PLAN
-----------------------------------------------------------------------------
-------------------------------------------------------------
 Nested Loop  (cost=94.21..15440.96 rows=25 width=22) (actual
time=0.711..11.865 rows=5 loops=1)
   -> Index Scan using users_username_key on users u  (cost=0.28..8.29
rows=1 width=14) (actual time=0.021..0.215 rows=1 loops=1)
         Index Cond: (username = 'fluca1978'::text)
   -> Bitmap Heap Scan on posts p  (cost=93.94..15432.42 rows=25 width=16)
(actual time=0.677..11.451 rows=5 loops=1)
         Recheck Cond: (author = u.pk)
         Filter: (daterange((CURRENT_DATE - 1), CURRENT_DATE) @>
(created_on)::date)
         Rows Removed by Filter: 4995
         Heap Blocks: exact=66
         -> Bitmap Index Scan on idx_posts_author  (cost=0.00..93.93
rows=5000 width=0) (actual time=0.628..0.671 rows=5000 loops=1)
               Index Cond: (author = u.pk)
 Planning Time: 0.630 ms
 Execution Time: 12.310 ms
(12 rows)
```

First of all, the query now takes only 12 milliseconds, compared to the 5 seconds of the previous execution plan. What has changed is that the posts table is now seeking to use the fresh index, and the bitmap built out of that index is then used to join the result of the index scan on the users' table. The date clause has been pushed as a check condition within the index scan of the posts table. Why has the plan changed so much? Clearly, the author column filters much more of the table tuples than the range of dates, so PostgreSQL decides to use this clause as a first approach to filtering the data. Since PostgreSQL expects no more than 5,000 tuples out of the author filter condition, and we know it is right about that, filtering on the created_on column after will be faster (if you remember, the created_on filter clause was causing 10,000 tuples compared to the 5,000 out of the author column).

It is also interesting to note that there is no longer any reason to keep around the index on the `created_on` column, since the preceding query plan is not using it anymore.

What is the space required by the indexes now? Again, it is quite simple to check:

```
forumdb=> SELECT pg_size_pretty( pg_relation_size( 'posts') ) AS
table_size,
        pg_size_pretty( pg_relation_size( 'idx_posts_date' ) ) AS
idx_date_size,
        pg_size_pretty( pg_relation_size( 'idx_posts_author' ) ) AS
idx_author_size;

 table_size | idx_date_size | idx_author_size
------------+---------------+-----------------
 482 MB     | 107 MB        | 107 MB
(1 row)
```

Both the indexes require the same space, but as we already said, we can drop the date-based index since it is no longer required. In fact, even with a specific date clause, the index is not used anymore:

```
forumdb=> EXPLAIN ANALYZE
    SELECT p.title, u.username
    FROM posts p
    JOIN users u ON u.pk = p.author
    WHERE u.username = 'fluca1978'
    AND   p.created_on::date = CURRENT_DATE -2;


----------------------------------------------------------------------------
----------------------------------------------------------------
 Nested Loop  (cost=94.21..15415.96 rows=25 width=22) (actual
time=0.851..6.264 rows=5 loops=1)
    -> Index Scan using users_username_key on users u  (cost=0.28..8.29
rows=1 width=14) (actual time=0.022..0.027 rows=1 loops=1)
          Index Cond: (username = 'fluca1978'::text)
    -> Bitmap Heap Scan on posts p  (cost=93.94..15407.42 rows=25 width=16)
(actual time=0.815..6.218 rows=5 loops=1)
          Recheck Cond: (author = u.pk)
          Filter: ((created_on)::date = (CURRENT_DATE - 2))
          Rows Removed by Filter: 4995
          Heap Blocks: exact=66
          -> Bitmap Index Scan on idx_posts_author  (cost=0.00..93.93
rows=5000 width=0) (actual time=0.755..0.756 rows=5000 loops=1)
                Index Cond: (author = u.pk)
```

Identifying unused indexes is important because it allows us to reclaim disk space and simplifies the management and data insertion: remember that every time the table changes, the index has to be updated, and this also requires extra resources, such as time and disk space.

Therefore, as you can see, it is really important to analyze the queries your applications execute the most and identify whether an index can help improve the execution speed, but also remember that an index has an extra cost in both space and maintenance, so don't abuse the use of indexes.

But how can you identify unused indexes without even knowing about the ongoing queries?

Luckily, PostgreSQL provides you with detailed information about the usage of every index: the special `pg_stat_user_indexes` view provides information about how many times an index has been used and how. For example, to get information about the indexes over the posts table, you can execute something such as the following:

```
forumdb=> SELECT indexrelname, idx_scan, idx_tup_read, idx_tup_fetch FROM
pg_stat_user_indexes WHERE relname = 'posts';
    indexrelname    | idx_scan | idx_tup_read | idx_tup_fetch
--------------------+----------+--------------+---------------
 posts_pkey         |        0 |            0 |             0
 idx_posts_date     |        3 |      5000002 |       5000002
 idx_posts_author   |       15 |        25010 |            10
(3 rows)
```

This tells us that `idx_posts_date` has been used 3 times, providing over 5 million tuples, while `idx_posts_author` has been used 15 times and also provided much fewer tuples (only 10), meaning it is very effective.

If the trend is confirmed, and `idx_posts_date` is seldomly used, you can safely drop it.

ANALYZE and how to update statistics

PostgreSQL exploits a statistical approach to evaluate different execution plans. This means that PostgreSQL does not know how many tuples there are in a table, but has a good approximation that allows the planner to compute the cost of the execution plan.

Statistics are not only related to the quantity (how many tuples) but also to the quality of the underlying data – for example, how many distinct values, which values are more frequent in a column, and so on. Thanks to the combination of all of this data, PostgreSQL is able to make a good decision.

There are times, however, when the quality of the statistical data is not good enough for PostgreSQL to choose the best plan, a problem commonly known as "out-of-date statistics." In fact, statistics are not updated in real time; rather, PostgreSQL keeps track of what is ongoing in every table in every database and summarizes the number of new tuples, updated ones, and deleted ones, as well as the quality of their data. It could happen that the statistics are not updated frequently enough or not at all for different reasons we are going to explain later, so the database administrator should always have a way to force PostgreSQL to start from scratch and "rebuild" the statistics.

The command that does this is ANALYZE.

ANALYZE accepts a table (and, optionally, a list of columns) and builds all the statistics for the specified table (or the specified columns only).

ANALYZE is a very intrusive command and requires a lot of I/O resources, so it is not a good idea to run it manually, and it is for that reason that the auto-analyze daemon is in charge of periodically updating the statistics when enough changes on a table happen.

> ANALYZE has nothing to do with the argument to EXPLAIN: EXPLAIN ANALYZE looks to the query plan and executes the query, while ANALYZE updates PostgreSQL statistics.

The synopsis for the ANALYZE command is the following:

```
ANALYZE [ ( option [, ...] ) ] [ table_and_columns [, ...] ]
```

Essentially, you are going to launch it against a single table, as follows:

```
forumdb=> \timing
forumdb=> ANALYZE posts;
ANALYZE
Time: 17674,106 ms  (00:17,674)
```

Please consider the time the preceding ANALYZE command required: 17 seconds to analyze a table with 5 million tuples. As already stated, ANALYZE is a very intrusive command and can lock a table (and the ongoing activity against it) for a long time.

> You can inspect the times required to execute commands and queries with the \timing psql special command, which will print a summary of the elapsed time after every statement. This is not a solid way to measure performances, only to get an idea of how much time a task is spending.

ANALYZE does not support a lot of options, mainly VERBOSE to display verbose output of what ANALYZE is doing, and SKIP_LOCKED, which makes ANALYZE skip a table if it cannot acquire the appropriate locks because there are other ongoing operations that have already acquired an incompatible lock.

Where does PostgreSQL store the statistics that ANALYZE collects? The pg_stats special catalog contains all the statistics used by the planner to determine the values and constraints to examine the attribute. For example, let's see what PostgreSQL knows about the author column of the posts table, and in particular, how many distinct values there are:

```
forumdb=> SELECT n_distinct
          FROM pg_stats
          WHERE attname = 'author' AND tablename = 'posts';
 n_distinct
------------
       1000
(1 row)
```

PostgreSQL knows that we have 1,000 different authors that have posted at least one post in our example database, as demonstrated by the EXPLAIN ANALYZE command in the previous section.

One bit of important information you can find in the pg_stats catalog is the most common values, correlated by the frequency that these values appear. Extracting this information requires a little more attention since both values and frequencies are stored as arrays, so the following query provides the most common values and frequency for the author column:

```
forumdb-> SELECT * FROM pg_stats;
 mcv  |       mcf
------+--------------
 2358 |     0.001039
 2972 |     0.001037
 2155 |     0.001035
 2794 | 0.0010336667
 2648 | 0.0010313333
 2649 | 0.0010286666
 2917 | 0.0010286666
 2629 | 0.0010276666
 2906 |     0.001027
 ...
```

What the preceding output means is that every author that appears in the mcv column – for instance, the author with primary key 2358 appears in the posts table with a frequency of 0.001039. This leads to the fact that this author appears 0.001039 multiplied by the number of tuples (5 million) times and produces 5,195 tuples, confirmed by a trivial counting:

```
forumdb=> SELECT count(*) FROM posts WHERE author = 2358;
 count
-------
  5000
```

Of course, remember that the information in pg_stats is an approximation of the real situation on the underlying table, so values can be slightly different but must be of the same order of magnitude to let the planner produce a good access method.

There is other information in pg_stats, such as the number of NULL values for a column, and so on.

In conclusion, PostgreSQL keeps track of the statistics of every column in every table; the statistics are updated by ANALYZE or the auto-analyze daemon so that the planner can always be trusted to have a good approximation of the quantity and quality of data that is stored in a table.

Auto-explain

Auto-explain is an extension that helps the database administrator get an idea of slow queries and their execution plan. Essentially, auto-explain triggers when a running query is slower than a specified threshold, and then dumps in the PostgreSQL logs (refer to Chapter 14, *Logging and Auditing*) the execution plan of the query.

In this way, the database administrator can get an insight into slow queries and their execution plan without having to re-execute these queries. Thanks to this, the database administrator can inspect the execution plans and decide if and where to apply indexes or perform a deeper analysis.

The auto-explain module is configured via a set of auto_explain parameter options that can be inserted in the PostgreSQL configuration (the postgresql.conf file), but you need to remember that in order to activate the module, you need to restart the cluster.

The auto-explain module can do pretty much the same things that a manual EXPLAIN command can do, including EXPLAIN ANALYZE, but it has to be properly configured.

In order to install and configure the module, let's start simple and add the following two settings to the cluster configuration in `postgresql.conf`:

```
session_preload_libraries = 'auto_explain'
auto_explain.log_min_duration = '500ms'
```

The first line tells PostgreSQL to load the library related to the auto-explain module, while the second instruments the module to triggers whenever a query takes longer than half of a second to conclude. Of course, you can raise the query duration or lower it depending on your needs.

With that configuration in place, it is now possible to execute quite a long query, as follows (assuming you have dropped/disabled the indexes created in the previous section):

```
forumdb=> \timing
forumdb=> SELECT count(*) -- p.title, u.username
FROM posts p
JOIN users u ON u.pk = p.author
WHERE u.username = 'fluca1978'
AND    daterange( CURRENT_DATE - 20, CURRENT_DATE ) @> p.created_on::date;
 count
-------
    80
(1 row)

Time: 6650,378 ms (00:06,650)
```

The query took 6 and a half seconds, enough time to trigger our auto-explain, and in fact, in the PostgreSQL logs, you can see the following:

```
$ sudo tail -f /postgres/12/log/postgresql.log
2020-04-08 18:36:15.901 CEST [79192] LOG:  duration: 5991.394 ms  plan:
        Query Text: SELECT count(*)
        FROM posts p
        JOIN users u ON u.pk = p.author
        WHERE u.username = 'fluca1978'
        AND    daterange( CURRENT_DATE - 20, CURRENT_DATE ) @>
p.created_on::date
        ;
        Finalize Aggregate  (cost=114848.33..114848.34 rows=1 width=8)
          -> Gather  (cost=114848.12..114848.33 rows=2 width=8)
              Workers Planned: 2
              -> Partial Aggregate  (cost=113848.12..113848.13 rows=1
width=8)
                    -> Hash Join  (cost=8.30..113848.09 rows=10 width=0)
                        Hash Cond: (p.author = u.pk)
                        -> Parallel Seq Scan on posts p
```

```
(cost=0.00..113812.33 rows=10417 width=4)
                                        Filter: (daterange((CURRENT_DATE - 20),
CURRENT_DATE) @> (created_on)::date)
                              ->  Hash  (cost=8.29..8.29 rows=1 width=4)
                                    Buckets: 1024  Batches: 1  Memory Usage:
9kB
                                    ->  Index Scan using users_username_key
on users u  (cost=0.28..8.29 rows=1 width=4)
                                          Index Cond: (username =
'fluca1978'::text)
```

That is exactly the output a normal `EXPLAIN` command would have produced for the same query.

The beauty of this approach is that you don't have to worry about or remember to execute `EXPLAIN` on queries or collected queries; you simply have to inspect the logs to find out the execution plan of slow queries. Once you have fixed queries such as the preceding, by creating indexes, for example, you can raise the threshold of auto-explain to catch slower queries and iterate the process again.

There is a full set of options you can configure to take greater advantage of the auto-explain module, but discussing all of them is beyond the scope of this section. However, a few interesting ones are discussed as follows. If you need to perform `EXPLAIN ANALYZE`, you can set the `auto_explain.log_analyze` parameter to `on`. As an example, enabling the parameter in your configuration file would look as follows:

```
session_preload_libraries = 'auto_explain'
auto_explain.log_min_duration = '500ms'
auto_explain.log_analyze = on
```

This will produce the following output in the logs whenever you execute the same query of the preceding example:

```
2020-04-08 09:15:55.046 CEST [88277] LOG:  duration: 5193.342 ms  plan:
        Query Text: SELECT count(*)
        FROM posts p
        JOIN users u ON u.pk = p.author
        WHERE u.username = 'fluca1978'
        AND   daterange( CURRENT_DATE - 20, CURRENT_DATE ) @>
p.created_on::date
        ;
        Finalize Aggregate  (cost=114848.33..114848.34 rows=1 width=8)
(actual time=5190.322..5190.323 rows=1 loops=1)
          ->  Gather  (cost=114848.12..114848.33 rows=2 width=8) (actual
time=5189.678..5193.226 rows=3 loops=1)
                Workers Planned: 2
```

```
                    Workers Launched: 2
                      -> Partial Aggregate  (cost=113848.12..113848.13 rows=1
    width=8) (actual time=4861.705..4861.712 rows=1 loops=3)
                            -> Hash Join  (cost=8.30..113848.09 rows=10 width=0)
    (actual time=2477.949..4861.639 rows=27 loops=3)
                                  Hash Cond: (p.author = u.pk)
                                  -> Parallel Seq Scan on posts p
    (cost=0.00..113812.33 rows=10417 width=4) (actual time=214.443..4761.128
    rows=26667 loops=3)
                                        Filter: (daterange((CURRENT_DATE - 20),
    CURRENT_DATE) @> (created_on)::date)
                                        Rows Removed by Filter: 1640000
                                  -> Hash  (cost=8.29..8.29 rows=1 width=4)
    (actual time=0.296..0.296 rows=1 loops=3)
                                        Buckets: 1024  Batches: 1  Memory Usage:
    9kB
                                        -> Index Scan using users_username_key
    on users u  (cost=0.28..8.29 rows=1 width=4) (actual time=0.109..0.144
    rows=1 loops=3)
                                              Index Cond: (username =
    'fluca1978'::text)
```

As you can see, the output now includes the same information as EXPLAIN ANALYZE would report. Another interesting feature of auto-explain is that it can log information about the trigger's execution, therefore providing a hint as to which triggers have fired and could be responsible for slowing down your query execution: the auto_explain.log_triggers Boolean parameter can be set for this purpose.

Lastly, you can change the output of the auto-explain information in the logs, choosing from plain text, XML, JSON, and YAML, exactly as you would do with the manual EXPLAIN command, and you can set your preferred log style via the auto_explain.log_format option.

Since PostgreSQL 13, you can also get information about WAL usage out from auto_explain, in a way that is similar to what EXPLAIN with the WAL option does. The auto_explain.log_wal special configuration parameter can be turned on to automatically log information about records and segments inserted in the write-ahead logs.

For other interesting features and tunables, refer to the official documentation of this module.

Summary

PostgreSQL provides very rich features for creating and managing indexes, both single column-based or multi-column-based, as well as multiple types of indexes that can be built.

Thanks to the EXPLAIN command, a database administrator can inspect a slow query and see how the optimizer has thought about what the best access to the underlying data is, and thanks to an understanding of how PostgreSQL works, the administrator can decide which indexes to create in order to tune the performances.

PostgreSQL also provides a rich set of statistics that is used to both extract the quality and the quantity of data within every table, therefore being able to generate an execution plan, and to monitor which indexes are used and when. Auto-explain is another useful module that can be used to silently monitor slow queries and execution plans and see how the cluster is performing without any need to manually execute every suspect statement.

It is important to emphasize that performance tuning is one of the most complex tasks in database administration, and that there is no one silver bullet or one-size-fits-all solution, so experience and a lot of practice are required. In the next chapter, we will look into how to do so, using logging and auditing.

References

- PostgreSQL official documentation about CREATE INDEX: https://www.postgresql.org/docs/12/sql-createindex.html
- PostgreSQL official documentation about pg_stats: https://www.postgresql.org/docs/12/view-pg-stats.html
- PostgreSQL official documentation about EXPLAIN: https://www.postgresql.org/docs/12/using-explain.html
- PostgreSQL official documentation about ANALYZE: https://www.postgresql.org/docs/12/sql-analyze.html
- Auto-explain official documentation: https://www.postgresql.org/docs/12/auto-explain.html

14

Logging and Auditing

PostgreSQL provides a very rich logging infrastructure. Being able to examine the log is a key skill for every database administrator—logs provide hints and information about what the cluster has done, what it is doing, and what happened in the past. This chapter will explain the basics about PostgreSQL log configuration, providing you with an explanation of how to examine logs using either manual approaches, such as reading every log line the cluster produces, or by using automated tools that can help in getting a complete overview of the cluster activity. Related to logging is the topic of auditing, which is the capability of tracking who did what to which data. Auditing is often enforced by government laws, rather than the needs of the database administrators. However, a good auditing system can also help administrators in identifying what happened in the database.

In this chapter, you will learn about the following topics:

- Introduction to logging
- Extracting information from logs using PgBadger
- Implementing auditing

Technical requirements

You will need to know the following:

- How to manage PostgreSQL configuration
- How to start, restart, and monitor PostgreSQL and interact with PGDATA files

You can find the code for this chapter at the following GitHub repository: https://github.com/PacktPublishing/Learn-PostgreSQL.

Introduction to logging

Like many other services and databases, PostgreSQL provides its own logging infrastructure so that the administrator can always inspect what the daemon process is doing and what the current status of the database system is. While logs are not vital for the data and database activities, they represent very important knowledge about what has happened or is happening in the whole system and represent an important clue by means of which an administrator can take action.

PostgreSQL has a very flexible and configurable log infrastructure that allows different logging configuration, rotation, archiving, and post-analysis.

Logs are stored in a textual form, so that they can be easily analyzed with common log analysis tools, including operating system utilities such as `grep(1)`, `sed(1)`, and text editors.

> The term "log," as used in this chapter, refers only to the system's textual logs, and not to the write-ahead logs that, on the other hand, are crucial in the database life cycle.

Usually, logs are contained in a specific sub-folder of the `PGDATA` directory, but as you will see in the following subsections, you are free to move logs to pretty much wherever you want in your operating system storage.

Every event that happens in the database is logged in a separate line of text within the logs, an important and useful aspect when you want to analyze logs with line-oriented tools such as the common Unix commands (for example, `grep(1)`). Of course, writing a huge amount of information into logs has drawbacks; it requires system resources and can fill the storage where the logs are placed. For this reason, it is important to manage the logging infrastructure according to the aim of the cluster, therefore logging only the minimum amount of information that can be used for post-analysis.

> Logs can quickly fill your disk storage if you don't configure them appropriately, and therefore you should be sure your cluster is not producing more logs than your system can handle.

Following the common Unix philosophy, PostgreSQL allows you to send logs to an external component named the syslog. The idea is that there could be, in your own infrastructure, a component or a machine that is responsible for collecting logs from all the available services, including databases, web servers, application servers, and so on. Therefore, you can redirect PostgreSQL logs to the same common `syslog` facility and get the cluster logs collected in the very same place as you already do for the other services. However, this is not always a good choice, and it is for this reason that PostgreSQL provides its own component, named the logging collector, to store logs.

In fact, under a heavy load, the syslog component could start to lose log entries, while the PostgreSQL logging collector has been designed explicitly to not lose a single piece of log information. Therefore, the logging collector shipped with PostgreSQL is usually the preferred way of keeping track of logs, so that you can be sure that once you have to analyze the logs, you have all the information the cluster has produced, without any missing bits.

PostgreSQL logging is configured via tunables contained in the main cluster configuration, namely the `postgresql.conf` file. In the following subsections, you will be introduced to the PostgreSQL logging configuration, and you will see how to tune your own log to match your needs.

Where to log

The first step in configuring the logging system is to decide where and how to store textual logs. The main parameter that controls the logging system is `log_destination`, which can assume one or more of the following values:

- `stderr` means the cluster logs will be sent to the standard error of the postmaster process, which commonly means they will appear on the console from which the cluster has been started.
- `syslog` means that the logs will be sent to an external syslog component.
- `csvlog` means that the logs will be produced as comma-separated values, useful for the automatic analysis of logs (more on this later).
- `eventlog` is a particular component available only on Microsoft Windows platforms that collects the logs of a whole bunch of services.

It is possible to set up the logging to send logs to multiple destinations, such as to the `stderr` and `csvlog` facilities at once, but most of the time you will choose a single destination and use that.

Another important setting of the logging infrastructure is `log_collector`, which is a Boolean value that switches on a process (named the logging collector) that captures all the logs sent to the standard error and stores them where you want.

To summarize, the two preceding parameters are somehow inter-dependent: you need to choose where to send the logs that PostgreSQL will always produce (`log_destination`), and in the case that you send them only (or also) to the standard error or to the `csvlog` facility, you can turn on a dedicated process (the `logging_collector` value) to catch any log entry and store it on disk. This means that your logging configuration will always be something like the following one:

```
log_destination = 'stderr'
logging_collector = on
```

Here, the first line tells the cluster to send the produced logs to the standard error, but from there to be managed and stored by a dedicated process named the logging collector.

In the rest of this section, we will concentrate on the configuration of the logging collector. The logging collector can be configured to place logs in the directory you desire, to name the log files as you wish, and to automatically rotate them. Log rotation is a quite common feature in every logging system and means that once a single log file has grown to a specified size, or when enough time has passed, the file log is closed and a new one (with a different name) is created. For example, you can decide to automatically rotate your log files once a single file becomes 100 MB or every 2 days: the first condition that happens triggers the rotation so that PostgreSQL produces a different log file at least every 2 days or every 100 MB of textual information.

Log rotation is useful because it allows you to produce smaller log files that can be constrained to a specific period of time and that are smaller than a single log file, therefore helping you select the right field to analyze depending on the time of the event you are interested in.

Once you have enabled the logging collector, you have to configure it so that it will store the logs as you want and where you want. In fact, you can use the following parameters to configure the logging collector process, by placing the right value for any of the following settings in the PostgreSQL configuration file:

- `log_directory`: This is a directory where individual log files must be stored. It can be a relative path, considered with regard to `PGDATA`, or an absolute path (that the process must be able to write into).
- `log_filename`: This is a single filename or a pattern to specify the name of every log file (within `log_directory`). The pattern can be specified following `strftime(3)` to format it with a date and time. For example, the value `postgresql-%Y-%m-%d.log` will produce a log filename with the date (respectively, year, month, and day), for example, `postgresql-2020-04-17.log`.
- `log_rotation_age`: This indicates how much time the log should wait before applying automatic log rotation. For example, `1d` means 1 day and specifies that the logs will be rotated once per day.
- `log_rotation_size`: This specifies the size of the log file before it is rotated to a new one.

With the preceding settings, our logging configuration within the `postgresql.conf` file will be as follows:

```
log_destination    = 'stderr'
logging_collector = on
log_directory      = 'log'
log_filename       = 'postgresql-%Y-%m-%d.log'
log_rotation_age  = '1d'
log_rotation_size = '100MB'
```

With the preceding settings, the cluster will produce a new log file every one day (or 100 MB of information) within the log directory (relative to `PGDATA`) using the logging collector, and every log file will have the indication of the year, month, and day it was created. In other words, things on disk will be as follows:

```
$ sudo ls -1 /postgres/12/log
postgresql-2020-04-13.log
postgresql-2020-04-15.log
postgresql-2020-04-16.log
...
```

When to log

It is important to decide when an event must be reported in the logs. There are a lot of options to control the triggering of a log action, specified by means of a threshold. The logging threshold can assume a mnemonic value that indicates the minimum value over which the log event will be inserted into the logs.

The most common values are, in order, info, notice, warning, error, log, fatal, and panic, with info being the minimum and fatal being the highest value.

This means that if you decide that warning is the threshold you want to accept as a minimum, every log event with a lower threshold (such as info and notice) will not be inserted into the logs. As you can see, the threshold increases as it moves toward error values such as fatal and panic, which are always logged automatically because they represent unrecoverable problems. There are also the lowest levels named debug1 through debug5 to get development information and inner details about the processes executions (that is, they are usually used when developing with PostgreSQL).

The cluster will therefore produce different log events at different times, and all with different levels of priority, which in turn will be inserted into the logs depending on the threshold you have configured.

In particular, there are two parameters that can be used to tune the log threshold: log_min_messages and client_min_messages.

The former, log_min_messages, decides the threshold of the logging system, while the latter decides the threshold of every new user connection. How are they different?

log_min_messages specifies what the cluster has to insert into the logs without any regard for incoming user connections, nor their settings. client_min_messages decides which log events the client has to report to the user during the connection. Both these settings can assume a value from the preceding list of thresholds.

A typical use case of a development or test environment could be the following one:

```
log_min_messages    = 'info'
client_min_messages = 'debug1'
```

With the preceding configuration, the cluster will log only info messages in the textual logs, which is something related to the normal execution of the processes, while incoming user connections will report more detailed messages such as development ones back to the user.

Setting thresholds is not the only way you can decide when to trigger log insertion: there are another couple of settings that can be used to take care of the duration of statements and utilities. For example, `log_min_duration_statement` inserts into the logs the textual representation of an SQL statement that has run for more than the specified amount of time (expressed in milliseconds). `log_autovacuum_min_duration` logs the autovacuum actions that took more than the specified amount of milliseconds.

Another very important parameter to decide when to log activity is `log_transaction_sample_rate`. This setting was introduced in PostgreSQL 12 and logs every statement within an explicit transaction with a sample rate. The sample rate is a value between 0 (no sampling) and 1 (full sampling), and therefore is a value of `0.5`:

```
log_transaction_sample_rate = 0.5
```

This will log every single statement from every explicit transaction out of two. As another example, consider setting it to a very low value as follows:

```
log_transaction_sample_rate = 0.01
```

Having a low value as in the preceding will log a transaction every 100 transactions. The aim of this parameter is to get a sample of a full transaction on a heavily loaded system, so as to help to get an idea of what transaction is executing without the need to log every single statement of every single transaction. As you can imagine, the sampling is not an absolute value, but rather an approximation of the number of transactions to wait before activating the logging.

As an example, assume there is a `log_transaction_sample_rate` of `0.5` (half of the transactions) and assume you repeatedly execute a transaction like the following:

```
BEGIN;
SELECT 'Transaction 1'; -- increase the number to identify the transaction
COMMIT;
```

The effect of the preceding simple transaction is to change, at every run, the counter in the message – with the preceding configuration, you will end up with something like the following in the logs:

```
2020-04-16 18:27:21 CEST [59924]: [4]
user=luca,db=forumdb,app=psql,client=192.168.222.1 LOG:  duration: 0.000 ms
statement: SELECT 'transaction 3';
2020-04-16 18:27:35 CEST [59924]: [7]
user=luca,db=forumdb,app=psql,client=192.168.222.1 LOG:  duration: 0.159 ms
statement: SELECT 'transaction 4';
2020-04-16 18:27:55 CEST [59924]: [10]
user=luca,db=forumdb,app=psql,client=192.168.222.1 LOG:  duration: 0.150 ms
```

```
statement: SELECT 'transaction 6';
2020-04-16 18:28:06 CEST [59924]: [13]
user=luca,db=forumdb,app=psql,client=192.168.222.1 LOG:  duration: 0.152 ms
statement: SELECT 'transaction 7';
```

As you can see, the system is not logging the even transactions but is trying to honor the logging of half of the transactions depending on how many transactions the system is executing and has executed. In other words, the sampling cannot be exactly determined in advance.

What to log

The quality of the information to log is configured with a rich set of parameters, usually a Boolean to turn a particular event to log on or off.

One very used and abused setting is `log_statement`: if turned on, it will log every statement executed against the cluster from every connection. This can be very useful because it allows you to reconstruct exactly what the database did and with which statements, but on the other hand, it can also be very dangerous. Logging every statement could make private or sensitive data available in the logs, which could, therefore, become available to unauthorized people. Moreover, logging all the statements could quickly fill up the log storage, in particular, if the cluster is under a heavy load and high concurrency.

 Usually, it is much more useful to configure the `log_min_duration_statement` setting to log only "slow" statements, instead of logging them all.

It is possible to fine-tune the category of statements to log via `log_statement`: the setting can value any of `off`, `ddl`, `mod`, or `all`. It is quite trivial to understand what `off` and `all` mean, while `ddl` means that all Data Definition Language statements (for example, `CREATE TABLE`, `ALTER TABLE`, and so on) are logged, while `mod` means that all data manipulation statements (for example, `INSERT`, `UPDATE`, `DELETE`, and `SELECT`) are logged.

Other useful logging tunables are `log_checkpoints`, `log_connections`, and `log_disconnections`, which can be turned on to log, respectively, when a checkpoint happens, and an incoming connection is established or closed.

The quality of the information in the log is also established by the `log_line_prefix` parameter. `log_line_prefix` is a pattern string that defines what to insert at the beginning of every logline, and therefore can be used to detail the event that is logged. The pattern is created with a few placeholders in the same way as `sprintf(3)`, and documenting every option here is out of the scope of the book. It does suffice to say that useful and common placeholders are as follows:

- `%a` represents the application name (for example, `psql`).
- `%u` represents the username connected to the cluster (role name).
- `%d` is the database where the event happened.
- `%p` is the operating system **process identifier (PID)**.
- `%h` represents the remote host from which the connection to the cluster has been established.
- `%l` is the session line number, an autoincrement counter that helps us to understand the ordering of every statement executed in an interactive session.
- `%t` is the timestamp at which the event happened.

For example, the following configuration will produce a logline that begins with the timestamp of the event, followed by the process identifier of the backend process, then the counter of the command within the session, and then the user, database, and application used to connect to the cluster from the remote host:

```
log_line_prefix = '%t [%p]: [%l] user=%u,db=%d,app=%a,client=%h '
```

The end result of the preceding configuration will be something like the following logline:

```
2020-04-16 11:49:32 CEST [97734]: [4-1]
user=luca,db=digikamdb,app=psql,client=192.168.222.1 LOG:   duration: 16.163
ms   statement: SELECT count(*) FROM get_images( 2019 );
```

You can have a look at how it has been built by referring to the single escaping sequences as shown here:

```
        %t                   [%p]   : [%l] user=%u,   db=%d,        app=%a,
client=%h
2020-04-16 11:49:32 CEST [97734]: [4]
user=luca,db=digikamdb,app=psql,client=192.168.222.1
```

Thanks to the configuration of the logline prefix, it is possible to tune log events in a way that automated applications can identify and analyze log events for you, or at least the administrator can have extra information about the event itself.

A complete example of logging configuration

With all the notions explained in the previous sections, it is possible to set up the logging infrastructure and see the result in action.

Instead of modifying the `postgresql.conf` main file in place, we are going to create a supplementary configuration file, named `logging.conf`, to be included in the main cluster configuration file. In this way, we will be able to switch back to the original logging configuration or replicate it on other systems by simply copying the new file.

The `logging.conf` file will have the following content:

```
$ sudo cat /postgres/12/logging.conf
log_destination             = 'stderr'
logging_collector           = on
log_directory               = 'log'
log_filename                = 'postgresql-%Y-%m-%d.log'
log_rotation_age            = '1d'
log_rotation_size           = '100MB'
log_line_prefix             = '%t [%p]: [%l] user=%u,db=%d,app=%a,client=%h
'
log_checkpoints             = on
log_connections             = on
log_disconnections          = on
log_min_duration_statement  = '100ms'
```

In order to activate the preceding settings, we need to insert, at the end of the `postgresql.conf` file, the following configuration line:

```
include_if_exists='logging.conf'
```

This will load and override the settings with the one coming from the `logging.conf` file (if found). Restart the cluster to be sure you have applied the changes, and you will start to see the new logging configuration in action:

```
$ sudo -u postgres pg_ctl -D /postgres/12 restart
```

In order to test the logging configuration, let's execute a few queries against the database so that it has to log some activity. For example, the following session has lasted less than a minute:

```
$ date
thu 16 apr 2020, 18.04.57, CEST

$ psql -U luca forumdb
psql (12.2 (Ubuntu 12.2-2.pgdg18.04+1))
```

```
SSL connection (protocol: TLSv1.3, cipher: TLS_AES_256_GCM_SHA384, bits:
256, compression: off)
Type "help" for help.

forumdb=> \timing
Timing is on.

forumdb=> SELECT CURRENT_DATE;
 current_date
--------------
 2020-04-16
(1 row)

Time: 1,479 ms

forumdb=> SELECT 1 + 1;
 ?column?
----------
        2
(1 row)

Time: 6,605 ms

forumdb=> SELECT count(*) FROM posts;
  count
---------
 5000000
(1 row)

Time: 1019,322 ms (00:01,019)
forumdb=> \q

$ date
thu 16 apr 2020, 18.05.09, CEST
```

What has been logged in the PostgreSQL logs? We can inspect the file
`postgresql-2020-04-16.log` within the `PGDATA/log` directory (respectively from
the `log_filename` and `log_directory` settings):

```
$ sudo tail /postgres/12/log/postgresql-2020-04-016.log

2020-04-16 18:05:00 CEST [29759]: [1]
user=[unknown],db=[unknown],app=[unknown],client=192.168.222.1 LOG:
connection received: host=192.168.222.1 port=45106

2020-04-16 18:05:00 CEST [29759]: [2]
user=luca,db=forumdb,app=[unknown],client=192.168.222.1 LOG:  connection
authorized: user=luca database=forumdb application_name=psql SSL enabled
```

```
(protocol=TLSv1.3, cipher=TLS_AES_256_GCM_SHA384, bits=256,
compression=off)

2020-04-16 18:05:05 CEST [29759]: [3]
user=luca,db=forumdb,app=psql,client=192.168.222.1 LOG:  duration: 1018.650
ms  statement: SELECT count(*) FROM posts;

2020-04-16 18:05:08 CEST [29759]: [4]
user=luca,db=forumdb,app=psql,client=192.168.222.1 LOG:  disconnection:
session time: 0:00:07.159 user=luca database=forumdb host=192.168.222.1
port=45106
```

There are four lines in the logs, each one starting with a timestamp. The first two lines report the connection request and establishment (the log_connections setting); it is possible to better read this information by removing part of the line prefix and getting to the event that has been logged:

```
2020-04-16 18:05:00 CEST [29759]: [1] ... LOG:  connection received ...
2020-04-16 18:05:00 CEST [29759]: [2] ... LOG:  connection authorized:
user=luca database=forumdb application_name=psql
```

The third line in the logs reports the execution of a slow statement: thanks to the log_min_duration_statement setting, PostgreSQL is logging every statement that requires more than 100 milliseconds, and in fact, only one of the three statements executed has taken more time:

```
2020-04-16 18:05:05 CEST [29759]: [3]
user=luca,db=forumdb,app=psql,client=192.168.222.1 LOG:  duration: 1018.650
ms  statement: SELECT count(*) FROM posts;
```

The preceding logline has not been mangled, because this is a typical use case for an administrator: the line reports a slow statement, which has been executed on the forumdb database by the user luca at time 18:05.

The fourth and last line in the logs is a disconnection notification:

```
2020-04-16 18:05:08 CEST [29759]: [4] ... disconnection: session time:
0:00:07.159 user=luca database=forumdb host=192.168.222.1 port=45106
```

After 7 seconds, the user `luca` disconnected from the database `forumdb`. Please note that the timestamp at the beginning of each log line corresponds to the output of the `date` command in the example session. It is also interesting to note that the logging reports exactly four events, but the connection produced much more. In fact, the session has performed three `SELECT` statements, but only one has been logged (the slowest one) and therefore the session counter has not been increased for the unlogged events, so the disconnection is event number 4.

Now that we have learned all about logging, we will move on to extracting information from the logs that are created, using a special tool called PgBadger.

Extracting information from logs – PgBadger

Thanks to the rich set of information that can be included in the logs, it is possible to automate log information analysis and extraction. There are several tools with this aim, and one of the most popular is PgBadger.

PgBadger is a self-contained Perl 5 application that carefully reads and extracts information from PostgreSQL logs, producing a web dashboard with a summary of all the information it has found in the logs. The aim of this application is to provide you with a more useful insight into the logs without having to manually search for specific information with low-level tools such as grep, awk, and text editors.

Using PgBadger is not mandatory; your cluster will work fine without it and you will be able to seek information and problems in the logs anyway. However, using PgBadger deeply simplifies log management and provides you with more useful hints about what your server has done.

It is important to note that using PgBadger, as well as performing any automated or manual log analysis, does not provide real-time information, but rather, a look at the past in the server activities.

In the following subsections, you will learn how to install and use PgBadger.

Installing PgBadger

PgBadger requires Perl 5 to be installed on the system it will run on, and that is the only dependency it has. You can run PgBadger on the same host the PostgreSQL cluster is running on, or on a remote system (as will be shown in a later subsection). In this section, we will assume PgBadger will be installed and executed on the very same machine the PostgreSQL cluster is running on.

The first step to get PgBadger installed is to download a recent version; for example, to get version 11.2, you can do the following:

```
$ wget https://github.com/darold/pgbadger/archive/v11.2.tar.gz
```

Once the download has completed, you have to extract the download archive as in the following snippet of code, and then run the build and installation process using `make`:

```
$ tar xzvf v11.2.tar.gz
...
$ cd pgbadger-11.2
$ perl Makefile.PL
Checking if your kit is complete...
Looks good
Generating a Unix-style Makefile
Writing Makefile for pgBadger
Writing MYMETA.yml and MYMETA.json

$ make
cp pgbadger blib/script/pgbadger
"/usr/local/bin/perl" -MExtUtils::MY -e 'MY->fixin(shift)' --
blib/script/pgbadger
echo "=head1 SYNOPSIS" > doc/synopsis.pod
./pgbadger --help >> doc/synopsis.pod
echo "=head1 DESCRIPTION" >> doc/synopsis.pod
sed -i.bak 's/ +$//g' doc/synopsis.pod
rm doc/synopsis.pod.bak
sed -i.bak '/^=head1 SYNOPSIS/,/^=head1 DESCRIPTION/d' doc/pgBadger.pod
sed -i.bak '4r doc/synopsis.pod' doc/pgBadger.pod
rm doc/pgBadger.pod.bak
Manifying 1 pod document
rm doc/synopsis.pod


$ sudo make install
echo "=head1 SYNOPSIS" > doc/synopsis.pod
./pgbadger --help >> doc/synopsis.pod
echo "=head1 DESCRIPTION" >> doc/synopsis.pod
sed -i.bak 's/ +$//g' doc/synopsis.pod
```

```
rm doc/synopsis.pod.bak
sed -i.bak '/^=head1 SYNOPSIS/,/^=head1 DESCRIPTION/d' doc/pgBadger.pod
sed -i.bak '4r doc/synopsis.pod' doc/pgBadger.pod
rm doc/pgBadger.pod.bak
Manifying 1 pod document
Installing /home/luca/perl5/man/man1/pgbadger.1p
Installing /home/luca/perl5/bin/pgbadger
Appending installation info to /home/luca/perl5/lib/perl5/amd64-freebsd-
thread-multi/perllocal.pod
rm doc/synopsis.pod
```

Ensure you have the binary script, pgbadger, in your PATH and test its functionality by running the script followed by the version argument as follows:

```
$ pgbadger --version
pgBadger version 11.2
```

If the program replies with the version number, everything should be fine and ready to be used.

Configuring PostgreSQL logging for PgBadger usage

PgBadger is not a magic tool: it cannot understand PostgreSQL logs unless they are produced in a suitable format. It is therefore important to configure PostgreSQL logging infrastructure in a suitable way for PgBadger.

The following is an example of the minimum set of configuration parameters that make PostgreSQL produce logs that PgBadger can mangle correctly:

```
log_destination = 'stderr'
logging_collector = on
log_directory       = 'log'
log_filename        = 'postgresql-%Y-%m-%d.pgbadger.log'
log_rotation_age    = '1d'
log_rotation_size   = '100MB'
log_min_duration_statement = 0
log_line_prefix = '%t [%p]: [%l] user=%u,db=%d,app=%a,client=%h '
log_checkpoints = on
log_connections = on
log_disconnections = on
log_lock_waits = on
log_temp_files = 0
log_autovacuum_min_duration = 0
log_error_verbosity = default
```

Most of the parameters have already been discussed in the previous sections. Please note that the filename for every log has changed to include a `pgbadger` suffix in order to discriminate older logs from PgBadger-compatible ones.

Please also note that we decided to include `log_min_duration_statement` with a zero value, which means PostgreSQL will log every statement that requires more than 0 milliseconds to run, which is short for "log all the statements." As already explained, this can lead to private data being included in the logs, so you should consider carefully whether you need such detail or not.

It is possible to store the preceding configuration parameter in a small configuration file. Let's call it `pgbadger.conf` and import it in the main cluster configuration as already seen, inserting at the end of `postgresql.conf` a line as follows:

```
include_if_exists='pgbadger.conf'
```

Once the server has been restarted to get the new log configuration, you can start using PgBadger.

Using PgBadger

Once PostgreSQL has begun producing logs, you can analyze the results with PgBadger. Before you run PgBadger, especially on a test system, you should generate (or wait for) some traffic and statements (as well as transactions), or the produced dashboard will be empty.

Before starting to use PgBadger, it is appropriate to create a location to store the reports and all the related stuff. This is not mandatory, but simplifies maintenance and archiving of reports later on when you could need to keep them. Let's create a directory, and let's assign the same Postgres user that runs the cluster the ownership of the directory (again, this is not mandatory but simplifies the workflow a little):

```
$ sudo mkdir /postgres/reports
$ sudo chown postgres:postgres /postgres/reports
```

It is now time to launch PgBadger for the first time:

```
$ sudo -u postgres pgbadger -o /postgres/reports/first_report.html
/postgres/12/log/postgresql-2020-04-17.pgbadger.log
[=========================>] Parsed 313252 bytes of 313252 (100.00%),
queries: 801, events: 34
LOG: Ok, generating html report..
```

The first argument, `-o`, specifies a filename where we want the report to be stored. PgBadger produces exactly one file for every run, so you need to change the filename if you want to generate another report without overwriting an existing report.
The second argument is the PostgreSQL log file to analyze.

The program runs for a few seconds, or minutes depending on the size of the log file, and reports some statistical information about what it found on the log file (in this example, 801 statements). You can check the generated report file quite easily:

> If you are going to analyze big log files, or many of them, you can use the parallel mode of PgBadger with the `-j` option followed by the number of parallel processes to spawn. For example, passing `-j 4` means that every log file will be divided into four parts, each one analyzed by a single process. Thanks to parallelism, you can exploit all the cores of your machine and get results faster for a large amount of logs.

```
$ ls /postgres/reports
first_report.html
```

If you point your web browser to the local file, you will see the report shown here. The report provides a glance at the cluster activity, including the number of statements, the time spent serving such statements, and graphs showing the statement traffic with regard to the period of time:

At the top of the web page, there is a menu bar that inludes several menus that allow you to look at different graphs and dashboards. For example, the **Connections** menu allows you to get information about how many concurrent connections you had, as shown in the example here:

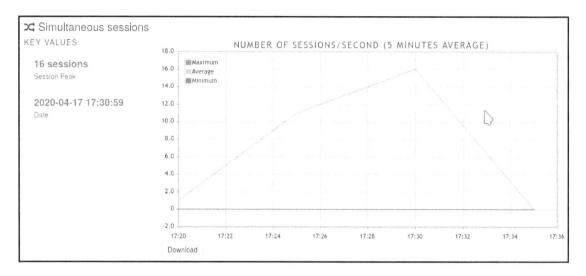

The **Queries** menu allows you to get an overview of the type and frequency of statements, as shown in the following screenshot, where the main percentage of queries has been of type **SELECT**:

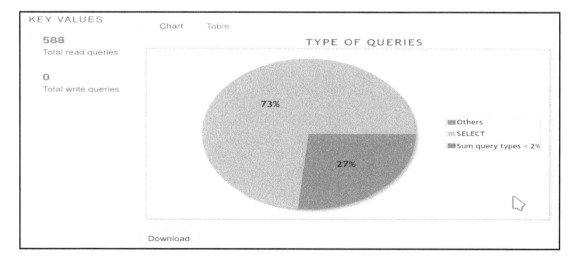

The **Top** menu allows to see the "top events," such as the slowest queries and the most time-consuming queries, shown respectively in the following screenshot:

Rank	Duration	Query
1	3s199ms	⏎SELECT *count* `*` FROM posts;
		[**Date**: 2020-04-17 17:23:42 - **Database**: forumdb - **User**: luca - **Remote**: 127.0.0.1 - **Application**: psql]

And the detailed version is shown in this screenshot:

Rank	Times executed	Total duration	Min duration	Max duration	Avg duration	Query
1	294 Details	4m56s	949ms	3s199ms	1s7ms	⏎SELECT *count* `*` FROM posts; Examples User(s) involved App(s) involved
2	294 Details	169ms	0ms	5ms	0ms	⏎SELECT *count* `*` FROM tags; Examples User(s) involved App(s) involved
3	73 Details	83ms	0ms	3ms	1ms	⏎ROLLBACK; Examples User(s) involved App(s) involved
4	73 Details	2ms	0ms	0ms	0ms	⏎BEGIN; Examples User(s) involved App(s) involved
5	67 Details	357ms	3ms	16ms	5ms	⏎TRUNCATE posts CASCADE; Examples User(s) involved App(s) involved

Discussing all the features and graphs of PgBadger is out of the scope of this book, so please see the official documentation for more details and a clear and accurate explanation of every single option.

Scheduling PgBadger

PgBadger can be used in a scheduled way so that it can produce accurate reports during a specific period of time. This is possible because PgBadger includes an incremental feature, with which the report is not overwritten every time but the program can produce a per-hour report and a per-week summary report.

This is handy because you can schedule `pgbadger` execution with, for example, `cron(1)` and forget about it. Let's first see how PgBadger can be run in incremental mode:

```
$ sudo -u postgres pgbadger -I --outdir /postgres/reports/
/postgres/12/log/postgresql-2020-04-17.pgbadger.log
[=========================>] Parsed 313252 bytes of 313252 (100.00%),
queries: 801, events: 34
LOG: Ok, generating HTML daily report into
/postgres/reports//2020/04/17/...
LOG: Ok, generating HTML weekly report into
/postgres/reports//2020/week-16/...
LOG: Ok, generating global index to access incremental reports...
```

The `-I` argument specifies the incremental mode, so PgBadger will produce separate files for the hourly and weekly reports. Please note that instead of specifying the output file, the `--outdir` option has been used to specify the directory to place the files in. Lastly, as usual, there is the log file to analyze.

The end result, as you can guess from the output of the program, is that a directory tree has been produced, something like the following:

```
/postgres/reports
   - index.html
   + 2020
      + 04
        + 17
          index.html
```

The main `index.html` file is the entry point for the whole incremental report. Then there is a tree that has a directory for the year (`2020`), the month (`04`), the day (`17`), and an `index.html` file for that day. The tree is therefore going to be expanded as more days come into play.

If you point your web browser to the main index file, you will see a calendar like the one in the following screenshot, where you can select the month and day to see the per-day report. The everyday report has the same structure as shown in the previous section:

Global Index - pgBadger

Year 2020

April

	Su	Mo	Tu	We	Th	Fr	Sa
14				01	02	03	04
15	05	06	07	08	09	10	11
16	12	13	14	15	16	17	18
17	19	20	21	22	23	24	25
18	26	27	28	29	30		

Thanks to the incremental approach, you can now schedule the execution in your own scheduler, for example, in `cron(1)` you can insert a line like the following:

```
59 23 * * * pgbadger -I --outdir /postgres/reports/
/postgres/12/log/postgresql-`date +'%Y-%m-%d'`.pgbadger.log
```

That is essentially the same command line as the preceding with the current date automatically computed. The preceding line will produce, at the end of every day, the report for the current day, so you will have the population of the report tree.

The previous crontab entry is just an example. Please consider wrapping everything in a robust script and testing the correctness of its execution.

Lastly, it is possible to run PgBadger from a remote host, so that you can dedicate a single machine to collect all the reports and information in a single place. In fact, PgBadger accepts a URI parameter that is the remote location of the log directory (or file) and can be accessed in either FTP or SSH (the recommended way).

As an example, the following represents the same command line as previously, which pulls in incremental mode the logs from a remote PostgreSQL host named `miguel`:

```
$ pgbadger -I --outdir /postgres/reports
ssh://postgres@miguel//postgres/12/log/postgresql-`date +'%Y-%m-
%d'`.pgbadger.log
[========================>] Parsed 313252 bytes of 313252 (100.00%),
queries: 841, events: 34
LOG: Ok, generating HTML daily report into /postgres/reports/2020/04/17/...
LOG: Ok, generating HTML weekly report into
/postgres/reportsy/2020/week-16/...
LOG: Ok, generating global index to access incremental reports...
```

Please note that the log file has been specified via an SSH URL. It is highly recommended to use a remote user that has access to the logs and perform an SSH key exchange to automate the login between the hosts.

Now that we know how to use logs, we will move on to another way of looking at tasks, auditing.

Implementing auditing

Auditing is the capability of performing introspection over an application or user session, in other words, to be able to reproduce, step by step, what the user or the application asked the cluster to do.

Auditing is slightly different from logging, as logging provides a simple way of saving whatever action of the user, but without providing an easy way to reconstruct the user or application interactions with the cluster. In fact, in a highly concurrent cluster, many actions made by different users will coexist in the logs in a mixed bunch of lines. Moreover, logging does not provide any particular logic on what it is storing, and therefore it becomes hard to find out what a user has done. This becomes even more true when the user or the application executes complex statements, in particular, statements where parameters and values are not explicitly provided.

As an example, consider the following simple section:

```
forumdb=> PREPARE my_query( text ) AS SELECT * FROM categories WHERE title
like $1;
PREPARE
forumdb=> EXECUTE my_query( 'PROGRAMMING%' );
 pk |        title        |            description
----+---------------------+------------------------------
```

```
   3 | PROGRAMMING LANGUAGES | All about programming languages
(1 row)
```

That will reveal, with verbose logging, the following:

```
LOG:  duration: 19.011 ms  statement: PREPARE my_query( text ) AS SELECT *
FROM categories WHERE title like $1;
LOG:  duration: 6.539 ms  statement: EXECUTE my_query( 'PROGRAMMING%' );
```

As you can see, in the logs, there is everything you need to reconstruct what the user has done, but that is not so simple. You have to understand that the two lines are related to each other and that the session from which the statements have been executed is the same. This is not always possible – especially if other queries are logged between the two lines you are interested in.

Moreover, it could happen that the logs do not report all the information you need – perhaps because you chose to not log statements that execute faster than a threshold.

Therefore, while you can use logging to perform auditing, that is not always the best choice. In this section, you will learn about the PgAudit extension, which was born to provide a reliable and easy-to-use auditing infrastructure. PgAudit exploits the excellent PostgreSQL logging facility, therefore, you need to configure your logging infrastructure in an appropriate way, as you will see in the next subsections.

Before we dig into the configuration and usage of PgAudit, there are some details and concepts that have to be explained. PgAudit can work in two different ways: auditing by session or by object. The former is a quick and simple way to audit a part (or a whole) session by a user or an application; the latter is a more complex and fine-grained way of logging actions related to specific database objects (for example, who deleted rows from that table?).

Auditing by session works by simply configuring the categories of statements to audit within a session. On the other hand, auditing by object requires you to configure individual database roles that, depending on their set of permissions, will trigger the auditing of specific actions. In the following subsections, you will see both ways used to audit.

Installing PgAudit

PgAudit is an extension that has to be built against the PostgreSQL version it will be installed on. The first step is to download the source code from the official GitHub repository:

```
$ wget https://github.com/pgaudit/pgaudit/archive/1.4.0.tar.gz
...
$ tar xzvf 1.4.0.tar.gz
...
```

Once you have downloaded and extracted the archive, you can enter the source directory and start compiling the extension by using make:

```
$ cd pgaudit-1.4.0
$ make USE_PGXS=1
cc -Wall -Wmissing-prototypes -Wpointer-arith -Wdeclaration-after-statement
-Werror=vla -Wendif-labels -Wmissing-format-attribute -Wformat-security -
fno-strict-aliasing -fwrapv -Wno-unused-command-line-argument -O2 -pipe  -
fstack-protector-strong -fno-strict-aliasing  -fPIC -DPIC -I. -I./ -
I/usr/local/include/postgresql/server -
I/usr/local/include/postgresql/internal  -I/usr/local/include -
I/usr/local/include -I/usr/local/include  -c -o pgaudit.o pgaudit.c
cc -Wall -Wmissing-prototypes -Wpointer-arith -Wdeclaration-after-statement
-Werror=vla -Wendif-labels -Wmissing-format-attribute -Wformat-security -
fno-strict-aliasing -fwrapv -Wno-unused-command-line-argument -O2 -pipe  -
fstack-protector-strong -fno-strict-aliasing  -fPIC -DPIC -shared -o
pgaudit.so pgaudit.o  -L/usr/local/lib   -L/usr/local/lib -lpthread -
L/usr/local/lib  -fstack-protector-strong   -L/usr/local/lib -Wl,--as-
needed -Wl,-R'/usr/local/lib'


$ sudo make USE_PGXS=1 install
/bin/mkdir -p '/usr/local/lib/postgresql'
/bin/mkdir -p '/usr/local/share/postgresql/extension'
/bin/mkdir -p '/usr/local/share/postgresql/extension'
/usr/bin/install -c -m 755  pgaudit.so
'/usr/local/lib/postgresql/pgaudit.so'
/usr/bin/install -c -m 644 .//pgaudit.control
'/usr/local/share/postgresql/extension/'
/usr/bin/install -c -m 644 .//pgaudit--1.4.sql
'/usr/local/share/postgresql/extension/'
```

Now that the extension is available among those of the cluster, you have to configure PostgreSQL to use PgAudit.

Configuring PostgreSQL to exploit PgAudit

PgAudit is an extension that needs to be loaded at server startup, therefore you have to change the main configuration file, `postgresql.conf`, to include the `pgaudit` library as follows:

```
shared_preload_libraries = 'pgaudit'
```

Then, restart your cluster to make the changes take effect:

```
$ sudo pg_ctl -D /postgres/12 restart
```

Since PgAudit is an extension, you have to enable it within the database you want to audit in order to activate it. For the sake of simplicity, let's enable it within our `forumdb` database (you need to connect as a database superuser):

```
forumdb=# CREATE EXTENSION pgaudit;
CREATE EXTENSION
```

It is now time to decide when and how to apply auditing.

Configuring PgAudit

PgAudit ships with a rich set of configuration parameters that allow you to specify exactly what to log, when, what to exclude from auditing, and so on. All configuration parameters live within the `pgaudit` namespace so that they will not clash with other existing settings with the same name.

The most important setting is `pgaudit.log`, which defines which statements and actions you want to audit. The parameter can assume any of the following values:

- `ALL` to audit every statement
- `NONE` to audit nothing at all
- `READ` to audit only `SELECT` and `COPY` statements
- `WRITE` to audit every statement that modifies data (`INSERT`, `UPDATE`, and `COPY`)
- `ROLE` to audit role changes or creation
- `DDL` to audit all the data-definition statements, and therefore any change to the database structure
- `FUNCTION` to audit all code execution, including `DO` blocks

- `MISC` to audit all the values not explicitly categorized above
- `MISC_SET` to audit all `SET` like commands

You are free to specify more than one setting at the same time by separating single names with a comma, for example:

```
pgaudit.log = 'WRITE,FUNCTION';
```

This function can be used to audit all data changes and code executions.

Another important configuration parameter is `pgaudit.log_level`, which specifies the log level that PgAudit will use to make the auditing messages appear in the logs. By default, this setting assumes the value log, but you can change it to any other log threshold except error ones (such as `ERROR`, `FATAL`, and `PANIC`).

In order to insert more details in the audit information, you will likely want to enable `pgaudit.log_parameter` to dump any query parameters (you will see an example later).

If you are going to configure PgAudit by object, you will set the `pgaudit.role` parameter as you will see later in this chapter.

Auditing by session

The first, and most simple to understand and try, way of using PgAudit is by session.

You don't have to manipulate any configuration files; you can set `pgaudit.log` directly in your interactive session and perform some actions to see what happens. As an example, suppose we want to audit any changes to the data:

```
forumdb=# SET pgaudit.log TO 'write, ddl';
SET
forumdb=# SELECT count(*) FROM categories;
 count
-------
     3
(1 row)

forumdb=# INSERT INTO categories( description, title ) VALUES( 'Fake', 'A
Malicious Category' );
INSERT 0 1

forumdb=# SELECT count(*) FROM categories;
 count
-------
```

```
     4
(1 row)
```

```
forumdb=# INSERT INTO categories( description, title ) VALUES( 'Fake2',
'Another Malicious Category' );
INSERT 0 1
```

> The `pgaudit.log` parameter can be set only by superusers, therefore if you want to try it dynamically in an interactive session, you need to connect as a database administrator. You can, of course, set for all users at a cluster-wide level setting the parameter in the `postgresql.conf` configuration file.

In the logs, PostgreSQL will write something like the following:

```
2020-04-20 18:40:03.582 CEST [25760] LOG:  AUDIT:
SESSION,1,1,WRITE,INSERT,,,"INSERT INTO categories( description, title )
VALUES( 'Fake', 'A Malicious Category' );",<not logged>
2020-04-20 18:41:32.326 CEST [25760] LOG:  AUDIT:
SESSION,2,1,WRITE,INSERT,,,"INSERT INTO categories( description, title )
VALUES( 'Fake2', 'Another Malicious Category' );",<not logged>
```

There are several details in such a logline, but before we examine the fields, please note that nothing has been written about the two SELECT statements: since we asked PgAudit to not audit READ queries, the SELECT statements have been discarded from auditing.

Please note that every audit line has a quite self-explanatory prefix, AUDIT, which makes it simple to understand whether the logline has been produced by PgAudit or by some other event internal to PostgreSQL.

Every line has the type of auditing – in the preceding, SESSION – and a counter that increments to indicate the chronological order in which statements have been audited. Then there is the category of statement that PgAudit recognizes – in the preceding, both are WRITE events – and then follows the complete statements that have been executed. There is room for other details, which will be discussed in further examples.

Let's move on with another example – consider the execution of a dynamically built query like the following one:

```
forumdb=# DO $$ BEGIN
EXECUTE 'TRUNCATE TABLE ' || 'tags CASCADE';
END $$;
NOTICE:  truncate cascades to table "j_posts_tags"
DO
```

Instead of executing a `TRUNCATE TABLE tags` statement, the statement has been built concatenating two strings. In the logs, PgAudit inserts a line as follows:

```
2020-04-20 18:46:45.640 CEST [25760] LOG:  AUDIT:
SESSION,3,1,WRITE,TRUNCATE TABLE,,,TRUNCATE TABLE tags CASCADE,<not logged>
```

Again, the line reports the auditing mode (`SESSION`), the auditing statement number (3), the category (`WRITE`), and the statement (`TRUNCATE TABLE`), as well as the fully executed statement. This last detail is important: if you execute the same statement without auditing, PostgreSQL logs will contain a line as follows:

```
2020-04-20 18:50:00.122 CEST [29365] LOG:  duration: 12.616 ms  statement:
DO $$ BEGIN
        EXECUTE 'TRUNCATE TABLE ' || 'tags CASCADE';
        END $$;
```

Here, you can see the logs have blindly copied the source statement, including string concatenation and newlines, making it difficult to read and search for.

Auditing by role

The auditing by role mechanism of PgAudit allows you to define in a very fine-grained way what events you are interested in auditing.

The idea is that you define a database role, and grant permissions related to the action you want to audit to the role. Once the role and its permissions are set, you inform PgAudit to audit by that role, which means PgAudit will report in the logs any action that matches the one granted to the auditing role without any regard to the role that has performed it.

The first step is therefore the creation of a role that is used only to specify which actions to audit, and therefore will not be used as an ordinary role for interactive sessions:

```
forumdb=# CREATE ROLE auditor WITH NOLOGIN;
CREATE ROLE
```

In order to specify which actions the role must audit, we simply have to GRANT those to the role. For example, assuming we want to audit all DELETE on every table and INSERT only on posts and categories, we have to grant the role with the following set of permissions:

```
forumdb=# GRANT DELETE ON ALL TABLES IN SCHEMA public TO auditor;
GRANT
forumdb=# GRANT INSERT ON posts TO auditor;
GRANT
forumdb=# GRANT INSERT ON categories TO auditor;
GRANT
```

Everything is now prepared for PgAudit to do its job, but it is fundamental that the auditing system knows that the auditor role has to be used, therefore we need either to configure pgaudit.role in the cluster configuration or in the current session. The former method is, of course, the right one to use with a production environment, while setting the configuration parameter in a single session is useful for testing purposes. Let's set the parameter in the session, as a database administrator, to test it in action:

```
forumdb=# SET pgaudit.role TO auditor;
SET
```

Now it is time to execute a few statements and see what the PgAudit stores in the cluster logs:

```
forumdb=# INSERT INTO categories( title, description ) VALUES( 'PgAudit',
'Topics related to auditing in PostgreSQL' );
INSERT 0 1

-- this will not be logged
forumdb=# INSERT INTO tags( tag ) VALUES( 'pgaudit' );
INSERT 0 1

forumdb=# DELETE FROM posts WHERE author NOT IN ( SELECT pk FROM users
WHERE username NOT IN ( 'fluca1978', 'sscotty71' ) );
DELETE
```

As you can imagine, PgAudit will log the first and last statement of the preceding example session: in fact, only those statements are related to tables and actions the auditor role has been granted. In the PostgreSQL logs, you will find something similar to the following lines:

```
2020-04-26 10:28:10.414 CEST [29550] LOG:  AUDIT:
OBJECT,1,1,WRITE,INSERT,TABLE,public.categories,"INSERT INTO categories(
title, description ) VALUES( 'PgAudit', 'Topics related to auditing in
PostgreSQL' );",<not logged>
2020-04-26 10:30:11.319 CEST [29550] LOG:  AUDIT:
```

```
OBJECT,2,1,WRITE,DELETE,TABLE,public.posts,"DELETE FROM posts WHERE author
NOT IN ( SELECT pk FROM users WHERE username NOT IN ( 'fluca1978',
'sscotty71' ) );",<not logged>
```

Please note that the tuple insertion against the tags table is missing: it has not been audited and logged because the auditor role does not include a specific GRANT permission for it.

Once our auditing role has been properly configured, we can save the configuration, modifying the configuration file, `postgresql.conf`, and setting the `pgaudit.role` tunable:

```
pgaudit.role = 'auditor'
```

As you can see, role-based auditing is much more flexible than session-only-based: while the latter allows you to specify only the categories of actions to audit, the former allows the fine-grained definition of exactly which statements to audit.

Summary

PostgreSQL provides a reliable and flexible infrastructure for logging that allows a database administrator to monitor what the cluster has done in the very near past. Thanks to its flexibility, the logs can be configured to be mangled by external tools for cluster analysis, such as PgBadger. Moreover, the same logging infrastructure can be exploited to perform auditing, a kind of introspection often required by local government laws.

In this chapter, you have learned how to configure the PostgreSQL logging system to match your needs, how to monitor your cluster by means of the web dashboards provided by PgBadger, and finally, how to perform auditing on your users and applications.

In the following chapter, you will learn how to back up your own cluster.

References

- The PgBadger official documentation available at `https://pgbadger.darold.net/documentation.html`
- The PostgreSQL log settings, official documentation available at `https://www.postgresql.org/docs/12/runtime-config-logging.html`
- PgAudit official website and documentation available at `https://www.pgaudit.org/`

15
Backup and Restore

It doesn't matter how solid your hardware and software is – sooner or later, you will need to go back in time to recover accidentally deleted or damaged data. That is the aim of backups – a safe copy that you can keep for a specific amount of time that allows you to recover from data loss. Being an enterprise-level database cluster, PostgreSQL provides a set of specific tools that allow a database administrator to take care of backups and restorations, and this chapter will show you all the main tools that you can exploit to be sure your data will last any accidental abuse.

Backup and restore isn't a very complex topic, but it's fundamental in any production system and requires careful planning. In fact, with a backup copy, you are holding another exact copy of your database just in case something nasty happens; this extra copy will consume resources, most notably storage space. Deciding how many extra copies, how frequently you grab them, and how long they must be kept is something that requires careful attention and is beyond the scope of this chapter. In this chapter, you will look at the main ways of performing a backup, either logically or physically, and all the provided tools that a PostgreSQL distribution provides so that you can manage backups.

In this chapter, we will cover the following topics:

- Introducing various types of backups and restores
- Exploring logical backups
- Exploring physical backups

Let's get started!

Technical requirements

You need to know about the following to complete this chapter:

- How to interact with command-line tools
- How to inspect your filesystem and the `PGDATA` directory

The code for this chapter can be found in the following GitHub repository: `https://github.com/PacktPublishing/Learn-PostgreSQL`.

Introducing various types of backups and restores

There are mainly two types of backups that apply to PostgreSQL: the **logical backup** (also known as a **cold backup**) and the **physical backup** (also known as a **hot backup**). Depending on the type of backup you choose, the restore process will differ accordingly. PostgreSQL ships will all the integrated tools to perform the classical logical backup, which in most cases suffices. However, PostgreSQL can easily be configured to support physical backups, which are useful when the size of the cluster becomes huge, as well as when you have particular needs, as you will discover later in this chapter.

But what is the difference between these two backup methods? As you can imagine, they both achieve the very same aim: allowing you to get a usable "copy" of your data to restore either on the same cluster or against another cluster. However, the difference between the two backup strategies come from the way data is extracted from the cluster.

A logical backup works as a database client that asks for all the data in a database, table by table, and stores the result in a storage system. It is like an application opening a transaction and performing a `SELECT` on every table, before saving the result on a disk file. Of course, it is much more complex than that, but this example gives you a simple idea of what happens under the hood.

This kind of backup is "logical" because it runs alongside other database connections and activities, as a dedicated client application, and relies on the database to provide data that is "logically" consistent. In fact, the backup is executed within a snapshot of the database so that all the foreign keys are consistent with each other.

The advantages of this backup strategy are that it is simple to implement since PostgreSQL provides all the software to perform a full backup, it is consistent, and it can be restored easily. However, this backup method also has a few drawbacks: being performed on a live system by means of a snapshot can slow down other concurrent database accesses, and it also requires the database to keep track of the ongoing backup process without trashing the snapshot as long as the backup is running. Moreover, the produced backup set is consistent at the time the backup has started; that is, if the backup requires a very long time to complete, data changes that occurred in the meantime will not be present in the backup (because it has to be consistent).

A physical backup, on the other hand, is not invasive of cluster operations: the backup requires a file-level copy of the `PGDATA` content – mainly the database file (`PGDATA/base`) and the WALs from the backup's start instance to the backup's end. The end result will be an inconsistent copy of the database that needs particular care to be restored properly. Essentially, the restore will proceed since the database has crashed and will redo all the transactions (extracted from the WALs) in order to achieve a consistent state.

This kind of backup is much more complex to set up, and while you can perform it on your own, as you will see in this chapter, several tools have emerged to help you perform this kind of backup in a more proficient and secure way. The main advantage of this kind of backup strategy is its less invasive nature – the database is not going to notice any particular activity related to the backup except for the storage I/O required to perform the file-level copy. Another important advantage of this backup strategy is that it allows for **point-in-time recovery** (**PITR**), which allows a database administrator to recover the database to any instance since the original backup. The main disadvantage of this is the complexity required to set up the backup.

There is another consideration to take into account here: logical backups are supposed to always work, regardless of the database version you are running (assuming you are running the latest version already) and, to some extent, regardless of whether the target database is PostgreSQL. On the other hand, physical backups will only work between the very same major versions of PostgreSQL instances. These are important considerations to take into account when dealing with backup and restores.

In the next section, you will learn how to perform both backup methods, as well as how to restore a backup. We'll start with logical backups.

Exploring logical backups

PostgreSQL ships with all the required tools to perform a logical backup and restore. Many operating systems, including FreeBSD and GNU/Debian, provide scripts and wrappers for the PostgreSQL backup and restore tools to ease the system administrator in scheduling backups and restores. Such scripts and wrappers will not be explained here. For more information, consider reading your operating system's PostgreSQL package documentation.

There are three main applications involved in backup and restore – `pg_dump`, `pg_dumpall`, and `pg_restore`. As you can imagine from their names, `pg_dump` and `pg_dumpall` are related to extracting (dumping) the content of a database, thus creating a backup, while `pg_restore` is their counterpart and allows you to restore an existing backup.

PostgreSQL does not require any special "backup" permissions. In order to perform a backup, the user must have all the required grants to access individual objects, such as tables and functions. The same applies to restoring a backup.

The `pg_dump` application is used to dump a single database within a cluster, `pg_dumpall` provides us with a handy way to dump all the cluster content, including roles and other intra-cluster objects, and `pg_restore` can handle the output of the former two applications to perform a restoration.

Remember that a backup is only good if it can be restored. `pg_dump` and `pg_dumpall` will not produce a corrupted backup, but your storage could accidentally damage your backup files, so to ensure you have a valid backup, you should always try to restore it on another machine or cluster.

All three commands can work locally or remotely on the cluster to backup or restore, which means you can use them from a remote backup machine or on the same server the cluster is running on. The applications follow the same parameter and variable conventions that `psql` does, so, for instance, you can specify the username that is going to perform the backup (or restore) via the `-U` command-line flag, as well as the remote host on which the cluster is running via `-h`, and so on. If no parameters are provided, the application assumes the cluster is running locally and connects to it via the current operating system user, just like `psql` does.

In the next subsections, you will learn how to back up and restore your own databases.

Dumping a single database

In order to dump – that is, to create a backup copy of – a database, you need to use the `pg_dump` command.

`pg_dump` allows three main backup formats to be used:

- **A plain text format**: Here, the backup is made of SQL statements that are reproducible.
- **A compressed format**: Here, the backup is automatically compressed to reduce storage space consumption.
- **A custom format**: This is more suitable for a selective restore by means of `pg_restore`.

By default, `pg_dump` uses plain text format, which produces SQL statements that can be used to rebuild the database structure and content, and outputs the backup directly to the standard output. This means that if you back up a database without using any particular option, you are going to see a long list of SQL statements:

```
$  pg_dump -U postgres forumdb
-- PostgreSQL database dump
...
SET client_encoding = 'UTF8';
SET standard_conforming_strings = on;
SELECT pg_catalog.set_config('search_path', '', false);
...
CREATE TABLE public.categories (
    pk integer NOT NULL,
    title text NOT NULL,
    description text
);

...
COPY public.categories (pk, title, description) FROM stdin;
1       DATABASE        Database related discussions
2       UNIX    Unix and Linux discussions
\.
...
```

As you can see, `pg_dump` has produced a set of ordered SQL statements that, if pushed to an interactive connection, allow you to rebuild not only the database structure (tables and functions) but also its content (data within tables), as well as permissions (grants and revokes) and other required objects. All lines beginning with a double dash are SQL comments that `pg_dump` has diligently placed to help you analyze and understand the database's backup content.

There are a few important things to note related to the backup content. The first is that `pg_dump` places a bunch of SET statements at the very beginning of the backup; such SET statements are not mandatory for the backup, but for restoring from this backup's content. In other words, the first few lines of the backup are not related to the content of the backup, but to how to use such a backup.

An important line among those SET statements is the following one, which has been introduced in recent versions of PostgreSQL:

```
SELECT pg_catalog.set_config('search_path', '', false);
```

Such lines remove the search_path variable, which is the list of schema names among those to search for an unqualified object. The effect of such a line is that every object that's created from the backup during a restore will not exploit any malicious code that could have tainted your environment and your search_path. The side effect of this, as will be shown later on, is that after restoration, the user will have an empty search path and will not be able to find any not fully qualified objects by their names.

Another important thing about the backup content is that pg_dump defaults to using COPY as a way to insert the data into single tables. COPY is a PostgreSQL command that acts like INSERT but allows for multiple tuples to be specified at once and, most notably, is optimized for bulk loading, resulting in a faster recovery. However, this can make the backup not portable across different database engines, so if your aim is to dump the database content in order to migrate it to another engine, you have to specify pg_dump to use regular INSERT statements by means of the --inserts command-line flag:

```
$ pg_dump -U postgres forumdb

...
INSERT INTO public.categories OVERRIDING SYSTEM VALUE VALUES (1,
'DATABASE', 'Database related discussions');
INSERT INTO public.categories OVERRIDING SYSTEM VALUE VALUES (2, 'UNIX',
'Unix and Linux discussions');
...
```

The entire content of the backup is the same, but this time, the tables are populated by standard INSERT statements. As you can imagine, the end result is a more portable but also longer (and therefore much heavier) backup content. However, note how, in the previous example, the INSERT statements did not include the list of columns every field value maps to; it is possible to get a fully portable set of INSERT statements by replacing the --inserts option with --column-inserts:

```
$ pg_dump -U postgres --column-inserts  forumdb
...
INSERT INTO public.categories (pk, title, description) OVERRIDING SYSTEM
VALUE VALUES (1, 'DATABASE', 'Database related discussions');
INSERT INTO public.categories (pk, title, description) OVERRIDING SYSTEM
VALUE VALUES (2, 'UNIX', 'Unix and Linux discussions');
...
```

Being able to dump the database content is useful, but being able to store such content in a file is much more useful and allows for restoration to occur at a later date. There are two main ways to save the output of pg_dump into a file. One requires that we redirect the output to a file, as shown in the following example:

```
$  pg_dump -U postgres --column-inserts  forumdb > backup_forumdb.sql
```

The other (suggested) way is to use the pg_dump -f option, which allows us to specify the filename that the content will be placed in. Here, the preceding command line can be rewritten as follows:

```
$  pg_dump -U postgres --column-inserts -f backup_forumdb.sql forumdb
```

This has the very same effect as producing the backup_forumdb.sql file, which contains the same SQL content that was shown in the previous examples.

pg_dump also allows for verbose output, which will print what the backup is performing while it is performing. The -v command-line flag enables this verbose output:

```
$  pg_dump -U postgres -f backup_forumdb.sql -v forumdb
pg_dump: last built-in OID is 16383
pg_dump: reading extensions
pg_dump: identifying extension members
pg_dump: reading schemas
pg_dump: reading user-defined tables
pg_dump: reading user-defined functions
pg_dump: reading user-defined types
pg_dump: reading procedural languages
...
...
```

Once you have your backup file ready, you can restore it easily. We'll learn how to do this in the next section.

Restoring a single database

If the backup you have produced is plain SQL, you don't need anything other than a database connection to restore it – after all, you will execute a bunch of statements in the correct order to recreate the database content.

pg_dump is smart enough to figure out the correct order by which tables and their content must be dumped to allow for foreign key recreation.

It is important to note that, by default, `pg_dump` does not issue, in its backup content, a `CREATE DATABASE` statement. In fact, let's say we produce a backup file as follows:

```
$ pg_dump -U postgres --column-inserts -f backup_forumdb.sql forumdb
```

The created `backup_forumd.sql` file will not include any instructions on how to create a new database. This can be handy, but also dangerous: this means that the restoration will happen within the database you are connected to.

Let's assume that we want to restore the database content to another local database that we are going to name `forumdb_test`. Here, the first step is to create a database, as follows:

```
$ psql -U postgres template1
psql (12.2)
Type "help" for help.

template1=# CREATE DATABASE forumdb_test WITH OWNER luca;
CREATE DATABASE
template1=# \q
```

Now, it is possible to connect to the target database and ask `psql` to execute the entire content of the backup file:

```
$ psql -U luca forumdb_test
psql (12.2)
Type "help" for help.

forumdb_test=> \i backup_forumdb.sql
```

You are going to see a list of command output codes such as `INSERT 0 1`, which means a single `INSERT` happened, as well as confirmation of the occurrence of `ALTER TABLE`, `GRANT`, and every other command the backup contains. Depending on the size of the backup, as well as the performance of the machine, the restoration could take a few seconds to minutes.

Once the restore has completed, it is possible to test whether the backup has been restored; for example, by querying a single table for its data:

```
forumdb_test=> SELECT * FROM tags;
ERROR:  relation "tags" does not exist
LINE 1: SELECT * FROM tags;
                      ^
```

Hold on – this does not mean that the backup and restore process didn't work properly! Remember that `pg_dump` has inserted an appropriate instruction to remove every entry from `search_path`, so `psql` doesn't know how to look up a table named `tags`, while it can regularly find a table with a fully qualified name such as `public.tags`:

```
forumdb_test=> SELECT * FROM public.tags;
 pk |   tag   | parent
----+---------+--------
 44 | pgaudit |
...
```

You can either close the connection and start it over to get a regularly set up version of `search_path` or set it manually in your current connection by means of `set_config()`, for example:

```
forumdb_test=> SELECT pg_catalog.set_config('search_path', 'public,
"$user"', false);
   set_config
-----------------
 public, "$user"
(1 row)

forumdb_test=> SELECT * FROM tags;
 pk |   tag   | parent
----+---------+--------
 44 | pgaudit |
...
```

As you can see, now, the connection works just fine.

It is also possible to perform a backup (and a restore) in the very same database. First of all, `pg_dump` must include a special option called `--create`, which instructs the application to issue `CREATE DATABASE` as the very first instruction for the restoration:

```
$ pg_dump -U postgres --column-inserts --create -f backup_forumdb.sql
forumdb
$ less backup_forumdb.sql
...
CREATE DATABASE forumdb WITH TEMPLATE = template0 ENCODING = 'UTF8'
LC_COLLATE = 'C' LC_CTYPE = 'C';
ALTER DATABASE forumdb OWNER TO luca;
\connect forumdb
...
```

As you can see, the output of pg_dump now includes the creation of the database, as well as the special \c command to connect immediately to such a database. In other words, launching this file through psql will restore the full content in the right database when the latter does not exist.

In order to test this, let's destroy our beloved database and restore it by means of initially connecting to template1:

```
$ psql -U postgres template1
psql (12.2)
Type "help" for help.

template1=# DROP DATABASE forumdb;
DROP DATABASE
template1=# \i backup_forumdb.sql
...
forumdb=#
```

Note how the Command Prompt has changed to reflect the fact that we are now connected to the restored forumdb database.

So, which version of dump and restoration should you use? If you are going to replicate the database in another cluster, for example, to migrate a staging database to production, you should include the --create option to let the database engine create the database for you. If you are migrating the database content to an existing database, then the --create option must not be present at all because there is no need to set up a database; this can be risky because you could restore objects to the wrong database, so you need to carefully check that you are connected into the right database before reloading the backup script.
If you are going to migrate the content of the database to another engine, such as another relational database, you should use options such as --inserts or --column-inserts to make the database backup more portable.

Limiting the amount of data to back up

pg_dump allows an extensive set of filters and flags to be used to limit the amount of data to back up. For example, you could decide to dump only the database schema without any data in it, and this can be achieved by means of the -s flag. On the other hand, you could already have the database schema in place, and you may only need the database content without any DDL statement. This can be achieved with the -a option. You can, of course, combine different pg_dump commands to get separate backups:

```
$ pg_dump -U postgres -s -f database_structure.sql forumdb
$ pg_dump -U postgres -a -f database_content.sql forumdb
```

You will end up with a file called `database_structure.sql` that contains all the different `CREATE TABLE` statements and another that contains only the `COPY` (or `INSERT` statements if you specified `--inserts`) statements.

You can also decide to limit your backup scope, either by schema or data, to a few tables by means of the `-t` command-line flag or, on the other hand, to exclude some tables by means of the `-T` parameter. For example, if we want to back up only the `users` table and `users_pk_seq` sequence, we can do the following:

```
$ pg_dump -U postgres  -f users.sql -t users -t user_pk_seq forumdb
```

The created `users.sql` file will contain only enough data to recreate the user-related stuff and nothing more. On the other hand, if we want to exclude the `users` table from the backup, we can do something similar to the following:

```
$  pg_dump -U postgres  -f users.sql -T users -T user_pk_seq forumdb
```

Of course, you can mix and match any option in a way that makes sense to you and, more importantly, allows you to restore exactly what you need. As an example, if you want to get all the data contained in the `posts` table and the table structure itself, you can do the following:

```
$  pg_dump -U postgres  -f posts.sql -t posts -a  forumdb
pg_dump: warning: there are circular foreign-key constraints on this table:
pg_dump:   posts
pg_dump: You might not be able to restore the dump without using --disable-
triggers or temporarily dropping the constraints.
pg_dump: Consider using a full dump instead of a --data-only dump to avoid
this problem.
```

`pg_dump` is smart enough to see that the table posts have different dependencies and foreign keys, so it warns you about the fact that your dump won't be able to restore all the content of the `posts` table. It is up to you to manage such dependencies in a correct way since you asked `pg_dump` to not perform a full backup (which, on the other hand, is always complete and consistent).

Dump formats and pg_restore

In the previous sections, you have only seen the plain SQL format for backups and restores, but `pg_dump` allows for more complex and smart formats. All formats except plain SQL must be used with `pg_restore` for restoration, and therefore are not suitable for manual editing.

Backup formats are specified by the -F command-line argument to pg_dump, which allows for one of the following values:

- c (custom) is the PostgreSQL-specific format within a single file archive.
- d (directory) is a PostgreSQL-specific format that's compressed where every object is split across different files within a directory.
- t (tar) is a .tar uncompressed format that, once extracted, results in the same layout as the one provided by the directory format.

Let's start with the first format: the custom single-file format. The command to back up a database resembles the one used for the plain SQL format, where you have to specify the output file, but this time, the file is not a plain text one:

```
$  pg_dump -U postgres -Fc --create -f backup_forumdb.backup forumdb
$ ls -lh backup_forumdb*
-rw-r--r--  1 luca  luca    24K May  2 17:20 backup_forumdb.backup
-rw-r--r--  1 luca  luca   168K May  2 17:04 backup_forumdb.sql
$ file backup_forumdb.backup
backup_forumdb.backup: PostgreSQL custom database dump - v1.14-0
```

The produced output file is smaller in size than the plain SQL one and can't be edited as text because it is binary. Many of the pg_dump command-line arguments apply the same to the custom formats, while others do not make sense at all. In any case, pg_dump is smart enough to know what to take into account and what to discard, so the following command lines will produce the same backup shown in the preceding example:

```
$ pg_dump -U postgres -Fc --create --inserts -f backup_forumdb.backup
forumdb
$ pg_dump -U postgres -Fc --create --column-inserts -f
backup_forumdb.backup forumdb
```

Once you have the custom backup, how can you restore the database content? Remember that custom backup formats require pg_restore to be used for a successful restoration. As we did previously, let's destroy our database again and restore it by means of pg_restore:

```
$ psql -U postgres -c 'DROP DATABASE forumdb;' template1
$ pg_restore -U postgres -C -d template1 backup_forumdb.backup
```

pg_restore runs silently and restores the specified database. The -C option indicates that pg_restore will recreate the database before restoring inside it. The -d option tells the program to connects to the template1 database first, issue a CREATE DATABASE, and then connect to the newly created database to continue the restore, similar to what the plain backup format did. Clearly, pg_restore requires a mandatory file to operate on; that is, the last argument specified on the command line.

It is interesting to note that pg_restore can produce a list of SQL statements that are going to be executed without actually executing them. The -f command-line option does this and allows you to store plain SQL in a file or inspect it before proceeding any further with the restoration:

```
$ pg_restore backup_forumdb.backup -f restore.sql
$ less restore.sql
--
-- PostgreSQL database dump
--

CREATE EXTENSION IF NOT EXISTS pgaudit WITH SCHEMA public;
...
```

As you can see, the content of the restore.sql file is plain SQL, similar to the output of a plain dump by means of pg_dump.

Another output format for pg_dump is the directory one, specified by means of the -Fd command-line flag. In this format, pg_dump creates a set of compressed files in a directory on disk; in this case, the -f command-line argument specifies the name of a directory instead of a single file. As an example, let's do a backup in a backup folder:

```
$ pg_dump -U postgres -Fd -f backup forumdb
$ ls -lh backup
total 56
-rw-r--r-- 1 luca  luca   202B May  2 17:42 3342.dat.gz
-rw-r--r-- 1 luca  luca    25B May  2 17:42 3344.dat.gz
-rw-r--r-- 1 luca  luca    25B May  2 17:42 3345.dat.gz
-rw-r--r-- 1 luca  luca    25B May  2 17:42 3346.dat.gz
-rw-r--r-- 1 luca  luca    25B May  2 17:42 3347.dat.gz
-rw-r--r-- 1 luca  luca    25B May  2 17:42 3348.dat.gz
-rw-r--r-- 1 luca  luca    39B May  2 17:42 3350.dat.gz
-rw-r--r-- 1 luca  luca   9.1K May  2 17:42 3352.dat.gz
-rw-r--r-- 1 luca  luca    14K May  2 17:42 toc.dat
```

The directory is created, if needed, and every database object is placed in a single compressed file. The `toc.dat` file represents a *Table Of Contents*, an index that tells `pg_restore` where to find any piece of data inside the directory. The following example shows how to destroy and restore the database by means of a backup directory:

```
$ psql -U postgres -c "DROP DATABASE forumdb;" template1
DROP DATABASE
$ pg_restore -C -d template1 -U postgres backup/
$ psql -U luca forumdb
psql (12.2)
Type "help" for help.

forumdb=> \q
```

The directory backup format is useful when the database grows in size since it can become a problem to store a single huge file that could overtake the filesystem's limitations.

The very last `pg_dump` format is the .tar one, which can be obtained by means of the `-Ft` command-line flag. The result is the creation of a `tar(1)` uncompressed archive that contains the same directory structure that we created in the previous example, but where every file is not compressed:

```
$ pg_dump -U postgres -Ft -f backup_forumdb.tar forumdb
$ tar tvf backup_forumdb.tar
-rw-------  0 postgres postgres 14637 May  2 17:47 toc.dat
-rw-------  0 postgres postgres   249 May  2 17:47 3342.dat
-rw-------  0 postgres postgres     5 May  2 17:47 3344.dat
-rw-------  0 postgres postgres     5 May  2 17:47 3345.dat
-rw-------  0 postgres postgres     5 May  2 17:47 3346.dat
-rw-------  0 postgres postgres     5 May  2 17:47 3347.dat
-rw-------  0 postgres postgres     5 May  2 17:47 3348.dat
-rw-------  0 postgres postgres    19 May  2 17:47 3350.dat
-rw-------  0 postgres postgres 63185 May  2 17:47 3352.dat
```

Next, we will look at running a selective restore, which will help you choose which elements of a database you want to restore.

Performing a selective restore

When performing a plain SQL database dump, you are allowed to manually edit the result, since it is plain text, and selectively remove parts you don't want to restore. With custom formats and `pg_restore`, you can do the very same thing, but you need to perform a few steps to do so.

First of all, you can always inspect the content of a binary dump by means of `pg_restore` and its `--list` option, which allows you to get a printed out index (**Table of Contents** or **TOC** for short) about the content of the backup. You need to specify, after the `--list` option, either the single file or directory that contains the backup to get the TOC printed:

```
$ pg_restore --list backup/
;
; Archive created at 2020-05-02 17:42:46 CEST
;     dbname: forumdb
;     TOC Entries: 56
;     Compression: -1
;     Dump Version: 1.14-0
;     Format: DIRECTORY
...
3360; 0 0 ACL - SCHEMA public postgres
2; 3079 34878 EXTENSION - pgaudit
3361; 0 0 COMMENT - EXTENSION pgaudit
218; 1255 34883 FUNCTION public f_load_data() luca
...
```

Lines beginning with a semicolon are comments, and as you can see, the first few lines that are printed out are a banner that describes the content of the backup, the date the backup was realized, the format (in this example, "directory"), and how many entries (objects) are in the backup.

Every line that is not a comment represents a database object or a single action that the restore process will perform. As an example, take a look at the following line:

```
218; 1255 34883 FUNCTION public f_load_data() luca
```

This indicates that the `f_load_data()` function will be restored by `luca` within the `public` schema.

Similarly, the following line means that the `public.tags` table will be created and assigned to the user `luca`:

```
211; 1259 34915 TABLE public tags luca
```

The following line means that the same table will be filled with the data:

```
3350; 0 34915 TABLE DATA public tags luca
```

Thanks to this table of contents, you can take control of the restoration process. In fact, if you move or delete lines from the table of contents, you can instruct `pg_restore` to change its execution. As an example, first, let's store the table of contents in a text file:

```
$  pg_restore --list backup/ > my_toc.txt
```

Now, edit the `my_toc.txt` file with your favorite editor and comment out the mentioned part as follows, by placing a semicolon as the first character of the line or by removing the line that fills the `tags` table:

```
;3350; 0 34915 TABLE DATA public tags luca
```

Now, save the `my_toc.txt` file. With that, it is possible to restore the database by means of `pg_restore`, but you have to instruct the program to follow your own table of contents and not the full and unmodified one that ships with the backup itself. To this aim, `pg_restore` allows the `-L` flag to be specified with the table of contents to use:

```
$ pg_restore -C -d template1 -U postgres -L my_toc.txt   backup/
$ psql -U luca forumdb
psql (12.2)
Type "help" for help.

forumdb=> SELECT count(*) FROM tags;
 count
-------
     0
(1 row)
```

As you can see, the table has been created, but it is empty. This is the result of using an ad hoc table of contents.

It is also possible to rearrange lines to make some objects be restored before others, but this is much more complicated, particularly when cross-references and dependencies between objects exist. Anyway, this is an incredibly flexible way to selectively decide what to restore and, moreover, create a different table of contents to restore the same format backup in different working sets.

Dumping a whole cluster

pg_dumpall is the tool to use to dump a full cluster. Here, pg_dumpall loops over all the databases available in the cluster and performs a single pg_dump on each, then dumps the specific objects that are at a cluster level, such as roles.

pg_dumpall works similarly to pg_dump, so pretty much all the concepts and options you have seen in the previous sections apply to pg_dumpall too. If you don't specify any output format and file, pg_dumpall prints all the required SQL statements on the standard output. Assuming you want to store the whole database content in a single SQL file, the following command line provides a full backup:

```
$ pg_dumpall -U postgres -f cluster.sql
```

The file can become large quickly, and this time, it begins by creating all the required roles:

```
$  less cluster.sql
...
CREATE ROLE auditor;
ALTER ROLE auditor WITH NOSUPERUSER INHERIT NOCREATEROLE NOCREATEDB NOLOGIN
NOREPLICATION NOBYPASSRLS;
CREATE ROLE book_authors;
ALTER ROLE book_authors WITH NOSUPERUSER INHERIT NOCREATEROLE NOCREATEDB
NOLOGIN NOREPLICATION NOBYPASSRLS;
CREATE ROLE enrico;
...
```

It then continues by restoring every single database, including template1. Then, all the databases are populated by means of SQL statements produced by single pg_dump runs.

pg_dumpall only produces an SQL script, so you need to restore your cluster by means of psql or an interactive connection. All the main options you can use with pg_dump that have been presented in the previous sections apply to pg_dumpall too.

Parallel backups

It is possible to use parallelization to speed up backups and restores. The basic idea is to have multiple processes (and database connections), each assigned a smaller task to perform, so that performing all the tasks in parallel will provide you with better performance, or at least lower times.

It is important to note that, often, it is not doing the backup faster that's the problem – rather, it's being able to perform the restoration as fast as you can. So, while it is possible to perform both backups and restoration in parallel mode, you will find restoration to be the most important one.

pg_dump allows you to specify the parallelism level via the -j command-line argument, to which you must assign a positive integer; that is, the number of parallel processes to start. pg_dump will then open parallel connections to the database in number equal to the parallelism, plus one connection to rule them all, and will force every connection to dump a separate table. This clearly means it does not make any sense to start more processes than the number of tables in your database that you need to back up.
Since all the processes will dump a single table, parallel mode is only available for the directory (-Fd) format where every table is stored in a separate file so that processes don't mix their writes together.

As an example, the following instruction will dump the database with three parallel jobs, thus opening four database connections:

```
$ pg_dump -U postgres -Fd -f backup_forumdb -v -j 3 forumdb
...
finished item 3350 TABLE DATA tags
dumping contents of table "public.categories"
pg_dump: finished item 3344 TABLE DATA foo
pg_dump: finished item 3342 TABLE DATA categories
```

The messages such as "finished items" are the single dumping processes that are completed as a single table, and will not be shown in the non-parallel verbose output of the pg_dump command. It is important to consider the number of connections opened by a parallel pg_dump: they are always done on every parallel job, plus one, to synchronize and manage the whole backup procedure. This means that in order to execute a parallel backup, you must ensure there are enough connections available against your database; otherwise, the backup will fail.

Another important aspect of parallel backups is that they could fail under concurrent circumstances. In fact, once pg_dump has started, the "master" process acquires light locks (shared locks) on every object the parallel processes are going to dump, while when started, every parallel process acquires an exclusive (heavy) lock on the object. This prevents the object (a table) from being destroyed before the parallel process has finished doing its work. However, between the acquisition of the first lock from the master process and the acquisition of the heavy lock from its spawned parallel process, another concurrent connection could try to acquire the lock on the table, resulting in a possible deadlock situation. To prevent this, the master pg_dump process will detect the dependency and abort the whole backup.

pg_restore does support parallel restoration too, by means of the same mnemonic -j command-line argument. The command will spawn the indicated number of processes involved in data loading, index creation, and all the other heavy and time-consuming operations.

Unlike pg_dump, pg_restore can work in parallel for both the directory format and the custom format. It is not simple to determine the number of parallel jobs to specify to pg_restore, but usually, this is the number of CPU cores, even if values slightly greater than that can produce a faster restoration.

As an example, the following command allows for parallel restoration of the backup we took previously (the first line drops the database for the restoration to succeed):

```
$ psql -U postgres -c "DROP DATABASE forumdb;" template1
$ pg_restore -C -d template1 -U postgres -j 2 -v backup_forumdb/
...
pg_restore: launching item 3370 ACL TABLE users
pg_restore: creating ACL "public.TABLE users"
pg_restore: finished item 3369 ACL TABLE tags
...
pg_restore: finished main parallel loop
```

Thanks to the verbose flag, it is clear how pg_restore has executed a parallel restoration of the data in the database. Messages such as "launching item" and "finished item" indicate when and on what object a parallel worker has been involved.

Backup automation

By combining pg_dump and pg_dumpall, it is quite easy to create automated backups, for example, to run every night or every day when the database system is not heavily used. Depending on the operating system you are using, it is possible to schedule such backups and have them be executed and rotated automatically.

If you're using Unix, for example, it is possible to schedule pg_dump via cron(1), as follows:

```
$ crontab -e
```

After doing this, you would add the following line:

```
30 23 * * * pg_dump -Fc -f /backup/forumdb,backup  -U postgres forumdb
```

This takes a full backup in custom format every day at 23:30. However, the preceding approach has a few drawbacks, such as managing already existing backups, dealing with newly added databases that require another line to be added to the crontab, and so on.

Thanks to the flexibility of PostgreSQL and its catalog, it is simple enough to develop a wrapper script that can handle backing up all the databases with ease. As a starting point, the following script performs a full backup of every database except for `template0`:

```
#!/bin/sh

BACKUP_ROOT=/backup

for database in $( psql -U postgres -A -t -c "SELECT datname FROM
pg_database WHERE datname <> 'template0'" template1 )
do
    backup_dir=$BACKUP_ROOT/$database/$(date +'%Y-%m-%d')
    if [ -d $backup_dir ]; then
        echo "Skipping backup $database, already done today!"
        continue
    fi

    mkdir -p $backup_dir
    pg_dump -U postgres -Fd -f $backup_dir $database
    echo "Backup $database into $backup_dir done!"
done
```

The idea is quite simple: the system queries the PostgreSQL catalog, `pg_database`, for every database that the cluster is serving, and for every database, it searches for a dedicated directory named after the database that contains a directory named after the current date. If the directory exists, the backup has already been done, so there is nothing to do but continue to the next database. Otherwise, the backup can be performed. Therefore, the system will back up the `forumdb` database to the `/backup/forumd/2020-05-03` directory one day, `/backup/forumb/2020-05-04` the next day, and so on. Due to this, it is simple to add the preceding script to your crontab and forget about adding new lines for new databases, as well as removing lines that correspond to deleted databases:

```
30 23 * * * my_backup_script.sh
```

Of course, the preceding script does not represent a complex backup system, but rather a starting point if you need a quick and flexible solution to perform an automated logical backup with tools your PostgreSQL cluster and operating system are offering. As already stated, many operating systems have already taken backing up a PostgreSQL cluster into account and offer already crafted scripts to help you solve this problem. A very good example of this kind of script is the `502.pgsql` script, which is shipped with the FreeBSD package of PostgreSQL.

Exploring physical backups

A physical backup is a low-level backup that's taken during the normal operations of the database cluster. Here, low-level means that the backup is somehow performed "externally" inside the backup cluster; that is, at the filesystem level.

As you already know from `Chapter 10`, *Users, Roles, and Database Security*, the database cluster requires both the data files contained in `PGDATA/base` and the **write-ahead logs (WALs)** contained in `PGDATA/wal`, as well as a few other files, to make the cluster work properly. The main concept, however, is that the data files and the WALs can make the cluster self-healing and recover from a crash. Hence, a physical backup performs a copy of all the cluster files and then, when the restore is required, it simulates a database crash and makes the cluster self-heal with the WALs in place.

The reason why physical backups are important is that they allow us to effectively clone a cluster, starting from the files it is made of. This means that, on one hand, you cannot restore a physically backed up cluster on a different PostgreSQL version, and on the other hand, that you need essentially no interaction at all with the cluster during the backup phase. The last point is particularly important: the physical backup can be taken pretty much in every moment without impacting the database with a hue transaction, which occurs in logical backups, and without interfering with the ongoing database activities such as client connections and queries. Is it true that the storage system – in particular, the filesystem – will be stressed during this kind of backup, but to the cluster, the backup is almost transparent.

It is fair to say that the cluster must be informed that the backup is starting to allow it to clearly mark that a backup is in progress inside the WALs, but apart from this "simple" action, the backup is totally outside the scope of the database cluster.

Moreover, physical backups allow you to choose the best tool that fits the low-level file copy. You are free to use any filesystem-specific command, such as `cp(1)`, `rsync(1)`, `tar(1)`, and so on; you can do the backup via a network by using any file copying mechanism provided by your operating system, and you can even develop your own tool. There are also a lot of backup solutions for PostgreSQL, including the authors' favorite, pgBackRest, so you are free to tailor your backup strategy to the tools that best fit your environment and requirements.

In the following subsections, you will learn how to perform a physical backup by means of a tool shipped with PostgreSQL: `pg_basebackup`. This tool has been developed as the primary tool for cloning a cluster, for example, as the starting point of a replicated system (replication will be shown in later chapters).

Please consider that, in any case, what `pg_basebackup` does is perform a set of steps that can be performed manually by any system administrator, so the tool is a convenient and well-tested way of performing a physical backup.

Performing a manual physical backup

The `pg_basebackup` tool performs either a local or remote database cluster clone that can be used as a backup. In order to work properly, the cluster that must be cloned must be set up accordingly. Since `pg_basebackup` "asks" PostgreSQL to provide the WALs, it is important that the target cluster has at least two WAL Sender processes active (WAL Sender processes are responsible for serving WALs over a client connection).

Therefore, the first step to perform on the database you want to back up is to check that the `max_wal_senders` configuration parameter (in the `postgresql.conf` file) has a value of 2 or greater:

```
max_wal_senders = 2
```

Another important setting is to allow `pg_basebackup` to perform a connection to the cluster: the tool will connect not as an ordinary client but as a "replication" client, and therefore the `pg_hba.conf` file must allow a rule that allows an administrative user to connect to the "replication" special database. Something similar to the following should work for a local backup:

```
host      replication     postgres   127.0.0.1/32   trust
```

Here, the user `postgres` is allowed to connect from the very same host to the special replication database without providing any authentication credentials.

> The replication word that's used as the database that the user `postgres` is going to connect to is not a real database; rather, it is a special keyword that tells the PostgreSQL host-based access system to accept connections marked with a replication purpose, like `pg_basebackup` is. In other words, there is no effective "replication" database on any PostgreSQL cluster.

Once the preceding settings have been put in place and the cluster has been restarted (due to the `max_wal_senders` option), it is possible to take a new backup.

Let's assume we want to perform the physical backup to store the result – that is, the backup itself – in the /backup/pg_backup directory. The following command will perform the backup:

```
$ sudo -u postgres pg_basebackup -D /backup/pg_backup -l 'My Physical
Backup' -v -h localhost -p 5432 -U postgres
pg_basebackup: checkpoint completed
pg_basebackup: write-ahead log start point: D/4C000028 on timeline 1
pg_basebackup: starting background WAL receiver
pg_basebackup: created temporary replication slot "pg_basebackup_19254"
pg_basebackup: write-ahead log end point: D/4C000138
pg_basebackup: waiting for background process to finish streaming ...
pg_basebackup: syncing data to disk ...
pg_basebackup: base backup completed
```

The -D flag specifies the directory that you want the backup to be stored in, which in this example is /backup/pg_backup. The -l optional flag allows you to provide a textual label to your backup, which can be used to inspect the backup to get some extra information about it. The -v flag enables verbose mode, which produces rich output about what the command is performing at every step. The other arguments are typical PostgreSQL libpq client flags that specify how to connect to the database so that it can be cloned, in this case by means of the user postgres on localhost at port 5432.

If you inspect the directory where the backup has been stored, you will see that it is effectively a clone of the PGDATA directory of the server you took the backup from, including its configuration files.

pg_verifybackup

A new tool, named pg_verifybackup, can be used in PostgreSQL 13 to verify the integrity of the backup that's done via pg_basebackup. At a glance, it works as follows:

```
$ sudo -u postgres pg_basebackup -D /backup/pg_backup -l 'My Physical
Backup' -v -h localhost -p 5432 -U postgres
...
$ sudo -u postgres pg_verifybackup /backup/pg_backup
backup successfully verified
```

The tool performs four main steps:

1. It evaluates the backup manifest to check if it is readable and contains valid backup information.
2. It scans the backup content to search for missing or modified data files (some configuration files are skipped in this step because the user could have changed them).
3. It compares all the data file checksums with the manifest values to ensure the files have not been corrupted.
4. By exploiting another utility, `pg_waldump`, it verifies that the WAL records that are needed in order to restore the backup are in place and readable.

Thanks to `pg_verifybackup`, you can be sure that your backup has not been damaged by a filesystem problem, a disk failure, or something else, and therefore you can resume from such a backup.

Starting the cloned cluster

`pg_basebackup` does a complete clone of the target cluster, including the configuration files. This means that the configuration of the cluster has not been "adapted" to where the clone is, including the data directory and the listening options (for example, the TCP/IP port). Therefore, you must be careful when starting the cloned cluster since it could clash with the original one, especially if the backup is performed locally (on the same machine). Here, you have the option of editing the configuration before attempting to start the backup cluster, changing the main settings on the command line, or having the backup on a remote host.

If you want to start the cloned cluster, assuming it has been kept local, as in the previous section, you can, for example, start it over with the following command-line settings:

```
$ sudo -u postgres pg_ctl -D /backup/pg_backup -o '-p 5433' start
waiting for server to start....
2020-05-10 11:31:51.685 CEST [39379] LOG:  starting PostgreSQL 12.2 on
amd64-portbld-freebsd12.0, compiled by FreeBSD clang version 6.0.1
(tags/RELEASE_601/final 335540) (based on LLVM 6.0.1), 64-bit
2020-05-10 11:31:51.686 CEST [39379] LOG:  listening on IPv6 address "::",
port 5433
2020-05-10 11:31:51.686 CEST [39379] LOG:  listening on IPv4 address
"0.0.0.0", port 5433
2020-05-10 11:31:51.688 CEST [39379] LOG:  listening on Unix socket
"/tmp/.s.PGSQL.5433"
2020-05-10 11:31:51.705 CEST [39379] LOG:  redirecting log output to
```

```
logging collector process
2020-05-10 11:31:51.705 CEST [39379] HINT:  Future log output will appear
in directory "log".
 done
server started
```

Here, the server has been started on the cloned PGDATA directory and TCP/IP port 5433. If you inspect the database cluster logs, you will see that the database has restored from a "forced crash"; that is, the cloned cluster did self-healing on its first startup:

```
$ sudo cat /backup/pg_backup/log/postgresql-2020-05-10.log
...
2020-05-10 11:31:51.711 CEST [39821] LOG:  database system was interrupted;
last known up at 2020-05-10 11:19:09 CEST
2020-05-10 11:31:52.050 CEST [39821] LOG:  redo starts at D/4C000028
2020-05-10 11:31:52.050 CEST [39821] LOG:  consistent recovery state
reached at D/4C000138
2020-05-10 11:31:52.050 CEST [39821] LOG:  redo done at D/4C000138
2020-05-10 11:31:52.203 CEST [39379] LOG:  database system is ready to
accept connections
```

Restoring from a physical backup

If you need to restore from a backup, you need to overwrite the original PGDATA directory with the cloned copy produced by pg_basebackup. This is a very risky operation because you will be losing all the content of the PGDATA directory and replacing it with the backup copy, which means the risk of errors occurring is high.

For that reason, instead of performing an online restoration, we suggest that you start a cloned cluster somewhere else, as shown in the previous section, so that you can extract the data you need to recover and restore only that data on the target cluster. For instance, you can start the cloned server, extract the data you need to recover by means of pg_dump, and restore it on the target cluster.

Of course, there are situations when you need to recover the entirety of the cluster, and therefore you need to do PGDATA overwriting, but even in such cases, we suggest that you use more advanced tools such as pgBackRest that drive and assist you in both the backup and restore part.

Physical backup and restoration are very powerful mechanisms, but they require you to deeply understand what is going on under the hood. So, take the time to experiment with them carefully so that you're ready to apply them in production.

Summary

In this chapter, we learned that PostgreSQL provides advanced tools so that we can perform backups and restorations. Backups are important because, even in a battle-tested and high-quality product such as PostgreSQL, things can go wrong: often, the users may accidentally damage their data, but other times, the hardware or the software could fail miserably. Being able to restore data, partially or fully, is therefore very important and every database administrator should carefully plan backup strategies.

We also learned that PostgreSQL ships with tools for both logical and physical backups. Logical backups are taken by means of reading the data from the database itself, by means of ordinary SQL interactions; physical backups are taken by means of cloning the PGDATA directory either by using operating system tools or PostgreSQL ad hoc solutions. Restoration is performed by specific tools in the case of logical backups, and by the database self-healing mechanism in the case of physical backups.

Finally, it is important to stress the concept that a backup alone is not valid until it is successfully restored, so to ensure that you will be able to recover your cluster, you need to test your backups as well.

Now that you can back up and restore your clusters, in the next chapter, we will look at configuration and monitoring.

Further reading

- PostgreSQL `pg_dump` tool official documentation: `https://www.postgresql.org/docs/12/app-pgdump.html`
- PostgreSQL `pg_dumpall` tool official documentation: `https://www.postgresql.org/docs/12/app-pg-dumpall.html`
- PostgreSQL `pg_restore` tool official documentation: `https://www.postgresql.org/docs/12/app-pgrestore.html`
- FreeBSD `502.pgsql` backup script: `https://www.freshports.org/databases/postgresql83-server/files/502.pgsql`
- PostgreSQL `pg_basebackup` tool official documentation: `https://www.postgresql.org/docs/12/app-pgbasebackup.html`
- PostgreSQL `pg_verifybackup` tool official documentation: `https://www.postgresql.org/docs/13/app-pgverifybackup.html`
- pgBackRest external tool for physical backups: `https://pgbackrest.org/`

16
Configuration and Monitoring

One of the duties of a database administrator is to configure the cluster so that it behaves well for the current workload and context. The configuration is not static: most of the time, you will find yourself making changes to the configuration, so it is important that you feel comfortable with inspecting and changing the cluster's configuration.

Another important task, partially related to configuration, is monitoring the cluster in order to understand how the system is actually behaving, and whether there are bottlenecks and problems to be solved. Such problems can sometimes be solved by making changes to the configuration of the cluster, by using different hardware (for example, increasing the available memory), and sometimes by fixing the applications that could be causing the bottleneck.

This chapter will show you how to manage and inspect the cluster configuration, generate it from scratch, find errors and mistakes, and how to interactively monitor the cluster's activity via the rich statistics subsystem. Finally, you will discover a very powerful and common extension, named `pg_stat_statements`, that allows you to monitor the cluster's activity with great detail and flexibility.

This chapter will cover the following topics:

- Cluster configuration
- Monitoring the cluster
- Advanced statistics with `pg_stat_statements`

Let's get started!

Technical requirements

You need to know about the following to complete this chapter:

- How to interact with configuration files within the `PGDATA` directory
- How to connect to your cluster as a database administrator
- How to execute SQL statements against the system catalogs

The code for this chapter can be found in this book's GitHub repository: `https://github.com/PacktPublishing/Learn-PostgreSQL`.

Cluster configuration

PostgreSQL is configured by means of a bunch of text files that contain directives and values used to bootstrap the cluster and get it running. We saw how configuration files are handled at the beginning of this book and throughout, whenever we needed to perform particular configurations, such as to manage logging. This section will revisit and explain how to configure a cluster in more detail.

There are two main configuration files that present the *starting point* for any configuration:

- `postgresql.conf` is the main cluster configuration file and contains all the data required to start the cluster, set up processes (as WAL senders) and logging, and configure how the cluster will accept connections (for example, on which TCP/IP address).
- `pg_hba.conf` is the file that's used to allow or deny the client connections to the cluster. It was explained extensively in `Chapter 3`, *Managing Users and Connections*, and is related to the users and roles authentication mechanisms.

There are other configuration files under the `PGDATA` directory, but they will not be discussed here. Moreover, you are free to create your own configuration files and plug them into the cluster, but these two files are the main ones you will work with. This section will mainly be dedicated to `postgresql.conf`, the default configuration file.

The `postgresql.conf` file is a text file, usually annotated with useful comments, that contains a set of configuration parameters. Each parameter is expressed in the form of `key = value`, where the **key** is the configuration parameter name and the **value** is the configuration value for the parameter. We saw a few configuration parameters in the previous chapters. For example, the `max_wal_senders = 2` configuration parameter sets the `max_wal_senders` configuration parameter to the value of `2`.

Each configuration parameter must be on a single line, and all lines starting with a # sign are comments, which will not be taken into account by the cluster. Comments are useful since they allow you to add extra information about your intentions regarding a specific configuration. For example, let's take a look at the following code snippet:

```
# set to 2 to allow pg_basebackup to work properly
max_wal_senders = 2
```

The previous example provides a clear hint about why the parameter has been configured as such. You are not required to place comments in your configuration file, but it is a very good habit to document what you are doing and why you are doing it.

If you don't want to configure a specific parameter, you can either delete the line of that parameter, making it disappear totally from the configuration file, or place a comment sign in front of it, transforming the line into a pure comment. It is important to note that if a parameter is not configured in the file because it is either missing or commented out, then it will take its default value. Every parameter has a default value, and you must look through the documentation to understand each default value.

Each configuration parameter will accept only a specific set of values that depend on the type of the configuration parameter itself. Mainly, you can encounter numeric values, string values, and lists (separated by a comma); there are also values that can be expressed with a measurement unit, such as time in 2ms (2 milliseconds).

Inspecting all the configuration parameters

You can inspect all the configuration parameters from a live system by issuing a query against the pg_settings special catalog. This catalog contains every setting that the current version of PostgreSQL will accept, along with default values, current values, and much more.

As an example, with the following query, you can gather information about every configuration parameter, including their short and long descriptions, default values, and current values:

```
forumdb=> SELECT name, setting || ' ' || unit AS current_value, short_desc,
extra_desc, min_val, max_val, reset_val FROM pg_settings;
...
name           | authentication_timeout
current_value  | 360 s
short_desc     | Sets the maximum allowed time to complete client
authentication.
extra_desc     |
```

```
min_val      | 1
max_val      | 600
reset_val    | 60
```

In the preceding example, the `authentication_timeout` setting has been set to 360 seconds, and its value can be tuned in the range of 1 second to 600 seconds; `reset_val` is the default value that the parameter will assume if it is not configured at all, and it is set to 60 seconds.

The `pg_settings` special catalog also contains other useful information, including the file and the line number from where a parameter has been loaded. This information can be used to quickly find where a configuration parameter has been set in the `postgresql.conf` file or another configuration file. As an example, the following query will show where each parameter has been loaded from:

```
forumdb=# SELECT name, setting AS current_value, sourcefile, sourceline,
pending_restart FROM pg_settings;
...
name            | log_destination
current_value   | stderr
sourcefile      | /postgres/12/logging.conf
sourceline      | 1
pending_restart | f
```

As you can see, the `log_destination` configuration parameter has been loaded from the `/postgres/12/logging.conf` file. This is a custom-defined file, starting at line 1 of that file. In other words, if you need to tune the `log_destination` setting, you need to edit the `/postgres/12/logging.conf` file at line 1.

 You need to have database superuser rights in order to gather extra information, such as the file location and line number.

There is another important piece of information that you can get out of the `pg_settings` special catalog: what happens if a parameter is changed at runtime? Depending on the nature of the parameter, changes can be applied immediately, or they can be delayed while waiting for a special event or even a cluster restart. The `pending_restart` column indicates whether the current parameter has changed from its boot time value and whether the value has been applied to the cluster. In the previous example, `pending_restart` is false, so the configuration you are seeing is effectively what is running on the cluster right now.

Finding configuration errors

PostgreSQL provides a very useful catalog called `pg_file_settings` that provides us with a glance at all the configuration parameters and the file they have been loaded from, thus also providing information about errors. The following query extracts all the information from the catalog, and the trimmed output gives us some important information:

```
forumdb=# SELECT name, setting, sourcefile, sourceline, applied, error FROM
pg_file_settings   ORDER BY name;
      name       | setting |           sourcefile           | sourceline |
applied | error
-----------------+---------+--------------------------------+------------+---
------+-------
 log_destination | stderr  | /postgres/12/postgresql.conf   |        420 | f
 |
 log_destination | stderr  | /postgres/12/logging.conf      |          1 | f
 |
 log_destination | stderr  | /postgres/12/pgbadger.conf     |          1 | t
 |
 ...
```

As you can see, the `log_destination` configuration parameter has been loaded multiple times: exactly three, from different source files. The applied one is the configuration settings from the `/postgres/12/pgbadger.conf` file, at line 1, as reported by the status of the `applied` column. This is not an error; instead, it's a possible contention in the configuration. In the case that the parameter contains an error, the `error` column provides a hint about the problem.

As an example of an error, consider the previous query again and look carefully at what changes are in the output:

```
forumdb=# SELECT name, setting, sourcefile, sourceline, applied, error FROM
pg_file_settings   ORDER BY name;
      name       | setting  |           sourcefile           | sourceline |
applied |           error
-----------------+----------+--------------------------------+------------+--
-------+---------------------------
 log_destination | stderr   | /postgres/12/postgresql.conf   |        420 | f
 |
 log_destination | stderr   | /postgres/12/logging.conf      |          1 | f
 |
 log_destination | stderror | /postgres/12/pgbadger.conf     |          1 | f
 | setting could not be applied
 ...
```

First of all, none of the settings have been applied. The last line has an `error` column that says that the setting has not been applied due to an invalid value. In fact, the value has been mistakenly written as `stderror` instead of the valid choice `stderr`.

Therefore, whenever you suspect that a setting you have placed in your configuration files has not been applied, look at `pg_file_settings` for hints about possible errors.

Nesting configuration files

In the example shown in the previous section, you saw how the same configuration parameter was defined in three different files. This is possible because PostgreSQL provides three main directives:

- `include_file`: Includes a single file in the configuration
- `include_dir`: Includes all the files contained in the specified directory
- `include_if_exists`: Includes a file only if it exists

The last directive is very handy because if an included file does not exist, PostgreSQL will throw an error, while with `include_if_exists`, the cluster will not warn you if the file to include has not been created. This is useful for provisioning, for example, where you can set up a main configuration file that includes multiple files and ships those files only to those systems that really require such a configuration.

Adding custom files to the configuration can quickly lead to multiple definitions of the same parameter. PostgreSQL handles multiple definitions in a very simple way: the last definition that's found wins over the previous ones. This is also true for a single configuration file; for example, if you place the following three lines in `postgresql.conf`, only the last one will be applied:

```
work_mem = 4096MB;
work_mem = 2048MB;
work_mem = 512MB;
```

With this duplicated configuration, the system will run with `work_mem` set to 512 megabytes. The special catalog, `pg_file_settings`, can help us find multiple definitions of the same configuration parameter, as explained in the previous section.

Configuration contexts

Each configuration parameter belongs to a so-called **context**, a group that defines *when* a change to the parameter can be applied. There are parameters that can be changed during the cluster's life cycle. However, others cannot and require the cluster to be restarted; the context of a configuration parameter helps the system administrator understand when changes will take effect.

Configuration contexts can be extracted from the `pg_settings` catalog, as shown in the following example:

```
forumdb=> SELECT distinct context FROM pg_settings ORDER BY context;
      context
-------------------
 backend
 internal
 postmaster
 sighup
 superuser
 superuser-backend
 user
(7 rows)
```

As you can see, the allowed configuration contexts are as follows:

- `internal`: The `internal` context means that the configuration value depends on the PostgreSQL source code and is established at compile time, so it cannot be changed unless you decide to compile it from scratch. For example, the size of every memory page is defined in the source code.
- `postmaster`: The `postmaster` configuration context means that the postmaster process is responsible for getting changes. In other words, the whole cluster (and its main process, postmaster) must be started over.
- `sighup`: The `sighup` context makes the cluster aware of changes by signaling it with a hang-up signal, typically a reload of the operating system service.
- `superuser-backend` and `backend`: The `backend` and `superuser-backend` contexts allow changes to be applied to client and administrator connections, respectively. Such changes will be perceivable from the very next connection of either type.
- `user`: The `user` and `superuser` contexts apply changes to the current client connection immediately, either by normal means or established by a database administrator, respectively.

Main configuration settings

PostgreSQL includes a lot of configuration options, and describing all of them here would require an entire book. Moreover, configuration depends on many different factors, including the cluster workload and the connection concurrency. Many parameters can imply different behaviors for other parameters. Therefore, it is not possible to provide a simple and effective step-by-step guide to configuration, but it is possible to provide some suggestions to help you start tuning your cluster.

In the following subsections, you will learn about the main configuration parameters, depending on the main category they belong to. Take your time to clearly understand what every setting does before applying a change, and keep in mind that the configuration contexts could prevent you from seeing immediate results.

WAL settings

WALs are fundamental for the cluster to work properly and to be able to recover from crashes. Therefore, settings related to WALs are vital for the cluster's life cycle.

The main settings are as follows:

- `fsync` tells the cluster to issue an operating system call of `fsync(2)` every time a COMMIT is performed; that is, every time something must be stored in the WAL segments.
- `wal_sync_method` tells PostgreSQL which effective `fsync(2)` system call to use.
- `synchronous_commit` tells PostgreSQL whether every COMMIT must be followed by an immediate and synchronous `fsync(2)` or whether the flush can be delayed a bit, depending on the value of `wal_writer_delay`.
- `checkpoint_timeout`, `checkpoint_completion_target`, and `max_wal_size` control checkpointing, as discussed in Chapter 11, *Transactions, MVCC, WALs, and Checkpoints*, when we explained transactions and WALs.

The `fsync` settings must be kept set to `on` because disabling this will make the cluster subject to data loss due to an incomplete communication and flushing data toward the filesystem. Never, ever, set this parameter to off.

`wal_sync_method` allows the administrator to configure a specific operating system call to sync dirty buffers. All the POSIX operating systems implement `fsync(2)`, but some of them provide special flavors that behave faster or better under heavy loads. It is possible to specify the exact name of the system call to use via `wal_sync_method`.

But how can you discover the best (or just the available) fsync(2) implementation that fits your operating system? You can launch the pg_test_fsync program on your machine to get a good guess about the possible methods you can use, as well as the best one. As an example, on a FreeBSD machine, the program provides the following output:

```
$ pg_test_fsync
5 seconds per test
O_DIRECT supported on this platform for open_datasync and open_sync.

Compare file sync methods using one 8kB write:
(in wal_sync_method preference order, except fdatasync is Linux's default)
        open_datasync                                   n/a
        fdatasync                          6845.727 ops/sec     146
usecs/op
        fsync                              3685.769 ops/sec     271
usecs/op
        fsync_writethrough                              n/a
        open_sync                          2521.228 ops/sec     397
usecs/op
. . .
```

You should compare the available options and choose the fastest one. So, in the preceding example, wal_sync_method = open_datasysnc is the best choice.

synchronous_commit is a boolean setting that's set to on by default set, meaning that every time a COMMIT is issued, an immediate fsync(2) will be performed. Setting this to off will not cause data consistency loss, instead improving performance since PostgreSQL will schedule an fsync with a small delay, measured by one to three times the value of wal_writer_delay. It is useful to turn this setting off when you are bulk loading data (for example, restoring a database from a backup), but for normal activities, you should keep it on.

Other important settings were explained in Chapter 10, *Users, Roles, and Database Security*, when we discussed transaction management, so please go back to that chapter for more details.

Memory-related settings

PostgreSQL exploits the volatile RAM memory of the system to cache the data coming from the permanent storage and to manage data that is going to be stored later on.

The main settings related to memory management are as follows:

- `shared_buffers` is the amount of memory PostgreSQL will use to cache data in memory.
- `work_mem` is the amount of memory PostgreSQL will provide, on-demand, to perform particular activities on data.
- `maintanance_work_mem` is the amount of memory PostgreSQL reserves for its internal operations.
- `wal_buffers` is the cache used for WAL segments.

`shared_buffers` is probably the most important setting here since it determines the total amount of memory PostgreSQL will use. This memory will be made exclusively available to PostgreSQL and its spawn processes; the memory will not be available to other services running on the same machine. Usually, you should start with a value that is between 25% and 45% of the total RAM your system has. Values that are too low will make PostgreSQL change data from the permanent storage on and off, while values that are too high will make PostgreSQL compete with the operating system's filesystem cache.

`work_mem` is the amount of memory that every connection can use to perform a particular data rearrangement, such as what's done in a SORT or a Hash Join. If more memory is required, PostgreSQL will exploit the permanent storage to swap data on and off; for example, converting a sort into a merge sort.

`maintanance_work_mem` establishes the amount of memory, per session, related to particularly intensive commands such as VACUUM and CREATE INDEX. Since only one of those commands can be active at any moment in a connection, you can raise the value depending on how many administrative connections you are supposed to serve.

`wal_buffers` is probably the easiest setting you can tune with regard to memory: it indicates how much memory to use for caching WAL segments. Since WAL segments are usually written in chunks of 16 megabytes, this is exactly the optimal value for such a setting.

Process information settings

PostgreSQL is a multi-process system, and it spawns a process for serving every incoming connection. There are a couple of settings that can help with monitoring, from the operating system's point of view, every PostgreSQL-related process:

- `update_process_title` makes every process report what it is doing; for example, what query it is executing when asked by operating system tools such as `ps(1)` and `top(1)`.
- `cluster_name` is a mnemonic name used to recognize the cluster that every process belongs to in the case that multiple clusters are running on the same machine.

It is worth noting that these settings could make the system work slower on certain operating systems, such as FreeBSD.

> Log-related settings were explained in detail in Chapter 14, *Logging and Auditing*, so they will not be discussed again here.

Networking-related settings

Usually, PostgreSQL listens on a TCP/IP address for incoming connections, which is specified by a bunch of network-related settings. The main settings for this are as follows:

- `listen_address` specifies the TCP/IP address to listen on.
- `port` specifies the TCP/IP port the postmaster will wait for incoming connections on.
- `max_connections` and `superuser_reserved_connections` specify the allowed incoming connections.
- `authentication_timeout` and `ssl` indicate the authentication timeout and encrypted mode.

`listen_address` can include multiple addresses, separated by a comma, in the case the server is multi-homed. It can even be specified by the special value `*` to indicate the server should listen on every available address. `port` specifies the TCP/IP port number, which is `5432` by default.

`max_connections` is the max allowance for incoming connections: no more connections will be allowed on the cluster if this threshold is reached. Since part of `max_connection` is made by `superuser_reserved_connections`, this is the number of connections by a system administrator that have been authorized.

`authentication_timeout` is the time before an authentication trial will expire, while `ssl` enables the server to handle SSL handshakes on connections (SSL will not be explained here).

Archive and replication settings

There are different archiving and replication settings that deal with how the cluster archives its WALs and communicates with other clusters as either a master or a slave. All the settings will be detailed in `Chapter 17`, *Physical Replication*, and `Chapter 18`, *Logical Replication*, and they are listed here at a glance:

- `wal_level` indicates how the information in the WALs will be used. This can be `minimal` (for a standalone system), `replica` (for a replicated system), or `logical` (for a logical replication).
- `archive_mode`, `archive_command`, and `archive_timeout` manage the archiving mode – that is, storing WALs to other locations for point-in-time recovery or replication.
- `primary_conninfo` and `primary_slot_name` is used if the secondary instrument in the cluster has to be replicated from a master to which it must connect, depending on the values of these parameters.
- `hot_standy`, when used on a replicating system, allows for read-only queries.
- `max_logical_replication_workers` and `max_wal_senders` are used to define how many process will manage replication.

These settings and other replication-related settings will be discussed in the chapters dedicated to physical and logical replication.

Vacuum andautovacuum-related settings

There are different settings that can be used to define and tune the vacuum and autovacuum settings. These were discussed in Chapter 10, *Users, Roles, and Database Security*.

Optimizer settings

The PostgreSQL optimizer is driven by a cost-based approach. It is possible to tune these costs, as discussed in Chapter 11, *Transactions, MVCC, WALs, and Checkpoints*, in the *Indexes and performances* section.

Statistics collector

PostgreSQL exploits the **statistics collector** to gather facts about what happened in the cluster, as you will learn later in this chapter in the *Monitoring the cluster* section.

Since collecting those numbers has a little runtime impact, it is possible to exclude the collection entirely or filter the statistics collector to gather only the facts you are truly interested in. The main settings for this are as follows:

- `track_activities` enables other processes to monitor the current command or query currently being executed.
- `track_counts` gathers counting information about tables and index usage.
- `track_functions` gather statistics about the use of functions and stored procedures.
- `track_io_timing` allows us to count the time spent in different input/output operations.
- `stat_temp_directory` is the (relative) directory name to use as temporary storage for statistics collection.

Modifying the configuration from a live system

It is possible to modify the cluster configuration from within a database connection by means of the `ALTER SYSTEM` command.

`ALTER SYSTEM` provides us with a SQL way to set a parameter value, and the parameter will be appended to a special file name, `postgresql.auto.conf`, which lives within the `PGDATA` directory. The `postgresql.auto.conf` file is loaded automatically at server boot or when a **reload** signal (HUP) is issued. Therefore, parameters contained in `postgresql.auto.conf` will take priority over those in `postgresql.conf` and the end result will be that the changes will be applied as if you have manually edited the `postgresql.conf` file.

`ALTER SYSTEM` can only be executed from a database administrator. For example, let's say you issue the following command:

```
forumdb=# ALTER SYSTEM SET archive_mode = 'on';
ALTER SYSTEM
```

The end result will be to have a `postgresql.auto.conf` file that looks as follows:

```
$ sudo cat /postgres/12/postgresql.auto.conf
# Do not edit this file manually!
# It will be overwritten by the ALTER SYSTEM command.
archive_mode = 'on'
```

As you can see, the changed parameter was placed in the file as you manually edited it. The file contains a warning banner about the fact that you should not edit it manually because the system will not take your changes into account and will overwrite its content.

It is also possible to specify `DEFAULT` as the value for an option, so that option will be removed from the `postgresql.conf.auto` file. It is also possible to use `RESET` to reset a setting to its default value, or to use `RESET ALL` to remove all the settings from `postgresql.auto.conf`.

Therefore, the following two inputs are equivalent and result in removing the changed settings from the `postgresql.auto.conf` file:

```
forumdb=# ALTER SYSTEM SET archive_mode TO DEFAULT;
ALTER SYSTEM
forumdb=# ALTER SYSTEM RESET archive_mode;
ALTER SYSTEM
```

The following input will remove *every* changed setting in `postgresql.auto.conf`:

```
forumdb=# ALTER SYSTEM RESET ALL;
ALTER SYSTEM
```

Configuration generators

Instead of starting from the annotated `postgresql.conf` file and tuning it by yourself, you can exploit an automated tuning system to get a starting point for the configuration.

A good configuration system is **PGConfig**, an online system where you can specify the main settings of the host serving your cluster, such as memory, hard disk type, concurrency, and so on. With those few details, as shown in the following screenshot, the system can produce different configurations, depending on the workload you are going to use the cluster for:

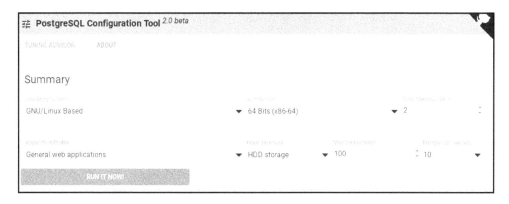

The following screenshot shows multiple configurations that you can use:

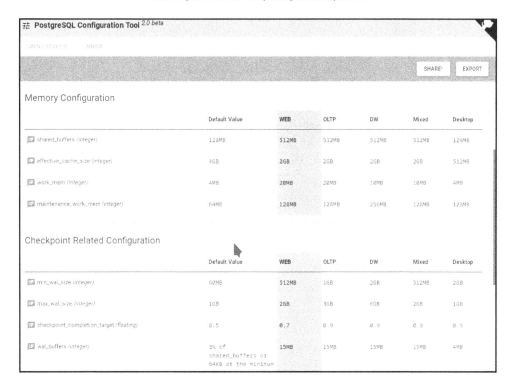

Once you have selected the configuration that best fits your workload, you can export such a configuration as a `postgresql.conf` file or as a set of `ALTER SYSTEM` statements to be executed as a SQL interactive script so that you can apply the configuration to your cluster:

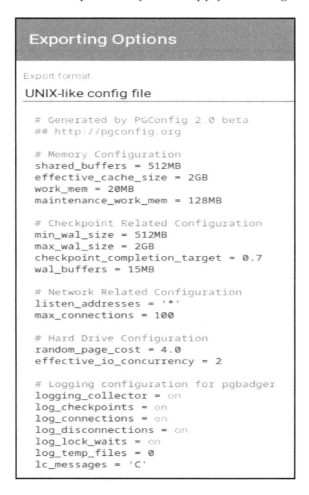

As you can see, the result is a bunch of configuration parameters that you can copy and paste into a "blank" configuration file. The idea is to start from this configuration and continue tuning on top of it.

The following screenshot shows the very same configuration by means of `ALTER SYSTEM` statements; that is, you can apply the configuration as a SQL script, depending on your needs:

```
Exporting Options

Export format
ALTER SYSTEM command

-- Generated by PGConfig 2.0 beta
---- http://pgconfig.org

-- Memory Configuration
ALTER SYSTEM SET shared_buffers TO '512MB';
ALTER SYSTEM SET effective_cache_size TO '2GB';
ALTER SYSTEM SET work_mem TO '20MB';
ALTER SYSTEM SET maintenance_work_mem TO '128MB';

-- Checkpoint Related Configuration
ALTER SYSTEM SET min_wal_size TO '512MB';
ALTER SYSTEM SET max_wal_size TO '2GB';
ALTER SYSTEM SET checkpoint_completion_target TO '0.7';
ALTER SYSTEM SET wal_buffers TO '15MB';

-- Network Related Configuration
ALTER SYSTEM SET listen_addresses TO '*';
ALTER SYSTEM SET max_connections TO '100';

-- Hard Drive Configuration
ALTER SYSTEM SET random_page_cost TO '4.0';
ALTER SYSTEM SET effective_io_concurrency TO '2';

-- Logging configuration for pgbadger
ALTER SYSTEM SET logging_collector TO 'on';
ALTER SYSTEM SET log_checkpoints TO 'on';
ALTER SYSTEM SET log_connections TO 'on';
ALTER SYSTEM SET log_disconnections TO 'on';
ALTER SYSTEM SET log_lock_waits TO 'on';
ALTER SYSTEM SET log_temp_files TO '0';
ALTER SYSTEM SET lc_messages TO 'C';
```

PGConfig is just one option you can use to get a customized configuration that you can start working on. Of course, there is no need to use it since PostgreSQL comes with a default configuration, and this configuration generator does not represent a "silver bullet" to provide you the optimal configuration for your cluster. In either case, you will need to tune and fix your parameters to optimize the cluster, depending on your needs, workload, and hardware.

In the next section, you are going to discover how to monitor your cluster, as well as how to discover bottlenecks and problems that can be fixed by tuning your queries or cluster configuration.

Monitoring the cluster

Monitoring the cluster allows you to understand what the cluster is doing at any given point in time and potentially act and react accordingly to avoid degradation in the performance and usability of databases. PostgreSQL provides a rich set of catalogs that allow a database administrator to monitor the overall activity by issuing only SQL statements and queries. You can also combine the results of the information coming from the catalog with other external monitoring tools, ranging from your operating system's tools to more complex ones such as Nagios.

In this section, we will have a look at the main PostgreSQL catalogs used to monitor and collect information about database activities. As you can imagine, only a database administrator can get complete information about overall cluster activities.

The cluster collects information about activities by means of the **statistic collector**, a dedicated process that is responsible for collecting, and therefore providing, information in a cluster-wide way. Statistics are not in real-time, even if you feel they are. This is because statistics are updated no more frequently than every 500 milliseconds by backend processes, assuming they are idle. Moreover, statistics within a transaction block are "frozen", meaning you cannot observe changes in the statistics unless your transaction has finished.

Statistics are kept across clean shutdowns and restarts of the cluster, but in the case of recovering from a crash, all the statistics are deleted and collection starts from scratch. There is also the possibility to manually reset the statistics for a specific database by invoking the `pg_stat_reset()` function as a database superuser.

Information about running queries

The `pg_stat_activity` catalog provides one tuple for every backend process active in the cluster, and therefore for every client connected. The following simple queries provide a detailed output:

```
forumdb=# SELECT usename, datname, client_addr, application_name,
          backend_start, query_start,
          state, backend_xid, query
   FROM pg_stat_activity;
...
-[ RECORD 4 ]----+-------------------------------------------------------
--------------------
usename           | luca
datname           | forumdb
client_addr       | 192.168.222.1
```

```
application_name | psql
backend_start    | 2020-05-13 16:42:50.9931+02
query_start      | 2020-05-13 16:44:20.601118+02
state            | idle
backend_xid      |
query            | INSERT INTO tags( tag ) SELECT 'A Fake Tag' FROM
generate_series( 1, 10000 );
```

As you can see, the user `luca` (ni the `username` field) was connected via `psql` (the `application_name` field) from a remote host (the `client_addr` field) and executed the `INSERT INTO` query called `tags` over the `forumdb` database. It is interesting to note the `state` field, which reports the status of the running query. In the preceding example, it says *idle*, meaning that the query is waiting for something else to happen, and may even be finished.

It is important to note that `pg_stat_activity` only reports the very last executed query from a session or connection. Remember that the catalog shows a tuple for every connected client and that the statistics are not updated until a new statement is executed.

Inspecting locks

The `pg_locks` special catalog provides a clear and detailed view of any locks that are acquired by different transactions and statements. The idea is that by inspecting this catalog, the system administrator can get a glance at possible bottlenecks and competition among transactions. It is useful to query this catalog by joining it with `pg_stat_activity` in order to get more detailed information about what is going on. The following is an example of a query and a partial result:

```
forumdb=# SELECT a.usename, a.application_name, a.datname, a.query,
         l.granted, l.mode
    FROM pg_locks l
    JOIN pg_stat_activity a ON a.pid = l.pid;
...
-[ RECORD 5 ]----+-----------------------------------------------------------
-
usename          | luca
application_name | psql
datname          | forumdb
query            | delete from tags;
granted          | t
mode             | RowExclusiveLock
...
```

```
-[ RECORD 9 ]----+------------------------------------------------------
-
usename          | luca
application_name | psql
datname          | forumdb
query            | insert into tags( tag ) values( 'FreeBSD' );
granted          | t
mode             | ExclusiveLock
```

There are two connections for the user `luca` to the `forumdb` database, and one connection has acquired a lock to delete tuples while the other is inserting tuples into the `tags` table. The `granted` column expresses whether the lock is acquired, so selecting only the non-granted locks is a good starting point to get advice on blocked queries. The `mode` column indicates what kind of lock the query is trying to acquire.

With these suggestions, and thanks again to an accurate join with `pg_stat_activity`, you can find blocked queries, as shown in the following example (this is a continuation of the same scenario depicted previously):

```
orumdb=# SELECT query, backend_start, xact_start, query_start,
         state_change, state,
         now()::time - state_change::time AS locked_since,
         pid, wait_event_type, wait_event
  FROM pg_stat_activity
  WHERE wait_event_type IS NOT NULL
  ORDER BY locked_since DESC;
...
-[ RECORD 6 ]---+---------------------------------------------
query           | insert into tags( tag ) values( 'FreeBSD' );
backend_start   | 2020-05-14 08:26:57.762887+02
xact_start      | 2020-05-14 08:27:00.017983+02
query_start     | 2020-05-14 08:27:14.745784+02
state_change    | 2020-05-14 08:27:14.775535+02
state           | idle in transaction
locked_since    | 00:07:33.411832
pid             | 60239
wait_event_type | Client
wait_event      | ClientRead
```

As you can see, the query has been waiting for 7 minutes and 33 seconds (the `locked_since` column), but the query is `idle in transaction` (the `state` column) and is waiting for input from a client (the `wait_event` and `wait_event_type` columns). In other words, the query is waiting for the user to complete (either `COMMIT` or `ROLLBACK`) the transaction.

Taking advantage of `pg_locks` can help you follow the evolution of transactions and their contention, as well as decide on how to terminate queries that are blocking other workloads.

Inspecting databases

You can get detailed information about the status of your databases by querying the `pg_stat_database` special catalog. This catalog provides information about commit and rolled back transactions, deadlocks, and conflicts. Please consider that deadlocks and rollbacks are a natural event in a database, but if you see the numbers grow quickly, this could mean there's been an application error or that there are clients who are trying to do things incorrectly in a database and thus are forced to roll back.

As an example, by using the following query, you can get details about your databases:

```
forumdb=# SELECT datname, xact_commit, xact_rollback, blks_read, conflicts,
deadlocks,
        tup_fetched, tup_inserted, tup_updated, tup_deleted, stats_reset
        FROM pg_stat_database;
...
-[ RECORD 6 ]-+------------------------------
datname       | forumdb_test
xact_commit   | 802
xact_rollback | 9
blks_read     | 1800
conflicts     | 0
deadlocks     | 0
tup_fetched   | 32977
tup_inserted  | 1391
tup_updated   | 46
tup_deleted   | 0
stats_reset   | 2020-05-02 17:37:20.226145+02
```

As you can see, the `forumdb` database doesn't have any conflicts or deadlocks, and the number of committed transactions (the `xact_commit` column) is much higher than the number of aborted transactions (the `xact_rollback` column). Therefore, we can assume that the database is fine and that the applications are issuing good queries.

The last column, `stats_reset`, is particularly important since it indicates whenever the statistics information for a database has been reset, meaning deleted. Knowing how much time has elapsed since the statistics have been reset helps in validating the database.

Inspecting tables and indexes

The `pg_stat_user_tables` and `pg_stat_user_indexes` special catalogs provide detailed information about the usage of a table or an index, such as the number of tuples, the number of reads and writes, and so on.

Regarding a specific table, the following query provides detailed information about the status of the memory for that table:

```
forumdb=# SELECT relname, seq_scan, idx_scan,
          n_tup_ins, n_tup_del, n_tup_upd, n_tup_hot_upd,
          n_live_tup, n_dead_tup,
          last_vacuum, last_autovacuum,
          last_analyze, last_autoanalyze
          FROM pg_stat_user_tables;
...
-[ RECORD 6 ]----+------------------------------
relname          | tags
seq_scan         | 5
idx_scan         | 0
n_tup_ins        | 10000
n_tup_del        | 0
n_tup_upd        | 0
n_tup_hot_upd    | 0
n_live_tup       | 10000
n_dead_tup       | 0
last_vacuum      | 2020-05-02 17:47:35.325376+02
last_autovacuum  | 2020-05-13 16:46:35.325376+02
last_analyze     | 2020-04-28 18:42:25.337372+02
last_autoanalyze | 2020-05-13 16:46:35.325376+02
```

The `last_vacuum`, `last_analyze`, `last_autovacuum`, and `last_autoanalyze` columns are particularly important to understand whether manual or automatic vacuuming and analysis ran on the table; this knowledge can be crucial to understanding whether the automatic daemons are working properly. The `n_live_tup` column reports the currently visible tuples, according to MVCC (see Chapter 10, *Users, Roles, and Database Security*), while the `n_dead_tup` column reports the number of no longer visible tuples that still occupy space but will be reclaimed by a manual or automatic vacuum.

The other columns are pretty much self-explanatory, with `seq_scan` and `idx_scan` being the number of times the table has been accessed in a sequential scan or by an index among those available; `n_tup_ins`, `n_tup_upd`, and `n_tup_del` provide information about how many tuples have been inserted as new and how many have been updated or deleted, respectively. The `n_tup_upd_hot` column reports the number of tuples that have been updated in place, instead of being created as new, by means of a mechanism called **Heap Only Tuple** (**HOT**).

The `pg_stat_user_indexes` special catalog provides detailed information about the usage of the available indexes. In particular, the `idx_scan`, `idx_tup_read`, and `idx_tup_fetch` fields specify the number of times the index has been used, how many index tuples have been read, and how many table tuples have been obtained thanks to the index. For more information, please see `Chapter 11`, *Transactions, MVCC, WALs, and Checkpoints*.

There are other, dual, catalogs whose names include "all" or "sys" to indicate they refer to all the available tables, including PostgreSQL internal tables, or to only the latter (system tables). Therefore, `pg_stat_all_tables` is the same as `pg_stat_user_tables` but also includes information about system tables, which is kept under `pg_stat_sys_tables`. The same applies to `pg_stat_all_indexes`; that is, the union of `pg_stat_user_indexes` and `pg_stat_sys_indexes`.

More statistics

PostgreSQL includes a very rich set of statistics-related catalogs, and not all of them can be described here due to space limitations.

Some of the most important ones to mention include the following:

- `pg_stat_replication`, `pg_stat_wal_receiver`, and `pg_stat_subscription` gather information about the replication of the cluster.
- `pg_stat_bgwriter` gets information about input/output.
- `pg_stat_archiver` gets information about how WALs are being archived.
- `pg_statio_user_tables`, `pg_statio_user_indexes`, and the related `pg_statio_all_tables` and `pg_statio_all_indexes` provide information about input/output at a table or index level, indicating the number of hits and misses from the buffer cache and reading new pages from storage.

You should take the time to become comfortable with all the statistics catalogs in order to be able to monitor your cluster with confidence.

In the next section, you are going to learn about a very handy extension that can help you manage your cluster and take control of cluster activities.

Advanced statistics with pg_stat_statements

While the PostgreSQL statistics collector is rich and mature, having to monitor connection activity can be a little tricky since the `pg_stat_activity` catalog does not provide historic information. For example, as we explained previously, there will be a single tuple with the last executed statement, so no history nor extended details will be provided.

The `pg_stat_statements` extension solves this problem by providing a single view that gives you a full history of executed statements, timing, and other little details that can come in very handy when doing introspection. Moreover, `pg_stat_statements` provides a count of how many times the same statement has been executed, resulting in important information that queries might need to pay attention to for optimization purposes.

In the following subsections, you will learn how to install this extension and use it.

Installing the pg_stat_statements extension

This extension is shipped with PostgreSQL, so the only thing you have to do is configure the database cluster to use it. Since `pg_stat_statements` requires a shared library, you need to configure the `shared_preload_libraries` setting of your configuration (the `postgrsql.conf` file) and restart the cluster.

The first step is to set the following in `postgresql.conf`:

```
shared_preload_libraries = 'pg_stat_statements'
```

Then, you need to restart the cluster.

`pg_stat_statements` collects information about all your clusters, but it will only export such information in the database you create the extension in, which in our example is the `forumdb` database:

```
$ psql -U postgres -c "CREATE EXTENSION pg_stat_statements;" forumdb
CREATE EXTENSION
```

The extension is now ready to be used.

Using pg_stat_statements

Once `pg_stat_statements` has been enabled, it will start collecting information. The runtime overhead of the extension is really minimal, so you can keep it enabled in production systems too.

Since `pg_stat_statements` collects data from the whole cluster, it is helpful to join the `pg_stat_statements` special view with other catalogs, such as `pg_database` and `pg_authid`, to gather information about the database and username a statement has been executed inside of, respectively. The following query provides an example of this:

```
forumdb=# SELECT auth.rolname,query, db.datname, calls, min_time, max_time
 FROM pg_stat_statements
       JOIN pg_authid auth ON auth.oid = userid
       JOIN pg_database db ON db.oid = dbid
 ORDER BY calls DESC;
...
rolname                | postgres
query                  | SELECT count(*) FROM posts WHERE last_edited_on >=
CURRENT_DATE - $1
datname                | forumdb
calls                  | 11
min_time               | 0.181
max_time               | 4.442902
```

The preceding example shows that the query has been executed 11 times since `pg_stat_statements` started collecting the data, and it required from `0.181` to `4.44` seconds to run. Depending on the frequency and timing of each query, it could be interesting to inspect and optimize the query by means of an index, for example.

The `pg_stat_statements` extension also provides fields related to block and shared buffer read and writes. This can be useful for inspecting the memory size of the database.

Resetting data collected from pg_stat_statements

It is possible, at any given time, to reset all the data that's been collected by the extension that's invoking the `pg_stat_statements_reset()` function as a database administrator. The function will erase all the data that's been collected and will allow the extension to collect new data from scratch. This can be useful when you want to test new configuration or hardware without having the collected data be biased due to old statistics:

```
forumdb=# SELECT pg_stat_statements_reset();
```

By default, `pg_stat_statements` data is kept across clean database shutdowns and restarts.

Tuning pg_stat_statements

The extension allows database administrators to limit the amount of data that's collected. In particular, you can tune the following parameters in your `postgresql.conf` configuration file:

- `pg_stat_statements.max` indicates the maximum number of individual queries to collect.
- `pg_stat_statements.save` is a boolean that indicates whether the content of the collected data must survive a clean system reboot. By default, this setting is `true`.
- `pg_stat_statements.track` allows you to specify the nesting level to track. With the `top` value, the extension will collect data about the query that was issued directly within clients and within tracking nested statements. This is triggered by the execution of other statements (for example, in function statements). With the value of `all`, the extension will trigger every statement and its descendants, while with `none`, no data will be collected about user statements.
- `pg_stat_statements.track_utility` tracks all statements that are not in `SELECT`, `INSERT`, `UPDATE`, `DELETE` – in other words, "non-ordinary" statements. By default, this setting is on.

Usually, you don't have to exploit these settings since `pg_stat_statements` comes already configured to track what most use cases need.

Summary

In this chapter you learned how PostgreSQL manages configuration through a main text file, `postgresql.conf`, that can be split into smaller pieces, depending on your needs. Every configuration option can be edited in the configuration file and can be inspected within the database thanks to dedicated system catalogs. This allows the database administrator to not only have a clear understanding of the currently running configuration but to also search for configuration errors and incorrectly loaded settings.

PostgreSQL also collects *statistics*; that is, runtime data that was gathered during the cluster's operational time. Those statistics can help an administrator understand what is going on, or what happened in the past, in the cluster. Thanks to a different set of catalogs, which was exposed in this chapter, you learned how to dig into the details of all the information that PostgreSQL has collected for you. Being able to track and analyze what single applications, users, and connections are doing in a specific moment against the cluster provides database administrators with a great way to fix bottlenecks and other problems, thus helping to improve the cluster experience.

Finally, you learned about the `pg_stat_statements` extension, thanks to which it is possible to collect historical data about query execution and timing so that it is possible to apply optimization and deep analysis of the cluster activity.

Now that you've understood how to configure and monitor your cluster, it is time to learn how to replicate this. The next chapter will show you how to perform physical replication by configuring the cluster appropriately.

Further Reading

- PostgreSQL 12 cluster configuration, official documentation: `https://www.postgresql.org/docs/12/runtime-config.html`
- PGConfig online configurator: `https://www.pgconfig.org/`
- PostgreSQL 12 statistics collector official documentation: `https://www.postgresql.org/docs/12/monitoring-stats.html`
- PostgreSQL 12 `pg_stat_statements` official documentation: `https://www.postgresql.org/docs/12/pgstatstatements.html`

Section 4: Replication

In this section, you will learn how replication in PostgreSQL can be used in multi-instance environments to provide high-availability, redundancy, and scalability.

This section contains the following chapters:

- *Chapter 17, Physical Replication*
- *Chapter 18, Logical Replication*

17
Physical Replication

When a database, after passing the development and testing phases, arrives in production, the first problem that the DBA must address is managing replicas. Replicas must be managed in real time and automatically updated. Replicas allow us to always have a copy of our data updated in real time on another machine. This machine can be placed in the same data center as our data or in a different one. This chapter differs from all that we have seen previously in that we will be talking about physical replication. In Postgres, starting from version 9.x, it is possible to have physical replication natively. We will talk about what physical replication means and we will see how to create a replica server and how to manage it. We will also see that there is the possibility of having synchronous or asynchronous replicas and that there can be multiple replicas of the same database, as well as the possibility of having replicas in a cascade.

In this chapter, we will return to the topic of WAL, something we have already discussed in `Chapter 11`, *Transactions, MVCC, WAL, and Checkpoints*. In order to execute the commands that will be shown in this chapter, we recommend installing a PostgreSQL server on a new machine or installing another instance of PostgreSQL on the same machine but on a different port. In the rest of the chapter, it will be presumed that you have two PostgreSQL installations available on different machines, to better simulate the situation of a real production environment.

In this chapter, we will talk about the following topics:

- Exploring basic concepts
- WAL archiving and PITR
- Managing streaming replication

Exploring basic concepts

In PostgreSQL, there are two kinds of replication techniques:

- **Asynchronous replication**: In asynchronous replication, the primary device (source) sends a continuous flow of data to the secondary one (target), without receiving any return code from the target. This type of copying has the advantage of speed, but it brings with it greater risks of data loss because the received data is not checked.

- **Synchronous replication**: In synchronous replication, a source sends the data to a target, that is, the second server; at this point, the server sends back a code to verify the correctness of the data. If the check is successful, the transfer is completed.

Both methods have advantages and disadvantages, and in the *Managing streaming replication* section of this chapter, we will analyze them.

WAL

Let's briefly summarize what we saw in the chapter on MVCC and WAL: in that chapter, we saw how PostgreSQL stores data on disk using WAL; as we saw in Chapter 11, *Transactions, MVCC, WAL, and Checkpoints*, WAL is mainly used in the event of a crash. After a crash, PostgreSQL retraces WAL segments and re-applies them to data starting from the last checkpoint; during the recovery time after a crash, the server puts itself in a recovery state mode. Here is a summary of the key information about WAL segments:

- The WAL size is fixed at 16 MB.
- By default, WAL files are deleted as soon as they are older than the latest checkpoint.
- We can maintain extra WAL segments using `wal_keep_segments`.
- WAL segments are stored in the `pg_wal` directory as shown here:

```
postgres@pg2:~/12/main/pg_wal$ ls -alh
totale 17M
drwx------ 3 postgres postgres 4,0K apr 18 20:10 .
drwx------ 19 postgres postgres 4,0K apr 19 14:13 ..
-rw------- 1 postgres postgres 16M apr 19 15:34
000000010000000000000001
drwx------ 2 postgres postgres 4,0K apr 18 20:10 archive_status
```

The wal_level directive

The `wal_level` directive sets what kind of information should be stored in WAL segments. The default value is `minimal`. With this value, all information that is stored in a WAL segment can support archiving and physical replication.

 For further information, see `https://www.postgresql.org/docs/12/runtime-config-wal.html#GUC-WAL-LEVEL`.

So, in this chapter, we will use the `wal_level=replica` value, which is the default value, and in the next chapter, we will use `wal_level=logical`. We have to remember that we need to restart the PostgreSQL server every time we change the `wal_level` parameter.

Preparing the environment setup for streaming replication

In this section, we will prepare the three servers that we need to proceed: the first one is the master server machine, the second one is the replica server, and the third one is the repository server of WAL segments. So, let's proceed with the installation of three virtual machines. For example, in the following examples, we will be using two Debian Linux virtual machines with `192.168.11.34` as the IP for the master server and `192.168.11.35` as the IP for the replication server, and `192.168.11.36` as the IP for the repository server. In this chapter, all the paths are referred to PostgreSQL 12 installed on Debian (for example, `/usr/lib/postgresql/12/bin/`); to have the correct path for version 13, in beta at the moment, we need to replace 12 with 13, for example, `/usr/lib/postgresql/13/bin/`:

1. For the master server, we will have the following output:

```
root@pg1# ip addr
enp1s0: <BROADCAST,MULTICAST,UP,LOWER_UP> mtu 1500 qdisc pfifo_fast
state UP group default qlen 1000
    link/ether 52:54:00:cb:79:5f brd ff:ff:ff:ff:ff:ff
    inet 192.168.12.34/24 brd 192.168.12.255 scope global dynamic
enp1s0
       valid_lft 3757sec preferred_lft 3757sec
    inet6 fe80::5054:ff:fecb:795f/64 scope link
       valid_lft forever preferred_lft forever
# su - postgres
postgres@pg1:$ psql
```

```
psql (12.2 (Debian 12.2-2.pgdg100+1))
Type "help" for help.
postgres=#
```

2. Similarly, for the replica server, we will have the following:

```
root@pg2# ip addr
2: enp1s0: <BROADCAST,MULTICAST,UP,LOWER_UP> mtu 1500 qdisc
pfifo_fast state UP group default qlen 1000
    link/ether 52:54:00:c9:14:a2 brd ff:ff:ff:ff:ff:ff
    inet 192.168.12.35/24 brd 192.168.12.255 scope global dynamic
enp1s0
    valid_lft 5308sec preferred_lft 5308sec
    inet6 fe80::5054:ff:fec9:14a2/64 scope link
    valid_lft forever preferred_lft forever
root@pg2:~# su -  postgres
postgres@pg2:$ psql
psql (12.2 (Debian 12.2-2.pgdg100+1))
Type "help" for help.
postgres=#
```

3. For the repository server, we will have the following:

```
root@pg3:~# ip addr
2: enp1s0: <BROADCAST,MULTICAST,UP,LOWER_UP> mtu 1500 qdisc
pfifo_fast state
UP group default qlen 1000
    link/ether 52:54:00:fa:3a:89 brd ff:ff:ff:ff:ff:ff
    inet 192.168.12.36/24 brd 192.168.12.255 scope global dynamic
enp1s0
        valid_lft 7027sec preferred_lft 7027sec
    inet6 fe80::5054:ff:fefa:3a89/64 scope link
        valid_lft forever preferred_lft forever
```

4. Let's check to see whether there is a connection between the two servers.

We will check for a connection from the master server to the replica server/repository server:

```
postgres@pg1:~$ ping 192.168.12.35
PING 192.168.12.35 (192.168.12.35) 56(84) bytes of data.
64 bytes from 192.168.12.35: icmp_seq=1 ttl=64 time=0.819 ms
64 bytes from 192.168.12.35: icmp_seq=2 ttl=64 time=0.889 ms

postgres@pg1:/root$ ping 192.168.12.36
PING 192.168.12.36 (192.168.12.36) 56(84) bytes of data.
64 bytes from 192.168.12.36: icmp_seq=1 ttl=64 time=0.497 ms
64 bytes from 192.168.12.36: icmp_seq=2 ttl=64 time=0.286 ms
```

We will check for a connection from the replica server to the master server/repository server:

```
postgres@pg2:/root$ ping 192.168.12.34
PING 192.168.12.34 (192.168.12.34) 56(84) bytes of data.
64 bytes from 192.168.12.34: icmp_seq=1 ttl=64 time=0.460 ms
64 bytes from 192.168.12.34: icmp_seq=2 ttl=64 time=0.435 ms

postgres@pg2:~$  ping 192.168.12.36
PING 192.168.12.36 (192.168.12.36) 56(84) bytes of data.
64 bytes from 192.168.12.36: icmp_seq=1 ttl=64 time=0.840 ms
64 bytes from 192.168.12.36: icmp_seq=2 ttl=64 time=1.28 ms
```

We will check for a connection from the repository server to the master server/replica server:

```
postgres@pg3:~$ ping 192.168.12.34
PING 192.168.12.34 (192.168.12.34) 56(84) bytes of data.
64 bytes from 192.168.12.34: icmp_seq=1 ttl=64 time=0.432 ms
64 bytes from 192.168.12.34: icmp_seq=2 ttl=64 time=0.954 ms
postgres@pg3:~$ ping 192.168.12.35
PING 192.168.12.35 (192.168.12.35) 56(84) bytes of data.
64 bytes from 192.168.12.35: icmp_seq=1 ttl=64 time=0.205 ms
64 bytes from 192.168.12.35: icmp_seq=2 ttl=64 time=1.15 ms
```

Now that everything is ready, let's start exploring the details of physical replication. In the next section, we will talk about WAL and **point-in-time recovery** (**PITR**), which are the building blocks of streaming replication.

Learning WAL archiving and PITR

In this section, we are going to look at the physical way of storing data in PostgreSQL in more detail. We will begin to process segment WAL manually and then move on to automatic modes, which make the work of the DBA much easier. Classic backups are made using the `pg_dump` command, and they are also called **logical backups**. What we want to do now is take a physical backup of the data; we want to obtain a continuous snapshot of our database. This technique offers the DBA the possibility to restore the database to any point in the past; this technique, called **PITR**, is widely used as a disaster recovery technique. This technique allows us to go back to our PostgreSQL cluster at an exact point in the past before a malicious event occurred (for example, a DROP of a table). This technique is often used by DBAs if we want to take a certain snapshot of a production environment back to a test environment.

PITR – the manual way

In the Chapter 14, *Backup and Restore* chapter, we discussed physical backups. In this section, we will resume the topics covered in that chapter to introduce physical replication. Let's suppose we have a new machine that we will call pg1. What we want to conduct is a continuous backup over time. On the pg1 machine, we have two directories, which we will call wallbackup, where we will store WAL segments, and databackup, where we will store the data.

The PITR technique works in this way:

1. WAL segments are copied to the walbackup directory (the wal archive).
2. A base backup is performed as a starting point for our archives.
3. During the recovery, the system starts from the last checkpoint before the recovery point that we want to obtain, and then it performs all operations up to the point in the past that we want to get to.

The WAL archive

Let's check out how the WAL archive works using the following steps:

1. Let's start by preparing the two directories, walbackup and databackup; they have to be accessible for writing by the postgres user:

```
root@pg1:/# mkdir walbackup
root@pg1:/# chown postgres.postgres walbackup/
root@pg1:/# mkdir /databackup
root@pg1:/# chown postgres.postgres databackup
root@pg1:/# chmod 0700 databackup/
```

2. Then, let's modify some rows of the postgresql.conf file. We will add these rows at the bottom of the file:

```
# Add settings for extensions here
archive_mode = on
archive_command = 'test ! -f /walbackup/%f && cp %p /walbackup/%f'
wal_level = replica
archive_timeout = 10 #optional
```

Now let's check out the values we have set here:

- `archive_mode = on`: With this directive, we tell PostgreSQL to make the archiving of `wal` possible.
- `archive_command = 'test ! -f /walbackup/%f && cp %p /walbackup/%f'`: With this directive, we tell PostgreSQL to copy the `wal` files to the `walbackup` directory.
- `wal_level = replica`: Replication is the default condition. In this way, PostgreSQL stores all the information in the WAL segments so that the cluster can be replicated physically.
- `archive_timeout = 10`: This forces the number of seconds after which PostgreSQL creates a new `wal` segment and also its copy. This is optional.

3. Once we have added these lines to the `postgresql.conf` file, we have to restart the service:

```
root@pg1:/# service postgresql stop
root@pg1:/# service postgresql start
```

4. Now, if we list the contents of the `walbackup` directory, we'll see that the system has started to archive WAL segments:

```
root@pg1:# ls -l /walbackup
-rw------- 1 postgres postgres 16777216 apr 26 17:37
000000010000000000000001
-rw------- 1 postgres postgres 16777216 apr 26 17:37
000000010000000000000002
-rw------- 1 postgres postgres 16777216 apr 26 17:37
000000010000000000000003
-rw------- 1 postgres postgres 16777216 apr 26 17:37
000000010000000000000004
-rw------- 1 postgres postgres 16777216 apr 26 17:38
000000010000000000000005
```

Thus, we have learned how to use the WAL archive.

Basebackup

To execute a basebackup, we use the `rysnc` command and we copy the data of the `$PGDATA` directory to the `databackup` directory. Be sure to remove `postmaster.pid` and `postmaster.opts` files. `$PGDATA` depends on the Linux distribution used, and in this example, we will refer to a Debian Linux server version. Let's now proceed with the basebackup process:

1. Let's open two shell windows; the first will be used to execute SQL commands using the `pgsql` environment, and the second will be used to launch the operating system commands.

2. The first thing to do is to inform the PostgreSQL server that we are starting the basebackup procedure. So, let's return to the first shell window and perform the following statement:

```
postgres=# SELECT pg_start_backup( 'MY_FIRST_PITR', true, false );
 pg_start_backup
-----------------
 0/A000028
(1 riga)
```

3. In the second shell window, we will execute the copy files commands:

```
root@pg1:/# sudo -u postgres rsync -a /var/lib/postgresql/12/main
/databackup/
root@pg1:/# sudo -u postgres rm /databackup/main/postmaster.pid
root@pg1:/# sudo -u postgres rm /databackup/main/postmaster.opts
```

4. Now, let's go back to the first window and run the following command:

```
postgres=# SELECT pg_stop_backup( false );
NOTICE: all required WAL segments have been archived
 pg_stop_backup
-------------------------------------------------------------------
--------
 (0/B000050,"START WAL LOCATION: 0/A000028 (file
000000010000000000000000A)+
 CHECKPOINT LOCATION: 0/A000060 +
 BACKUP METHOD: streamed +
 BACKUP FROM: master +
 START TIME: 2020-04-26 17:49:30 CEST +
 LABEL: MY_FIRST_PITR +
 START TIMELINE: 1 +
 ","")
(1 row)
```

At this point, the basebackup procedure is finished.

5. To do some tests, we then populate our new database with the test data available on GitHub:

```
postgres=# \i /tmp/setup_00-forum-database.sql
```

6. Let's try a test query on the categories table:

```
forumdb=# select * from categories;
 pk | title                 | description
----+-----------------------+-------------------------------
 1  | Database              | Database related discussions
 2  | Unix                  | Unix and Linux discussions
 3  | Programming Languages | All about programming languages
 4  | Database              | Database related discussions
 5  | Unix                  | Unix and Linux discussions
 6  | Programming Languages | All about programming languages
(6 rows)
```

7. Now let's check the last transaction in the database:

```
forumdb=# SELECT txid_current(), current_timestamp;
 txid_current | current_timestamp
--------------+----------------------------
 499          | 2020-04-26 17:59:36.88933+02
(1 row)
```

8. Let's insert a new record:

```
forumdb=# insert into categories (title,description) values
('BSD','Unix BSD discussions');
INSERT 0 1
```

9. Let's check again what the last transaction is:

```
forumdb=# SELECT txid_current(), current_timestamp;
 txid_current | current_timestamp
--------------+----------------------------
 499          | 2020-04-26 17:59:36.88933+02
(1 riga)
```

10. Check the data in the categories table:

```
forumdb=# select * from categories;
 pk | title                 | description
----+-----------------------+-------------------------------
 1 | Database              | Database related discussions
```

```
2 | Unix                    | Unix and Linux discussions
3 | Programming Languages   | All about programming languages
4 | Database                | Database related discussions
5 | Unix                    | Unix and Linux discussions
6 | Programming Languages   | All about programming languages
7 | BSD                     | Unix BSD discussions
(7 rows)
```

Thus, we have executed the basebackup process and found it to be effective.

Recovery

In this section, we will see how to make a recovery. Starting from the data copied to the data backup directory, we want to start a new instance of PostgreSQL on port 5433 with the cluster as it was at transaction 498.

We have to do some things to make the recovery possible:

1. Create an empty recovery.signal file in the /databackup /main directory.
2. Insert the pg_hba.conf, pg_ident.conf, and postgresql.conf configuration files in the /databackup /main directory.
3. Modify the postgresql.conf file by adding these lines at the bottom of the file:

```
#--------------------------------------------------------------------
------------
# CUSTOMIZED OPTIONS
#--------------------------------------------------------------------
------------
# Add settings for extensions here
#----- PATH AND PORT OPTIONS
data_directory = '/databackup/main' # use data in another directory
 # (change requires restart)
hba_file = '/databackup/main//pg_hba.conf' # host-based
authentication file
 # (change requires restart)
ident_file = '/databackup/main/pg_ident.conf' # ident configuration
file
 # (change requires restart)
port = 5433 # (change requires restart)
#--- PITR OPTIONS -----
restore_command = 'cp /walbackup/%f "%p"'
recovery_target_xid = 498
```

The `restore_command` option tells PostgreSQL where to go to pick up WAL segments and the `recovery_target_xid = 498` option tells PostgreSQL at which transaction the recovery procedure should stop.

4. Now, as a root user, let's perform the following command:

```
# sudo -u postgres /usr/lib/postgresql/12/bin/pg_ctl -D
/databackup/main/ start
```

The system will begin the recovery and, as we can verify from the log file, the recovery will end exactly at transaction `498`:

```
2020-04-26 19:15:33.279 CEST [6294] LOG: starting point-in-time
recovery to XID 498
2020-04-26 19:15:33.299 CEST [6294] LOG: restored log file
"000000010000000000000000D" from archive
2020-04-26 19:15:33.375 CEST [6294] LOG: redo starts at 0/D0007E8
2020-04-26 19:15:33.396 CEST [6294] LOG: restored log file
"000000010000000000000000E" from archive
2020-04-26 19:15:33.487 CEST [6294] LOG: restored log file
"000000010000000000000000F" from archive
2020-04-26 19:15:33.576 CEST [6294] LOG: restored log file
"0000000100000000000000010" from archive
2020-04-26 19:15:33.643 CEST [6294] LOG: consistent recovery state
reached at 0/10000088
2020-04-26 19:15:33.644 CEST [6293] LOG: database system is ready
to accept read only connections
LOG: recovery stopping after commit of transaction 498, time
2020-04-26 17:58:0
5.255896+02
2020-04-26 19:15:33.645 CEST [6294] LOG: recovery has paused
2020-04-26 19:15:33.645 CEST [6294] LOG: recovery has paused
2020-04-26 19:15:33.645 CEST [6294] HINT: Execute
pg_wal_replay_resume() to continue
```

5. Let's go back to the first shell and execute the following SQL command:

```
forumdb=# select * from categories;
pk  | title                 | description
----+-----------------------+--------------------------------
  1 | Database              | Database related discussions
  2 | Unix                  | Unix and Linux discussions
  3 | Programming Languages | All about programming languages
  4 | Database              | Database related discussions
  5 | Unix                  | Unix and Linux discussions
  6 | Programming Languages | All about programming languages
(6 righe)
```

As we can see, the data on our system is data from prior to the `2020-04-26 17:58: 0` timestamp, as set in the recovery configuration. At this point, the system only accepts read-only transactions so that we can verify that everything is OK.

6. If we try to execute a write statement, PostgreSQL will return an error, as seen here:

```
forumdb=# create table my_table(id integer);
ERROR: cannot execute CREATE TABLE in a read-only transaction
```

7. To make write operations possible, we must execute the following command:

```
forumdb=# select pg_wal_replay_resume();
pg_wal_replay_resume
----------------------

(1 riga)
```

8. Now it is possible to perform write operations on our database, as shown here:

```
forumdb=# create table my_table(id integer);
CREATE TABLE
```

The `pg_wal_replay_resume` command puts the system in full read/write mode and deletes `recovery.signal` files; now it is possible to use our cluster in read/write mode. Now that we have understood well how WAL and PITR work, in the next section, we will talk about streaming replication.

Managing streaming replication

In this section, we will talk about replication. Why do we have to have replicas? The problem with PITR is that recovery often takes a long time before a server can be restored:

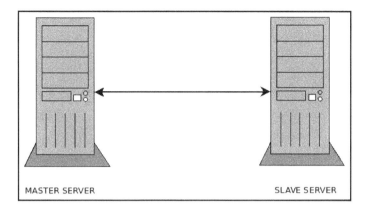

In a production environment, you often need to be able to restore the production environment as quickly as possible after a system crash. In order do this, we have to use the streaming replication technique. To make this possible, we need at least two servers, one master server and one slave server. The master server performs all the operations that will be requested by the application programs; the slave server will be available only for read operations and will have the data copied in real time.

Basic concept

The idea behind streaming replication is to copy the WAL files from the master server to another (slave) server. The slave server will be in a state of continuous recovery and it continuously executes the WAL that is passed by the master machine; in this way, the slave machine binarily replicates the data of the master machine through the WAL.

In a classic PITR situation, WAL segments are saved somewhere by the master and then they are taken by the recovery machine using manual scripts:

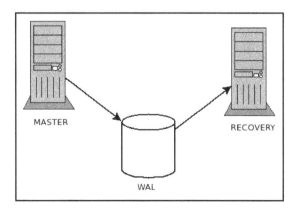

In a streaming replication context, a communication channel will be open between the slave and master, and the master will send the WAL segments through it:

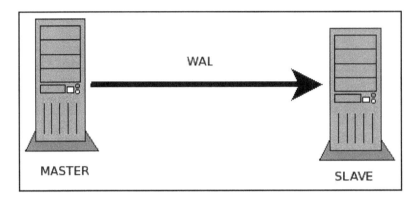

The slave server will receive the WAL segments and rerun them, remaining in a permanent recovery state.

We will now look at how to perform asynchronous physical replication. The technique is very similar to PITR.

Replication environment

Let's prepare our `develop` environment. We need two servers: the first one will be called `pg1` and its IP will be `192.168.12.34`; the second one will be called `pg2` and its IP will be `192.168.12.35`. Let's take a look at the preparatory steps for physical replication.

On the master server, we need to do the following:

1. The first thing we have to do is modify `listen_address` so that it is listening to the network. If we set `listen_address = '*'`, PostgreSQL will be listening to any IP; otherwise, we can specify a list of IP addresses separated by commas. This change requires a restart of the PostgreSQL service.

2. We need to create a new user that is able to perform the replication:

    ```
    CREATE USER replicarole WITH REPLICATION ENCRYPTED PASSWORD
    'SuperSecret';
    ```

3. We have to modify the `pg_hba.conf` file so that from the slave machine with the user `replicarole`, it is possible to reach the master machine:

```
host replication replicarole 192.168.12.35/32 md5
```

4. To make this configuration active, we need to run a reload of the PostgreSQL server. For example, we can run the following:

```
postgres=# select pg_reload_conf();
 pg_reload_conf
----------------
 t
(1 riga)
```

5. On the slave server, we have to turn off the PostgreSQL service, destroy the `PGDATA` directory, and remake it, this time empty and with the right permissions. To do this, we can use these statements:

```
root@pg1:/# systemctl stop postgresql
root@pg1:/# cd /var/lib/postgresql/12/
root@pg1:/# rm -rf main
root@pg1:/# mkdir main
root@pg1:/# chown postgres.postgres main
root@pg1:/# chmod 0700 main
```

All the paths used in this example are valid for Debian-based distributions; for other distributions, please consult the respective official documentation.

The wal_keep_segments option

From what we have understood, physical replication is done through the transfer of WAL segments. Now suppose for a moment that the slave server goes down for some reason. How does the master behave? When the slave server becomes functional again, will it realign itself with the master node or not? These are questions we need to ask ourselves if we want our replication system to work correctly.

The `postgresql.conf` directive that tells PostgreSQL how many WAL segments to keep on disk is called `wal_keep_segments`; by default, `wal_keep_segments` is set to zero. This means that PostgreSQL will not store any extra WAL segments as buffers. This means that if the slave machine (standby) goes down, then it will no longer be able to realign itself when it comes back up. This happens because in the time it takes the slave to get back up, it is possible that the master machine has produced and deleted new WAL segments. The first way to overcome this problem is to set the `wal_keep_segments` directive to a value greater than zero in `postgresql.conf`. For example, if we set a value of `wal_keep_segments = 100`, this means that at least 100 files of WAL segments will be present in the `pg_wal` folder, for a total occupied disk space of 100 * 16 MB = 1.6 GB.

In this case, the master always keeps these extra WAL segments and, if the slave should go down, then it will only be able to realign itself once back up if the master has produced a number of WAL segments less than `wal_keep_segments`.

This solution offers a static buffer in that you can store old WAL segments and offers a save anchor that is shorter than the time taken by the master to produce a number of WAL segments greater than `wal_keep_segments`. This solution is a static solution; it also has the disadvantage that the space occupied on disk is always equal to `wal_keep_segments` * 16 MB, even when it is no longer necessary to keep WAL segments on the master server because they have already been processed by the replica server.

The slot way

In PostgreSQL 12, there is another approach that can be used to solve the problem of storing WAL segments: the slot technique. Through the slot technique, we can tell PostgreSQL to keep all the WAL segments on the master until they have been transferred to the replica servers. In this way, we have dynamic, variable, and fully automated management of the number of WAL segments that the master server must keep as a buffer. This is a very easy way to manage our physical replicas and it is the way we will focus on in this book.

The instruction we need to perform on PostgreSQL to create a new slot is as follows:

```
postgres=#  SELECT * FROM pg_create_physical_replication_slot('master');
 slot_name | lsn
-----------+-----
 master    |
(1 row)
```

The instruction we need to perform on PostgreSQL to drop a slot is this:

```
postgres=# select pg_drop_replication_slot('master');
 pg_drop_replication_slot
---------------------------

(1 riga)
```

Later on in this chapter, we will look more at these instructions.

The pg_basebackup command

In the PITR section, we talked about the base backup; it is a hot backup that is performed as a starting base on which to then perform all the WAL segments. In that section, we made the backup base using a combination of commands:

- pg_start_backup
- rsync
- pg_stop_backup

There is another command called pg_basebackup that implements the procedures just described almost automatically. It requires a configuration line in the pg_hba.conf file to allow one connection to the source database of the type shown previously:

```
host replication replicarole 192.168.12.35/32 md5
```

It is also necessary that the max_wal_senders value is at least 2. It is a very useful command for the DBA because it allows us to do everything we need to do with a single instruction. We will use and better explain this command in the next section, where we will implement our first asynchronous physical replication.

For further information about the pg_basebackup command, please refer to https://www.PostgreSQL.org/docs/12/app-pgbasebackup.html.

Asynchronous replication

We now have all the building blocks necessary for easily and quickly making our first asynchronous physical replication. By default, in PostgreSQL, physical replication is asynchronous. Let's now start with the replication technique. By following the steps from the previous sections of this chapter, we already have a master server ready to be connected to the slave server, and we have the slave ready to receive information from the master. The slave server will now have the PostgreSQL service turned off and the PGDATA data folder created, empty, and with the right permissions:

1. Let's go inside the PGDATA directory as the system postgres user:

   ```
   root@pg2:# su - postgres
   postgres@pg2:~$ cd /var/lib/PostgreSQL/12/main
   ```

2. Now let's run the pg_basebackup command with the right options. This command will execute the base_backup command from the master machine to the slave machine and prepare the slave machine to receive and execute the received WAL segments, causing the slave server to remain in a state of permanent recovery:

   ```
   postgres@pg2:~/12/main$ pg_basebackup -h pg1 -U replicarole -p 5432
   -D /var/lib/PostgreSQL/12/main -Fp -Xs -P -R -S master
   Password:
   32725/32725 kB (100%), 1/1 tablespace
   postgres@pg2:~/12/main$
   ```

   The password that we have to insert is the password of the replicarole user; in our case, this is SuperSecret.

   Let's analyze this command in more detail:

   - -h: With this option, we see the host that we want the slave to connect to.
   - -U: This is the user created on the master server used for replication.
   - -p: This is the port where the master server is listening.
   - -D: This is the PGDATA value on the slave server.
   - -Fp: This performs a backup on the slave, maintaining the same data structure present on the master.
   - -Xs: This opens a second connection to the master server and starts the transfer of the WAL segments at the same time as the backup is performed.
   - -P: This shows the progress of the backup.
   - -S: This is the slotname created on the master server.

3. Create the `standby.signal` file and add the connection settings to the `PostgreSQL.auto.conf` file:

```
postgres@pg2:~/12/main$ cat postgresql.auto.conf
# Do not edit this file manually!
# It will be overwritten by the ALTER SYSTEM command.
primary_conninfo = 'user=replicarole password=SuperSecret host=pg1
port=5432 sslmode=prefer sslcompression=0 gssencmode=prefer
krbsrvname=postgres target_session_attrs=any'
primary_slot_name = 'master'
```

4. Now let's start the PostgreSQL service on the slave machine and physical replication should work. As the root user, let's execute the following:

```
root@pg2:/var/lib/postgresql/12# systemctl start postgresql
```

As we can see from the PostgreSQL log file (`/var/log/postgresql/postgresql-12-main.log`), the slave machine started in `stand_by` mode and in read-only mode:

```
2020-05-01 17:09:59.072 CEST [1422] LOG:  entering standby mode
2020-05-01 17:09:59.079 CEST [1422] LOG:  redo starts at 0/22000060
2020-05-01 17:09:59.079 CEST [1422] LOG:  consistent recovery state
reached at 0/23000060
2020-05-01 17:09:59.080 CEST [1421] LOG:  database system is ready
to accept read only connections
2020-05-01 17:09:59.080 CEST [1422] LOG:  invalid record length at
0/23000060: wanted 24, got 0
2020-05-01 17:09:59.099 CEST [1426] LOG:  started streaming WAL
from primary at 0/23000000 on timeline 1
```

5. Let's connect to the replica server and try to see whether everything has been replicated:

```
postgres=# \l
  List of databases
    Name     |  Owner   | Encoding |   Collate   |    Ctype    |
Access privileges
-----------+----------+----------+-------------+-------------+-----
-------------------
 forumdb    | postgres | UTF8     | it_IT.UTF-8 | it_IT.UTF-8 |
 postgres   | postgres | UTF8     | it_IT.UTF-8 | it_IT.UTF-8 |
 template0  | postgres | UTF8     | it_IT.UTF-8 | it_IT.UTF-8 |
=c/postgres +
            |          |          |             |             |
postgres=CTc/postgres
 template1  | postgres | UTF8     | it_IT.UTF-8 | it_IT.UTF-8 |
```

```
                   =c/postgres    +
                       |               |              |                |                   |
                   postgres=CTc/postgres
                   (4 rows)
```

6. Let's go inside the `forumdb` database and try to create another table:

```
forumdb=# create table test_table (id integer);
ERROR:  cannot execute CREATE TABLE in a read-only transaction
forumdb=#
```

As we can see, the server is now in read-only mode.

Replica monitoring

After successfully installing our first asynchronous replica server, let's look at how we can monitor the health of our replica. PostgreSQL offers us a view through which we can monitor the status of replicas in real time; its name is `pg_stat_replication`. This view must be queried by connecting to the master node.

For example, if we connect to the main node, we can see the following:

```
postgres=# select * from pg_stat_replication ;
-[ RECORD 1 ]----+-----------------------------
pid               | 1435
usesysid          | 16471
usename           | replicarole
application_name  | 12/main
client_addr       | 192.168.12.35
client_hostname   |
client_port       | 41306
backend_start     | 2020-05-01 17:09:59.085132+02
backend_xmin      |
state             | streaming
sent_lsn          | 0/23000060
write_lsn         | 0/23000060
flush_lsn         | 0/23000060
replay_lsn        | 0/23000060
write_lag         |
flush_lag         |
replay_lag        |
sync_priority     | 0
sync_state        | async
reply_time        | 2020-05-01 17:23:19.723054+02
```

Using this view, we have a lot of information that we need in order to know whether our `stand_by` server is in excellent health.

For example, we can see that the last reply message received from the replica server is `2020-05-01 17: 23: 19.723054 + 02`, and we can see, thanks to the difference between the `sent_lsn` value and the `replay_lsn` value, that our replication server is perfectly aligned. For further information about `pg_stat_replication`, please refer to the official documentation (`https://www.postgresql.org/docs/12/monitoring-stats.html#PG-STAT-REPLICATION-VIEW`).

Cascading replication

We have explored how to create an asynchronous replica starting from a master server. However, in some cases, we may need multiple replicas, and the simplest way to do this is to hook a second replica machine to the master machine with the procedure we have just seen. This procedure, however, could increase the load on the master machine, so PostgreSQL offers an alternative to it: cascading physical replication. The scheme we want to achieve is this:

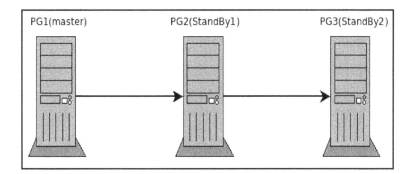

In order to make our example work, we will use a third machine called `pg3`.

The machines will have the following IPs:

1. `master` (pg1): IP `192.168.12.34`
2. `standby1` (pg2): IP `192.168.12.35`
3. `standby2` (pg3): IP `192.168.12.36`

1. In a similar way to what we did before, let's configure the pg2 machine (standby1) so that it can receive requests from the pg3 machine. We have to add this line to the pg_hba.conf file:

```
IPv4 local connections:
host replication replicarole 192.168.12.36/32 md5
```

2. Now, we have to reload the PostgreSQL service:

```
systemctl reload postgresql
```

3. On the pg2 machine, let's execute the following SQL command:

```
SELECT * FROM pg_create_physical_replication_slot('standby1');
```

4. As before, we have created a reference slot for cascade replication. Now let's go to the pg3 machine and turn off the PostgreSQL service:

```
# systemctl stop postgresql
```

5. Let's delete the contents of the /var/lib/postgresql/12/main directory:

```
rm -rf /var/lib/postgresql/12/main/*
```

6. As a PostgreSQL user, let's perform the basebackup procedure:

```
postgres@pg3:~$ pg_basebackup -h pg2 -U replicarole -p 5432 -D
/var/lib/PostgreSQL/12/main -Fp -Xs -P -R -S standby1
Password:
32743/32743 kB (100%), 1/1 tablespace
```

7. At this point, we can restart the PostgreSQL service. As the root user, let's execute the following:

```
# systemctl start postgresql
```

At this point, we are done! If we query the pg_stat_replication view on the standby1 server, we will see that a second replica will exist. Now our system has two replicas and we have achieved the goal that we set ourselves.

8. This is pg_stat_replication on the master server (pg1):

```
postgres=# select * from pg_stat_replication ;
-[ RECORD 1 ]----+-----------------------------
pid              | 834
usesysid         | 16471
usename          | replicarole
```

```
application_name | pg2
client_addr      | 192.168.12.35
client_hostname  |
client_port      | 34934
backend_start    | 2020-05-02 16:03:48.527444+02
backend_xmin     |
state            | streaming
sent_lsn         | 0/35000060
write_lsn        | 0/35000060
flush_lsn        | 0/35000060
replay_lsn       | 0/35000060
write_lag        |
flush_lag        |
replay_lag       |
sync_priority    | 1
sync_state       | async
reply_time       | 2020-05-02 16:41:14.1749+02
```

9. This is pg_stat_replication on the standby1 server (pg2):

```
postgres=# select * from pg_stat_replication;
-[ RECORD 1 ]----+----------------------------
pid              | 850
usesysid         | 16471
usename          | replicarole
application_name | 12/main
client_addr      | 192.168.12.36
client_hostname  |
client_port      | 55276
backend_start    | 2020-05-02 16:14:50.648676+02
backend_xmin     |
state            | streaming
sent_lsn         | 0/35000060
write_lsn        | 0/35000060
flush_lsn        | 0/35000060
replay_lsn       | 0/35000060
write_lag        |
flush_lag        |
replay_lag       |
sync_priority    | 0
sync_state       | async
reply_time       | 2020-05-02 16:43:55.007117+02
```

Thus, we have learned how cascade replication works.

Synchronous replication

So far, we have talked about asynchronous replication; this means that the master server passes information to the standby server without being sure that the standby server has replicated the data. In asynchronous replication, the master server does not wait for the slave server to actually replicate the data. In synchronous replication, when the master performs a commit, all the slaves replicated synchronously commit. In synchronous replication, after the execution of the commit, we are sure that the data is replicated on the master and on all the replicas. When we want to achieve synchronous replication, it is good practice to have all identical machines and a good network connection between the machines; otherwise, performance can become slow.

PostgreSQL settings

Starting with what has been done for asynchronous replication and simply changing some settings, it is possible to change from asynchronous replication to synchronous replication.

Master server

On the master server, we have to check whether the `synchronous_commit` parameter is set to `on`. Now, `synchronous_commit = on` is the default value on a new PostgreSQL installation.

After setting this parameter, we must add the `synchronous_standby_names` parameter, listing the names of all standby servers that will replicate the data synchronously. We can also use the `'*'` wildcard, thus indicating to PostgreSQL that each standby server can potentially have a synchronous replica. For example, to transform the master of the previous example so that it can support asynchronous replication for the `pg2` server, we have to write this:

```
synchronous_standby_names = 'pg2'
synchronous_commit = on
```

After this, we need to restart our server:

```
# systemctl restart postgresql
```

Standby server

On the standby server, we have to add a parameter to the connection string to the master so that the master knows from whom the reply request comes. We need to edit the postgresql.auto.conf file; it is currently as follows:

```
# Do not edit this file manually!
# It will be overwritten by the ALTER SYSTEM command.
primary_conninfo = 'user=replicarole password=SuperSecret host=pg1
port=5432 sslmode=prefer sslcompression=0 gssencmode=prefer
krbsrvname=postgres target_session_attrs=any'
primary_slot_name = 'master
```

We need to change it to the following:

```
# Do not edit this file manually!
# It will be overwritten by the ALTER SYSTEM command.
primary_conninfo = 'user=replicarole password=SuperSecret host=pg1
port=5432 sslmode=prefer sslcompression=0 gssencmode=prefer
krbsrvname=postgres target_session_attrs=any application_name=pg2'
primary_slot_name = 'master
```

We have added the application_name=pg2 option.

After doing this, let's restart the standby server. Now if we get back on the master server and re-check the g_stat_replication view, we will see this result:

```
postgres=# select * from pg_stat_replication ;
-[ RECORD 1 ]----+------------------------------
pid              | 571
usesysid         | 16471
usename          | replicarole
application_name | pg2
client_addr      | 192.168.12.35
client_hostname  |
client_port      | 45714
backend_start    | 2020-05-01 21:14:53.856502+02
backend_xmin     |
state            | streaming
sent_lsn         | 0/31000060
write_lsn        | 0/31000060
flush_lsn        | 0/31000060
replay_lsn       | 0/31000060
```

```
write_lag       |
flush_lag       |
replay_lag      |
sync_priority   | 1
sync_state      | sync
reply_time      | 2020-05-01 21:23:03.619776+02
```

As shown here, the master server and standby servers are replicated in a synchronous way and `sync_state=sync`.

Summary

In this chapter, we introduced the concept of physical replication. We started by reviewing and deepening our knowledge of WAL segments from previous chapters. We have introduced, seen, and configured an asynchronous physical replica and a synchronous physical replica. We looked at the difference between the two modes and we saw how easy it is to switch from one mode to another. We then explored some useful tools for monitoring replicas and checking their good health.

In the next chapter, we use the concepts that we have discussed in this chapter to address the topic of logical replication.

References

- https://www.postgresql.org/docs/12/runtime-config-wal.html#GUC-WAL-LEVEL
- https://www.postgresql.org/docs/12/app-pgbasebackup.html
- https://www.postgresql.org/docs/12/monitoring-stats.html#PG-STAT-REPLICATION-VIEW
- https://www.postgresql.org/docs/12/runtime-config-replication.html
- https://www.postgresql.org/docs/12/high-availability.html

18
Logical Replication

In the previous chapter, we talked about WAL segments and physical replication in synchronous, asynchronous, and cascading modes. In this chapter, we will cover the topic of logical replication. We will look at how to perform a logical replica, how a logical replication is different from a physical replication, and when it's better to use logical replication instead of physical replication. We'll also see that logical replication can be used to make a PostgreSQL hot upgrade. This chapter is intended to be just an introduction to logical replication; for further information, refer to more advanced texts, such as *Mastering PostgreSQL 12*, Hans-Jürgen Schönig, Packt Publishing.

This chapter covers the following topics:

- Understanding basic concepts
- Exploring logical replication setup

Understanding basic concepts

Logical replication is a method that we can use to replicate data based on the concept of identity replication. `REPLICA IDENTITY` is a parameter present in table management commands (such as `CREATE TABLE` and `ALTER TABLE`); this parameter is used by PostgreSQL to obtain additional information within WAL segments to recognize which tuples have been eliminated and which tuples have been updated. The `REPLICA IDENTITY` parameter can take four values:

- `DEFAULT`
- `USING INDEX index_name`
- `FULL`
- `NOTHING`

The concept behind logical replication is to pass the logic of the commands executed on the master machine to the server and not the exact copy of the blocks to be replicated byte by byte. At the heart of logical replication, there is a reverse engineering process that, starting from the WAL segments and using a logical decoding process, is able to extrapolate the original SQL commands and pass them on to the replication machine using a logical decoding process.

Let's analyze for a moment a flowchart that shows how PostgreSQL internally executes queries:

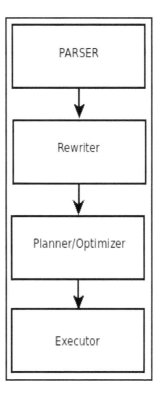

As we can see, a query, before being executed, requires several internal steps; this is because the system tries to execute the query in the best possible way according to the conditions prevailing at that moment in the database. Now suppose we want to replicate the data logically; at this point, we have two possibilities in front of us:

- We can capture commands before they get to the parser and transfer these commands to a second machine.
- We can try, in some way, to get the queries that are already parsed.

The first method is implemented by systems designed prior to native logical replication, which was based on triggers; an example of the application of this method can be found on Slony (`https://www.slony.info/`).

The second way is used in logical replication.

In logical replication, we are going to take the commands to be sent to the replica server within the WAL segments. The problem is that within the WAL segments, we have a physical representation of the data. In other words, within the WAL segments, the data is ready to be sent or archived to make physical copies, not to make logical copies.

In PostgreSQL 12, however, as we said earlier, we have the possibility of also having logical replications, and the way in which it realizes them is the following:

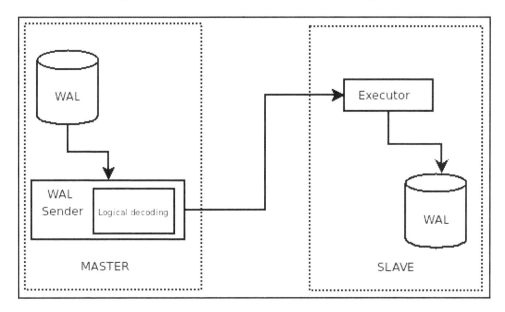

As we can see from the diagram, using a reverse engineering process, instructions are retrieved from the WAL segments, and these instructions are ready to be processed by the executor of the replica server without any parsing action. This second method is much faster than the first method. The first method was the only one that could be used for PostgreSQL versions prior to 9.4; starting from 9.4, there is an extension called `pglogical`, and starting from version 10.x, the logical replica is native.

Comparing logical replication and physical replication

Let's now examine how a logical replica differs from a physical replica:

- One of the positive characteristics of physical replicas is their speed. However, a distinct disadvantage is that we have to replicate all the databases in the cluster. Using a physical replica, it is not possible to replicate a single database belonging to an instance of PostgreSQL, and it is not possible to replicate only some tables of a database. Logical replication is a little bit slower than physical replication, but in using logical replication, we can decide which databases we want to replicate within a cluster and/or which tables we want to replicate within a single database.

- Another thing we need to keep in mind if we use physical replicas is that physical replication is only possible if the two servers have the same version of PostgreSQL. With logical replication, since the logical instruction to be executed is passed to the slave machine, it is also possible to perform replications between different versions of PostgreSQL.

- One last thing to consider in favor of physical replication is that, with the exception of operations on temporary and unlogged tables, all other operations are replicated. In a logical replication, only **data manipulation language** (**DML**) operations are replicated, and **data definition language** (**DDL**) operations such as `ALTER operation TABLE` are not replicated.

 Note: Because we can do replications between different versions of PostgreSQL, logical replication is a tool that can be used to perform PostgreSQL hot upgrades.

Logical replication is based on the concept that after being processed through a logical decoding process, WAL segments are made available through a publication mechanism. The master will then start a publication process and the replica will start a subscription process that, by connecting to the master's publication, is able to pass the decoded instructions directly to the query executor of the replica machine.

Exploring logical replication setup

Let's explore now how to perform logical replication. In this section, we will prepare the environment we need to be able to perform our logical replication.

Logical replication environment settings

Suppose we have two machines that we will call pg1 and pg2. We must remember to set our internal DNS, or the /etc hosts file, so that pg1 can reach pg2; for example, for the pg1 server, the master server will have an IP of 192.168.122.20, and for the pg2 server, the replica server will have an IP of 192.168.122.36.

First of all, let's check whether there is a connection between the two servers:

```
pg1:~$ ping pg2
PING pg2.pgtraining.com (192.168.122.36) 56(84) bytes of data.
64 bytes from pg2.pgtraining.com (192.168.122.36): icmp_seq=1 ttl=64
time=0.893 ms
64 bytes from pg2.pgtraining.com (192.168.122.36): icmp_seq=2 ttl=64
time=0.639 ms

pg2:~$ ping pg1
64 bytes from pg1.pgtraining.com (192.168.122.20): icmp_seq=1 ttl=64
time=1.40 ms
64 bytes from pg1.pgtraining.com (192.168.122.20): icmp_seq=2 ttl=64
time=1.33 ms
```

As shown here, there is a connection between the two servers.

The replica role

In order to perform a logical replication, as we have already done in the previous chapter when we talked about physical replication, we need a database user with replication permissions. So, let's create the following user on both servers:

```
postgres=# CREATE USER replicarole WITH REPLICATION ENCRYPTED PASSWORD
'SuperSecret';
```

This user will be used to manage logical replication.

Master server – postgresql.conf

Now we will modify the `postgresql.conf` file on both servers; this is to ensure that the two servers are listening on port `5432` for network interfaces. We will then modify some other values to try to optimize the logical replication procedure:

1. To the `postgresql.conf` file, we add the following line to the end of the file:

   ```
   # Add settings for extensions here
   listen_addresses = '*'
   wal_level = logical
   max_replication_slots = 10
   max_wal_senders = 10
   ```

   Now let's look at each parameter:

   - `listen addresses = '*'`: In this way, we make PostgreSQL listen on port `5432` on all network interfaces. We could also simply add the IP address of the interface where we want the PostgreSQL service to listen.
   - `wal level = logical`: We changed the value from `replica` (default) to `logical`; in this way, PostgreSQL, in addition to all the information present in the `wal level = replica` model, will add more information so that it can make the reverse engineering process possible. With `wal level = logical`, we make logical replication possible.
   - `max_replication_slots = 10`: This value must be set as at least one for each subscriber plus some for the initialization of the tables.
   - `max_wal_senders = 10`: This value must be set to a number at least equal to one for each replication slot plus those necessary for physical replication.

2. After setting these values, let's restart the master PostgreSQL server:

   ```
   # systemctl restart postgresql
   ```

3. Once that is done, we will run this command from the shell:

   ```
   # netstat -an | grep 5432
   tcp 0 0 0.0.0.0:5432 0.0.0.0:* LISTEN
   tcp6 0 0 :::5432 :::* LISTEN
   unix 2 [ ACC ] STREAM LISTENING 19910
   /var/run/postgresql/.s.PGSQL.5432
   ```

As we can see, PostgreSQL is now listening to all the network interfaces available on the server.

Replica server – postgresql.conf

When it comes to the slave server, the changes to `postgresql.conf` are as follows:

```
# Add settings for extensions here
listen_addresses = '*'
wal_level = logical
max_logical_replication_workers = 4
max_worker_processes = 10
```

As we can see, the values of `listen_addresses` and `wal_level` are identical to the master; here we don't have the values for `max_replication_slots` and `max_wal_senders`, but we have the values for the following:

- `max_logical_replication_workers`: This parameter must be set to one per subscription, plus some values to consider for table synchronizations.
- `max_worker_processes`: This must be set to at least one for each replication worker plus one.

Here, as we did with the master, let's restart the PostgreSQL server:

```
# systemctl restart postgresql
```

Once restarted, run this command from the shell:

```
# netstat -an | grep 5432
tcp 0 0 0.0.0.0:5432 0.0.0.0:* LISTEN
tcp6 0 0 :::5432 :::* LISTEN
unix 2 [ ACC ] STREAM LISTENING 19910 /var/run/postgresql/.s.PGSQL.5432
```

As we can see, PostgreSQL is now listening to all the network interfaces available on the server.

The pg_hba.conf file

Let's now configure this file on the master server so that it is possible to connect the slave machine and the master machine, using the user to replicate them. On the master machine, we set the following:

```
# IPv4 local connections:
host all all 127.0.0.1/32 md5
host all replicarole 192.168.122.36/32 md5
```

This allows the user to replicate them on the replica machine to query the master server. To activate the change, it is necessary to reload the master server:

```
# systemctl reload postgresql
```

Logical replication setup

At this point, we have everything ready to begin preparing our logical replica:

1. Let's go to the master machine and create our database:

```
db_source# create database db_source;
CREATE DATABASE
db_source=# \c db_source
You are now connected to database "db_source" as user "postgres"
```

2. Let's now create a table, t1, making sure that it has the primary key:

```
db_source=# create table t1 (id integer not null primary key, name varchar(64));
```

3. Now let's give the REPLICAROLE user SELECT permissions:

```
db_source=# SELECT ON ALL TABLES IN SCHEMA public TO replicarole;
```

4. Now let's create the publication on the master machine, where we are going to indicate the list of tables that we want to replicate on the slave machine. We can also indicate all the tables, as in our example:

```
db_source=# CREATE PUBLICATION all_tables_pub FOR ALL TABLES;
CREATE PUBLICATION
```

5. At this point, we go to the slave machine and create a new database:

```
postgres=# create database db_destination;
CREATE DATABASE
postgres=# \c db_destination
You are now connected to database "db_destination" as user "postgres".
```

6. We recreate the exact structure of the table that we created in the master machine:

```
create table t1 (id integer not null primary key, name varchar(64));
```

7. After this, we have to set the subscription so that the data from the publication is replicated on the slave machine:

```
db_destination=# CREATE SUBSCRIPTION sub_all_tables CONNECTION
'user=replicarole password=SuperSecret host=pg1 port=5432
dbname=db_source' PUBLICATION all_tables_pub;
NOTICE: created replication slot "sub_all_tables" on publisher
CREATE SUBSCRIPTION
```

Now our logical replication setup is complete.

8. We can try to insert some data into the master server:

```
db_source=# insert into t1 values(1,'Linux'),(2,'FreeBSD');
INSERT 0
```

9. As we can see here, the same data has been replicated on the replica server:

```
db_destination=# select * from t1;
 id | name
----+---------
  1 | Linux
  2 | FreeBSD
(2 rows)
```

Thus, we have successfully prepared our logical replica. We will now learn how to monitor it in the next section.

Monitoring logical replication

Just as it does for physical replication, PostgreSQL provides the necessary tools also for monitoring logical replication.

For logical replication, we must query the pg_stat_replication table, which is the same table used for monitoring physical replication, as we can see here:

```
db_source=# select * from pg_stat_replication ;
-[ RECORD 1 ]----+-----------------------------
pid              | 1311
usesysid         | 16384
usename          | replicarole
application_name | sub_all_tables
client_addr      | 192.168.122.36
client_hostname  |
client_port      | 45910
backend_start    | 2020-05-23 10:57:31.487134+02
```

```
backend_xmin       |
state              | streaming
sent_lsn           | 0/19F0D18
write_lsn          | 0/19F0D18
flush_lsn          | 0/19F0D18
replay_lsn         | 0/19F0D18
write_lag          |
flush_lag          |
replay_lag         |
sync_priority      | 0
sync_state         | async
reply_time         | 2020-05-23 11:25:42.417701+02
```

The information shown by this query is the same as what we saw in the case of physical replication, but we know this information refers to a logical replica. This query must be performed on the master server (pg1).

If we run the same query on the replica machine (pg2), we do not get any results, as we can see here:

```
db_destination=# select * from pg_stat_replication ;
(0 rows)
```

There are also two other catalog tables that we can query for more information about publications and subscriptions. Say that, on the master server, we perform this:

```
db_source=# select * from pg_publication;
-[ RECORD 1 ]+---------------
oid          | 16410
pubname      | all_tables_pub
pubowner     | 10
puballtables | t
pubinsert    | t
pubupdate    | t
pubdelete    | t
pubtruncate  | t
```

If we do that, we get information about all publications created in the database. For more information about this, consult the official documentation: https://www.postgresql.org/docs/12/catalog-pg-publication.html.

Similarly, say we run this other query on the replica server:

```
db_destination=# select * from pg_subscription;
 -[ RECORD 1 ]---+-----------------------------------------------------------
-----------------
 oid             | 16409
```

```
subdbid         | 16393
subname         | sub_all_tables
subowner        | 10
subenabled      | t
subconninfo     | user=replicarole password=SuperSecret host=pg1 port=5432
dbname=db_source
subslotname     | sub_all_tables
subsynccommit   | off
subpublications | {all_tables_pub}
```

We then have information about all subscriptions created in the database. For more information about this, consult the official documentation: `https://www.postgresql.org/docs/10/catalog-pg-subscription.html`.

Comparing physical replication and logical replication

At this point in the discussion, let's consider the main differences between physical replication and logical replication. Physical replication creates by definition a physical copy; it binarily replicates all the content of the master on the replica server that passes through the WAL. Logical replication, on the other hand, only replicates the instructions, that is, the statements that we must give to the replica server, to have a copy of the data on the master server.

Simulating on test versus bloating elimination

Physical replication, with the exception of unlogged tables, makes an identical copy of the master on the replica server. Physical replication copies absolutely everything; so, because the copy is physical at the page level, we copy not just the data but also the bloating associated with it. Sometimes this can be useful, for example, if we want to simulate the exact behavior of the production server in our test environment.

> We can use physical replication to simulate the exact behavior of the production server in our test environment.

Logical replication, however, through a reverse engineering mechanism, passes the queries to be executed directly to the query executor of the slave machine. For example, if I want to get a copy of my database starting with a low bloating percentage, I can perform a logical replica on a second machine and the second machine will initially begin with a very clean starting point. This is because all data will be passed in a non-physical but logical way to the second server.

We can use logical replication to replicate our database in a new server with a very clean starting situation.

Read-only versus write allowed

In the previous chapter, we saw that we can access a physical replication server only using read operations and that write operations are not allowed. We have also seen that physical replication replicates any type of operations, both DML operations and DDL operations. Using logical replication, we can also access write operations on the replica server, but, in a logical replica, only DML operations are replicated to the replica server; the DDL operations are not replicated. Let's conduct some tests and see what happens. In the following examples, the master server will always be called pg1 and the server with logical replication will always be called pg2.

This is our initial situation on the pg1 server:

```
db_source=# select * from t1;
 id | name
----+---------
  1 | Linux
  2 | FreeBSD
(2 rows)
```

This is our initial situation on the pg2 server:

```
db_destination=# select * from t1;
 id |  name
----+---------
  1 | Linux
  2 | FreeBSD
(2 rows)
```

Let's insert a record on the `pg2` server:

```
db_destination=# insert into t1 values (3,'OpenBSD');
INSERT 0 1
```

This is now the situation on the `pg2` server:

```
db_destination=# select * from t1;
 id | name
----+---------
  1 | Linux
  2 | FreeBSD
  3 | OpenBSD
(3 rows)
```

On the `pg1` server, we still have the following:

```
db_source=# select * from t1;
 id | name
----+---------
  1 | Linux
  2 | FreeBSD
(2 rows)
```

 The logical replica allows write operations on the replica server.

Let's see what happens if we add one record to the `pg1` server:

```
db_source=#  insert into t1 values(4,'Minix');
 INSERT 0 1
```

The situation on the `pg1` server is as follows:

```
db_source=# select * from t1;
 id  | name
----+---------
  1  | Linux
  2  | FreeBSD
  4  | Minix
 (3 rows)
```

The situation on the `pg2` server is as follows:

```
db_destination=# select * from t1;
 id | name
```

```
----+---------
 1  | Linux
 2  | FreeBSD
 3  | OpenBSD
 4  | Minix
(4 rows)
```

As we can see, the values have been inserted in the table of the master server pg1 and replicated through the logical replica on the pg2 server. Let's now see what happens if we try to insert a record with a key value already inserted on the pg2 server. For example, let's try to insert this record:

```
db_source=# insert into t1 values(3,'Windows');
INSERT 0 1
```

The situation on the pg1 server is now this:

```
db_source=# select * from t1;
 id | name
----+---------
 1  | Linux
 2  | FreeBSD
 4  | Minix
 3  | Windows
(4 rows)
```

However, the situation on the pg2 server is now this:

```
db_destination=# select * from t1;
 id | name
----+---------
 1  | Linux
 2  | FreeBSD
 3  | OpenBSD
 4  | Minix
(4 rows)
```

No record has been inserted on the pg2 server. If we examine the postgresql.log file of the pg2 replica server, we can see that there is this error:

```
20-05-23 17:58:53.698 CEST [431] LOG: background worker "logical
replication worker" (PID 3936) exited with exit code 1
2020-05-23 17:58:58.711 CEST [3938] LOG: logical replication apply worker
for subscription "sub_all_tables" has started
2020-05-23 17:58:58.732 CEST [3938] ERROR: duplicate key value violates
unique constraint "t1_pkey"
2020-05-23 17:58:58.732 CEST [3938] DETAIL: Key (id)=(3) already exists.
```

If we examine the log of the pg1 master server, we see that there are these messages:

```
2020-05-23 17:59:54.074 CEST [4548] replicarole@db_source LOG:  logical
decoding found consistent point at 0/19FE398
2020-05-23 17:59:54.074 CEST [4548] replicarole@db_source DETAIL:  There
are no running transactions.
2020-05-23 17:59:59.103 CEST [4549] replicarole@db_source LOG:  starting
logical decoding for slot "sub_all_tables"
2020-05-23 17:59:59.103 CEST [4549] replicarole@db_source DETAIL:
Streaming transactions committing after 0/19FE3D0, reading WAL
from 0/19FE398.
```

The duplicate key error on the replica server has the effect of causing the message described here on the master server.

So, now if we try to add another record on the master server, this record will not be inserted on the replica server. Say we tried on the pg1 server to perform this statement:

```
db_source=# insert into t1 values(5,'Unix');
INSERT 0 1
```

We would then have this on the pg1 server:

```
db_source=# select * from t1;
 id | name
----+----------
 1  | Linux
 2  | FreeBSD
 4  | Minix
 3  | Windows
 5  | Unix
(5 rows)
```

In the replica pg2 server, though, we would still have this:

```
db_destination=# select * from t1;
 id | name
----+----------
 1  | Linux
 2  | FreeBSD
 3  | OpenBSD
 4  | Minix
(4 rows)
```

From now on, logical replication no longer replicates data.

If we want to write records on the replica server, we have to make sure that these records do not conflict with the records on the master server.

A simple way to realign our replica server is to drop the subscription, truncate the table, and make the subscription again:

```
db_destination=# drop subscription sub_all_tables ;
NOTICE: dropped replication slot "sub_all_tables" on publisher
DROP SUBSCRIPTION
db_destination=# truncate t1;
TRUNCATE TABLE
db_destination=# CREATE SUBSCRIPTION sub_all_tables CONNECTION
'user=replicarole password=SuperSecret host=pg1 port=5432 dbname=db_source'
PUBLICATION all_tables_pub;
NOTICE: created replication slot "sub_all_tables" on publisher
CREATE SUBSCRIPTION
```

Now if we check both servers, the master server and the replica server will have all data aligned. On the pg1 server, we have the following:

```
db_source=# select * from t1;
 id | name
----+---------
 1  | Linux
 2  | FreeBSD
 4  | Minix
 3  | Windows
 5  | Unix
(5 rows)
```

On the replica pg2 server, we have this:

```
db_destination=# select * from t1;
 id | name
----+---------
 1  | Linux
 2  | FreeBSD
 4  | Minix
 3  | Windows
 5  | Unix
(5 rows)
```

DDL commands

In the previous section, we said that logical replication does not replicate DDL commands, but what happens if we apply a DDL statement on a master server that is already replicated using the logical replication? DDL commands are as follows:

- CREATE
- ALTER
- DROP
- RENAME
- TRUNCATE
- COMMENT

Suppose now we want to add a field on the t1 table of the master server, pg1:

```
db_source=# alter table t1 add description varchar(64);
ALTER TABLE
```

The situation on the pg1 server is now as follows:

```
db_source=# \d t1
Table "public.t1"
Column       | Type                   | Collation | Nullable | Default
-------------+------------------------+-----------+----------+---------
id           | integer                |           | not null |
name         | character varying(64)  |           |          |
description  | character varying(64)  |           |          |
Indexes:
 "t1_pkey" PRIMARY KEY, btree (id)
Publications:
 "all_tables_pub"
```

The situation on the replica server is the same as it was before:

```
db_destination=# \d t1
 Table "public.t1"
 Column  | Type                  | Collation | Nullable | Default
 --------+-----------------------+-----------+----------+---------
 id      | integer               |           | not null |
 name    | character varying(64) |           |          |
 Indexes:
 "t1_pkey" PRIMARY KEY, btree (id)
```

Let's try now to make a DML command on the pg1 server. Some examples of DML commands follow:

- INSERT
- DELETE
- UPDATE

For example, say we tried to delete a record from the t1 table of the pg1 server:

```
db_source=# delete from t1 where id=5;
DELETE 1
```

On the pg1 server, we would have the following:

```
db_source=# select * from t1;
 id | name     | description
----+----------+-------------
 1  | Linux    |
 2  | FreeBSD  |
 4  | Minix    |
 3  | Windows  |
(4 rows)
```

On the pg2 server, though, we would still have this:

```
db_destination=# select * from t1;
 id | name
----+---------
 1  | Linux
 2  | FreeBSD
 4  | Minix
 3  | Windows
 5  | Unix
(5 rows)
```

If we examine postgresql.log on the pg2 server, we'll see this:

```
2020-05-23 18:27:02.693 CEST [4231] ERROR: logical replication target
relation "public.t1" is missing some replicated columns
2020-05-23 18:27:02.696 CEST [431] LOG: background worker "logical
replication worker" (PID 4231) exited with exit code 1
```

The logical replication does not work anymore because the logical replication target relation `public.t1` is missing some replicated columns, as the server log said. If we want to solve this problem, we must execute the DDL on the replica server:

```
db_destination=# alter table t1 add description varchar(64);
ALTER TABLE
```

Now if we check the records on the `pg2` server, we have the same records that are present on the `pg1` server:

```
db_destination=# select * from t1;
 id | name     | description
----+---------+-------------
  1 | Linux    |
  2 | FreeBSD  |
  4 | Minix    |
  3 | Windows  |
(4 rows)
```

DDL commands must always be replicated on the replica servers.

Disabling logical replication

In the previous section, we used the DROP SUBSCRIPTION command to drop a subscription. There may be cases where we cannot use this command directly. For example, suppose that the master server becomes unreachable and we need to drop the subscription on the replica server. If we try to execute a DROP SUBSCRIPTION command, we will get the following response:

```
db_destination=# drop subscription sub_all_tables ;
ERROR: could not connect to publisher when attempting to drop the
replication slot "sub_all_tables"
DETAIL: The error was: could not connect to server: Connection refused
 Is the server running on host "pg1" (192.168.122.20) and accepting
 TCP/IP connections on port 5432?
HINT: Use ALTER SUBSCRIPTION ... SET (slot_name = NONE) to disassociate the
subscription from the slot.
```

PostgreSQL suggests using `ALTER SUBSCRIPTION ... SET (slot_name = NONE)` to disassociate the subscription from the slot. The problem is that we cannot execute this command before having disabled the subscription. In fact, if we try to perform the command suggested by PostgreSQL now, we will get this:

```
db_destination=# alter subscription sub_all_tables SET (slot_name = NONE);
ERROR: cannot set slot_name = NONE for enabled subscription
```

The correct steps that we have to execute are as follows:

1. Disable the subscription.
2. Set `slot_name` to `NONE`.
3. Drop the subscription.

We have to perform the following three statements:

```
db_destination=# alter subscription sub_all_tables disable;
ALTER SUBSCRIPTION
db_destination=# alter subscription sub_all_tables SET (slot_name = NONE);
ALTER SUBSCRIPTION
db_destination=# drop subscription sub_all_tables ;
DROP SUBSCRIPTION
```

These are the correct steps if we want to drop a subscription when the master server becomes unreachable. We can also use the `ALTER SUBSCRIPTION sub_name DISABLE` command to detach the subscription from the publication and the `ALTER SUBSCRIPTION sub_name ENABLE` command to re-attach the subscription to the publication.

Summary

In this chapter, we discussed logical replication. We have seen that logical replication is based on a concept of reverse engineering that starts with the analysis of WAL segments to extract the logical commands that have to be passed to a replica server. We saw that logical replication is useful when we want to replicate parts of databases and when we want to make hot migrations between different versions of PostgreSQL. Logical replication makes this possible because it does not binarily replicate data but rather extracts the logical DML commands from WAL, which are then replicated on the replica server. We have seen how to make a logical replica in practice and have addressed some of the issues that can occur when we work with logical replication.

In the next chapter, we'll talk about useful tools and useful extensions. We will see which tools are best to make life easier for a PostgreSQL DBA.

References

- Slony website: `https://www.slony.info`
- `https://www.postgresql.org/docs/10/logical-replication.html`
- `https://www.postgresql.org/docs/12/catalog-pg-publication.html`
- `https://www.postgresql.org/docs/10/catalog-pg-subscription.html`

Section 5: The PostegreSQL Ecosystem

5

In this section, you will be presented with useful tools and extensions that can make your PostgreSQL experience better and more professional. You will also learn about various features offered in the next version of PostgreSQL – PostgreSQL 13.

This section contains the following chapters:

- *Chapter 19, Useful Tools and Extensions*
- *Chapter 20, Towards PostgreSQL 13*

19
Useful Tools and Extensions

This chapter is to be considered as an appendix to the book. In this chapter, we will talk about some tools and extensions that allow a DBA to maximize the efficiency of their work by minimizing the effort needed to complete it.

We will talk about these extensions:

- `pg_trgm`
- Foreign data wrappers and the `postgres_fdw` extension
- `btree_gin`

These are some of the official extensions for PostgreSQL. A site that can be very useful for finding extensions available for PostgreSQL is `https://pgxn.org/`.

In addition to extensions, we will also talk about useful tools for the PostgreSQL DBA. There are dozens of tools available for PostgreSQL, but in this chapter, we will talk about `pgbackrest`, a powerful tool useful to manage disaster recovery and **point-in-time recovery** (**PITR**).

The first one is very useful for managing continuous backup and the second one is an out-of-the-box tool that can help us to better monitor our PostgreSQL server. This chapter is intended to be just a quick overview of some of the most useful PostgreSQL extensions.

The following topics will be covered here:

- Exploring the `pg_trgm` extension
- Using foreign data wrappers and the `postgres_fdw` extension
- Exploring the `btree_gin` extension
- Managing the `pgbackrest` tool

Exploring the pg_trgm extension

In previous chapters, we talked about query optimization and indexing. When we talked about indexing, we learned how to make our queries faster through the use of indices. However, B-tree indices do not index all types of operations. Now let's consider textual data types (`char`, `varchar`, or `text`). We have seen that the B-tree, using the `varchar_pattern_ops` opclass, is able to index `like` operations only as regards the `'search%'` type queries, but it is not able to index queries with a `where` condition of the `'%search'` or `'search%'` type:

1. Before diving into our example, let's do `set enable_seqscan to 'off'` in order to force PostgreSQL to use any index if it exists. We need to do this because, in our example case, PostgreSQL would always use sequential scanning, because we have less data in our table and all data that is present in the table is stored on a single page:

   ```
   db_source=# set enable_seqscan to 'off';
   SET
   ```

2. In our database, we now have these records in the `t1` table:

   ```
   db_source=# select * from t1;
    id | name    | description
   ----+---------+-------------
    1  | Linux   |
    2  | FreeBSD |
    4  | Minix   |
    3  | Windows |
   (4 rows)
   ```

3. Let's create a B-tree index with the `varchar` opclass in order to check whether PostgreSQL uses index access to the table when we perform a query with the `like` operator:

   ```
   db_source=# create index db_source_name_btree on t1 using
   btree(name varchar_pattern_ops);
   CREATE INDEX
   ```

4. Let's now perform some `like` queries:

   - Here's a `like` query using a `'search%'` predicate:

     ```
     db_source=# explain analyze select * from t1 where name like
     'Li%';
     QUERY PLAN
     ```

```
-------------------------------------------------------------
-----------------------------
 Index Scan using db_source_name_btree on t1 (cost=0.13..8.15
rows=1 width=104) (actual time=0.331..0.335 rows=1 loops=1)
 Index Cond: (((name)::text ~>=~ 'Li'::text) AND ((name)::text ~<~
'Lj'::text))
 Filter: ((name)::text ~~ 'Li%'::text)
 Planning Time: 2.102 ms
 Execution Time: 0.413 ms
(5 rows)
```

- Here's a `like` query using a `'%search'` predicate:

```
db_source=# explain analyze select * from t1 where name like
'%Li';
 QUERY PLAN
-------------------------------------------------------------
-----------------------------
 Seq Scan on t1 (cost=10000000000.00..10000000001.05 rows=1
width=104) (actual time=34.406..34.406 rows=0 loops=1)
 Filter: ((name)::text ~~ '%Li'::text)
 Rows Removed by Filter: 4
 Planning Time: 0.115 ms
 JIT:
 Functions: 2
 Options: Inlining true, Optimization true, Expressions true,
Deforming true
 Timing: Generation 0.900 ms, Inlining 11.458 ms, Optimization
15.832 ms, Emission 6.777 ms, Total 34.967 ms
 Execution Time: 35.442 ms
(9 rows)
```

- Here's an `ilike` query using a `'%search'` predicate:

```
db_source=# explain analyze select * from t1 where name ilike
'%Li';

 QUERY PLAN
-------------------------------------------------------------
-----------------------------
 Seq Scan on t1 (cost=10000000000.00..10000000001.05 rows=1
width=104) (actual time=28.874..28.874 rows=0 loops=1)
 Filter: ((name)::text ~~* '%Li'::text)
 Rows Removed by Filter: 4
 Planning Time: 0.136 ms
 JIT:
 Functions: 2
 Options: Inlining true, Optimization true, Expressions true,
```

```
Deforming true
 Timing: Generation 1.048 ms, Inlining 8.150 ms, Optimization
14.527 ms, Emission 6.054 ms, Total 29.778 ms
 Execution Time: 29.975 ms
(9 rows)
```

As we can see, only in the first case did PostgreSQL use an index approach. In the other cases, PostgreSQL used a sequence scan; to improve this kind of search, we can use the pg_trgm extension. pg_tgrm is an official extension and is included in the official PostgreSQL contribs package. When we use this extension, PostgreSQL splits every word into a set of trigrams and makes a GIST or GIN index on it. For example, if we consider a word such as dog, its set of trigrams consists of d, do, og, and dog. Let's look at how this works in practice:

1. First of all, let's install the extension:

```
db_source=# create extension pg_trgm;
CREATE EXTENSION
```

2. Now we can create a GIN or GIST index using the opclass trigram. For example, let's create a GIN index using the gin_trgm_ops opclass:

```
db_source=# create index db_source_name_gin on t1 using gin
(name gin_trgm_ops);
CREATE INDEX
```

3. Now let's perform our ilike query:

```
db_source=# explain analyze select * from t1 where name ilike
'%Li';
QUERY PLAN
----------------------------------------------------------------
----------------------------------
 Bitmap Heap Scan on t1 (cost=8.00..12.01 rows=1 width=104) (actual
time=0.062..0.062 rows=0 loops=1)
 Recheck Cond: ((name)::text ~~* '%Li'::text)
 -> Bitmap Index Scan on db_source_name_gin (cost=0.00..8.00 rows=1
width=0) (actual time=0.048..0.048 rows=0 loops=1)
 Index Cond: ((name)::text ~~* '%Li'::text)
 Planning Time: 5.786 ms
 Execution Time: 0.152 ms
(6 rows)
```

As can be seen here, PostgreSQL is now able to create an index access using a `like` query. The same thing happens for all types of `like` and `ilike` queries; the `pg_trgm` extension solves the access index to tables for this type of query. For further information about the `pg_trgm` extension, see `https://www.postgresql.org/docs/12/pgtrgm.html`. Now, we will move on to the next extension, `postgres_fdw`.

Using foreign data wrappers and the postgres_fdw extension

This section will provide a brief introduction to what foreign data wrappers are. Foreign data wrappers allow us to access data that is hosted on an external database as if it was kept on a normal local table. We can connect PostgreSQL to various data sources, we can connect PostgreSQL to another PostgreSQL server, or we can connect PostgreSQL to another data source that can be relational or non-relational. Once the foreign data wrapper is connected, PostgreSQL is able to read the remote table as if it were local. There are foreign data wrappers for well-known databases such as Oracle and MySQL, and there are foreign data wrappers for lesser-known systems. A complete list of foreign data wrappers available for PostgreSQL is available at `https://wiki.postgresql.org/wiki/Foreign_data_wrappers`.

In this section, we will consider an example using the `postgresql_fdw` foreign data wrapper, which is used to connect a PostgreSQL server to another PostgreSQL server.

Our starting situation is with two servers. We have one server called `pg1` server with an IP address of `192.168.12.34` and a second server called `pg2` with an IP address of `192.168.12.35`. In the `pg1` server, there is a `db1` database with a `t1` table, and on the `pg2` server, there is a `db2` database with a `t2` table. Our goal will be to connect the `pg2` server to the `pg1` server and make it possible to query the `t2` table, as if it were local, from the `pg1` server:

1. Our starting situation on the `pg1` server is as follows:

```
db1=# select * from t1;
 id | name
----+-------
  1 | Linux
```

2. Our starting situation on the `pg2` server is as follows:

```
db2=# select * from t2;
 id | name
```

```
----+------
 1  | Unix
```

3. Let's start with the installation of the `postgres_fdw` extension on the `pg1` server:

```
forumdb=# create extension postgres_fdw ;
CREATE EXTENSION
```

Suppose that on the `pg2` server, `pg_hba.conf` is configured as follows:

```
host    all             all             192.168.12.0/24
trust
```

4. Now we have to create the connection between the two servers:

```
db1=# CREATE SERVER remote_pg2 FOREIGN DATA WRAPPER postgres_fdw
OPTIONS (host '192.168.12.36', dbname 'db2');
CREATE SERVER
```

5. Now, we have to write a user map between the two servers:

```
db1=#  CREATE USER MAPPING FOR CURRENT_USER SERVER remote_pg2
OPTIONS (user 'postgres', password '');
CREATE USER MAPPING
```

6. With the following statement, we say to PostgreSQL that the current user of the `pg1` server, which in this case is the `postgres` user, will be mapped to the `postgres` user of the remote server:

```
db1=# create foreign table f_t2 (id integer, name varchar(64))
SERVER remote_pg2 OPTIONS (schema_name 'public', table_name 't2');
CREATE FOREIGN TABLE
```

7. Now we can query the `f_t2` table as if it were a local table:

```
db1# select * from f_t2;
 id | name
----+------
 1  | Unix
```

As we can see in the preceding example, we can query a foreign table as if it were on the local server. For further information, please see `https://www.postgresql.org/docs/12/postgres-fdw.html`. With this, we have learned all about that extension. We will now move on to the next one, which is the `btree_gin` extension.

Exploring the btree_gin extension

Before we start talking about the actual extension, let's spend a moment on the concept of GIN. **GIN** is an acronym for **Generalized Inverted Index**. In this chapter, we assume that you know what an inverted index is. Using GIN libraries, it is possible to build indices for different data types; it is also possible to create B-tree-type indices with the use of the GIN library. The `btree_gin` extension can index the following data types: `int2`, `int4`, `int8`, `float4`, `float8`, `timestamp with time zone`, `timestamp without time zone`, `time with time zone`, `time without time zone`, `date`, `interval`, `oid`, `money`, `char`, `varchar`, `text`, `bytea`, `bit`, `varbit`, `macaddr`, `macaddr8`, `inet`, `cidr`, `uuid`, `name`, `bool`, `bpchar`, and `enum types`.

The question we need to ask is, *When should we use* `tree_gin` *indices* *instead of default b-tree indices?* As the structure of the GIN index is constructed, it is useful when we are dealing with fields with many records but low cardinality; in this case, the `btree_gin` index will be much smaller than the classic b-tree index. For those coming from the Oracle world, there are bitmap indices in Oracle that are used for this purpose.

 In PostgreSQL, b-tree GIN indices can be used like Oracle bitmap indices.

We will now see an example of this:

1. Let's start by installing the extension:

   ```
   db1=# create extension btree_gin;
   CREATE EXTENSION
   ```

2. Now let's create an example table as follows:

   ```
   db1=# create table users (id serial not null primary key,name
   varchar(64) ,surname varchar(64),
    sex char(1));
   CREATE TABLE
   ```

3. Now let's populate it with some random data:

```
db1=# insert into users (name,surname) select
'name_'||generate_series(1,10000)::text,'surname_'||generate_series
(1,10000);
INSERT 0 10000
db1=# update users set sex = case when id%2 = 0 then 'M' else 'F'
end;
UPDATE 10000
```

In the `users` table, there are now 10,000 records, as shown here:

```
db1=# select count(*) from users;
 count
-------
 10000
(1 riga)
```

All the records with an even value for the `id` field will have `sex=F` and all records with an odd value for `id` will have `sex = M`, as we can see here:

```
db1=# select * from users limit 4;
 id | name   | surname   | sex
----+--------+-----------+-----
 1  | name_1 | surname_1 | F
 2  | name_2 | surname_2 | M
 3  | name_3 | surname_3 | F
 4  | name_4 | surname_4 | M
(4 rows)
```

So, we have 5,000 records with `sex='F'`, as shown here:

```
db1=# select count(*) from users where sex='F';
 count
-------
 5000
(1 row)
```

We have 5,000 records with `sex='M'`:

```
db1=# select count(*) from users where sex='M';
 count
-------
 5000
(1 row)
```

4. Now that we have set up our environment, let's try to create two indices for the sex field, the first one being of the btree type:

```
db1=# create index sex_btree on users using btree (sex);
CREATE INDEX
```

5. We'll also create another index of the btree_gin type:

```
db1=# create index sex_gin on users using gin (sex);
CREATE INDEX
```

We will see that the GIN index is much smaller than the b-tree index. The b-tree index will be of this size:

```
db1=# select pg_size_pretty(pg_relation_size('sex_btree'));
 pg_size_pretty
----------------
 240 kB
(1 row)
```

The GIN index will be of this size:

```
db1=# select pg_size_pretty(pg_relation_size('sex_gin'));
 pg_size_pretty
----------------
 32 kB
(1 row)
```

 We can use B-tree GIN indices to reduce index size for fields with very low granularity.

We have looked at the GIN index and various other extensions that can be used in PostgreSQL. We will now move on to the pgbackrest tool, which helps in disaster management.

Managing the pgbackrest tool

In `Chapter 18`, *Logical Replication*, we talked about disaster recovery and PITR, and we saw how to conduct them programmatically. In the real world, a DBA has to manage multiple PostgreSQL servers and it is useful to have some tools to make life easier. The open source world offers us a lot of solutions to address disaster recovery in an easy way. Some of these tools are listed here:

- WAL-E
- `pgbarman`
- OmniPITR

There are many others, and at `https://wiki.postgresql.org/wiki/Binary_Replication_Tools`, you can find a good comparison of them all.

In this section, we will give a nod to `pgbackrest`. The `pgbackrest` tool is a tool for PostgreSQL disaster recovery and PITR, and it has been designed for heavy load servers. Its official URL is `https://pgbackrest.org/`.

These are some of the features of the tool:

- It supports parallel backup and parallel restore.
- It can make full base backups, incremental backups, or differential backups.
- We can choose to make local operations or remote operations.
- We can choose our policy retention for backups and archive expiration.
- It supports a backup resume.
- It is possible to make streaming compression and create checksums.
- For a restore procedure, we can use the delta restore feature.
- It is possible to use parallel WAL archiving.
- It supports tablespaces and links.
- It is supported by S3.

Basic concepts

`pgbackrest` has the concept of the stanza, and it can also use an external repository:

- A stanza is a configuration of a remote server for backup. It is a set of targets to be backed up. A stanza configuration can contain multiple servers, in which case the first (`pg1`) is the master and the others are considered standby servers.
- A repository is local or remote storage (SSH) to which backups are saved; it can be encrypted. A repository can contain multiple definitions, but only the first one (`repo1`) is currently supported.
- It is important to have a public key exchange between users who use `pgbackrest`. The simplest thing to do is to have public keys exchanged between the Postgres user of the PostgreSQL server and the Postgres user of the server where the `pgbackrest` repository is present.

Environment setting

Before starting and testing our `pgbackrest` tool, we must make sure we have what we need to start working. We need the following:

- A running PostgreSQL server
- A new server where we will install and config the `pgbackrest` tool

In this scenario, we will continue to use our `pg1` PostgreSQL server with `ip=192.168.12.35`. We add another server called `pgdr` with an IP address of `192.168.12.37`.

The exchange of public keys

We will now see how to exchange public keys before we install `pgbackrest`:

1. First of all, let's create an `ssh` key for the Postgres user on both servers. As a PostgreSQL user, let's execute the following:

```
postgres@pgdr:~$ ssh-keygen -t rsa -b 4096
Generating public/private rsa key pair.
Enter file in which to save the key
(/var/lib/postgresql/.ssh/id_rsa):
Created directory '/var/lib/postgresql/.ssh'.
Enter passphrase (empty for no passphrase):
Enter same passphrase again:
```

```
Your identification has been saved in
/var/lib/postgresql/.ssh/id_rsa.
Your public key has been saved in
/var/lib/postgresql/.ssh/id_rsa.pub.
The key fingerprint is:
SHA256:PUT6Z7BTH83CNKyR3JLDLkbHjOTiYmE4OuO3Zrbwra0 postgres@pgdr
The key's randomart image is:
+---[RSA 4096]----+
|          .o* =o  |
|      .    +o %oo+ |
|     o o o.+o.=+ o|
|   . o o =o+o. o  |
|   +  o S.*.o .   |
| . o . .    =     |
|  o .             |
|   +++            |
|   +E+o           |
+----[SHA256]-----+
```

When we execute the `ssh-keygen` command, we have to make sure to press only the *Enter* key at the request of the passphrase.

2. Now, on both servers, we will have two files in the `/var/lib/postgresql/.ssh/` directory:

```
postgres@pg1:~/.ssh$ ls -l
totale 8
-rw------- 1 postgres postgres 3381 mag 30 11:41 id_rsa
-rw-r--r-- 1 postgres postgres 738 mag 30 11:41 id_rsa.pub
```

3. Now let's copy the public key of the `pg1` server inside the `authorized_keys` part of the `pgdr` server and vice versa. For the `pg1` server, as a Postgres user, we have to execute this:

```
postgres@pg1:$ cd $HOME/.ssh
postgres@pg1:~/.ssh$ touch authorized_keys
postgres@pg1:~/.ssh$ chmod 640 authorized_keys
```

4. Now let's go on the `pgdr1` server and show the contents of the `id_rsa.pub` file:

```
postgres@pgdr:~/.ssh$ cat id_rsa.pub
ssh-rsa
AAAAB3NzaC1yc2EAAAADAQABAAACAQC/evaUYDH1AzyajSfEPuuolHFT6GaLOI3lrbN
CYNyr3hgFzyr62jjoHYRofpX7iTGch/nLqvY9bqFHNxyKeNw96LS+aOHzF6JpgIg5Hg
xuIy3GbX8bF2p1tVek7yhEN/VTUIDTQM33w1vJ57YdyW2LbstVl0jn8cLjBmn9eLH7a
CjBGtRlSKZfNPbd26vBqZeV2nr0lBK5kAunalSRI8vZ9OWAEjqC4BoYSVL1Q+VyJrf2
```

```
QD17YVv2uCYBzGwIrsOG8tQBy3jEmnhOROraqKc6pr4AoSOIHiUAhgsvWf1Zo/ysBT2
oOdu+Vey0wJNv300wJWGAGqj3RdnrRe/3grTwJ2ZLnfck5FbFWji9wQuLND2vACztnV
RV/DWynLtIf9sia82PQgB4+xsh9yojrrOHN5QCOTKP/4x3ANTqqfM+nX6r5iBcFwQ92
nRzN1TwLopY9d+PlESb5l/PLX62FAE+YiYRECiyJIc/QMKjy34CkLu9sHI2E+i5Hpxb
AUk/9vjKCpSIDsnGHgCaom4QyPhUkItxHwQOCCy9Qwh/cKreGD+9bOYl0XvhZexywyJ
jM4vZZ0XEgf36vWvFJed1TjLtlRntuqQjxpOLso+Yqb9mWorPtC9+ypLK1Faoiee8TS
aR917L/rV0OgpogGdyClReOqlPSIxQQqVHqKWEFzXQ2w== postgres@pgdr
```

5. Next, we copy and paste this content onto the `authorized_keys` part of the `pg1` server. After this operation, our `authorized_keys` file will be as follows:

```
postgres@pgdr:~/.ssh$ cat id_rsa.pub
ssh-rsa
AAAAB3NzaC1yc2EAAAADAQABAAACAQC/evaUYDH1AzyajSfEPuuolHFT6GaLOI3lrbN
CYNyr3hgFzyr62jjoHYRofpX7iTGch/nLqvY9bqFHNxyKeNw96LS+aOHzF6JpgIg5Hg
xuIy3GbX8bF2p1tVek7yhEN/VTUIDTQM33w1vJ57YdyW2LbstVl0jn8cLjBmn9eLH7a
CjBGtRlSKZfNPbd26vBqZeV2nr0lBK5kAunalSRI8vZ9OWAEjqC4BoYSVL1Q+VyJrf2
QD17YVv2uCYBzGwIrsOG8tQBy3jEmnhOROraqKc6pr4AoSOIHiUAhgsvWf1Zo/ysBT2
oOdu+Vey0wJNv300wJWGAGqj3RdnrRe/3grTwJ2ZLnfck5FbFWji9wQuLND2vACztnV
RV/DWynLtIf9sia82PQgB4+xsh9yojrrOHN5QCOTKP/4x3ANTqqfM+nX6r5iBcFwQ92
nRzN1TwLopY9d+PlESb5l/PLX62FAE+YiYRECiyJIc/QMKjy34CkLu9sHI2E+i5Hpxb
AUk/9vjKCpSIDsnGHgCaom4QyPhUkItxHwQOCCy9Qwh/cKreGD+9bOYl0XvhZexywyJ
jM4vZZ0XEgf36vWvFJed1TjLtlRntuqQjxpOLso+Yqb9mWorPtC9+ypLK1Faoiee8TS
aR917L/rV0OgpogGdyClReOqlPSIxQQqVHqKWEFzXQ2w== postgres@pgdr
```

6. Now we can connect ourselves using the `ssh` command from the `pgdr` server to the `pg1` server without using a password:

```
postgres@pgdr:~/.ssh$ ssh postgres@192.168.12.35
The authenticity of host '192.168.12.35 (192.168.12.35)' can't be
established.
ECDSA key fingerprint is
SHA256:ewMgh+/TdX0CsfrMErYRKxyliqvoIUwmZ8P7apADqJ4.
Are you sure you want to continue connecting (yes/no)? yes
Warning: Permanently added '192.168.12.35' (ECDSA) to the list of
known hosts.
Linux pg1 4.19.0-9-amd64 #1 SMP Debian 4.19.118-2 (2020-04-29)
x86_64The programs included with the Debian GNU/Linux system are
free software;the exact distribution terms for each program are
described in the
individual files in /usr/share/doc/*/copyright.Debian GNU/Linux
comes with ABSOLUTELY NO WARRANTY, to the extentpermitted by
applicable law.
Last login: Sat May 30 11:53:04 2020 from 192.168.12.37
postgres@pg1:~$
```

If we connect for a second time from the `phdr` host to the `pg1` host, no question will be shown.

7. We now have to perform the same operations for the `pg1` host, and after we do this, also on the `pgdr1` server, we will have an `authorized_key` file with this content:

```
postgres@pg1:~/.ssh$ cat id_rsa.pub
ssh-rsa
AAAAB3NzaC1yc2EAAAADAQABAAACAQC9QAHzwQkAzFEDxhjpfIkL1I+lX0sVdeGrJHx
k08IKcBX40rrTC97FhjYJgAGai+MMe3JC5mHaY2FCTbM9iT66ai57kkqPUCPR6tnT6W
0zHtybgriboa/RDsLYhzsqhkezDnMMOZsvSeYJEcTW4M0KKO1HqlTQMSxXQNzyAvCPT
k9vdlT8I+KJsx/tGnYXs4mh/A2HrEQJ6IPMbd7ZKE/wgwp+wT7hEQreLec12F+www8T
wJTXT9xvERdDJCrqNdj0+XRhl0YmlpWOOAo2cjCSdWwVwAXPwiFUlICT9djNkQ5JuWa
GRFjp8PYi6Y+Meqm9pmGLCknMJzn5pGnC9fJL5Nhe3PQ1vkg6L8Fi3yh6P3JzAp7D4V
Z0YS627kZ8h8EC06ZSCNq2QQRHyGoiYBhYiS0qSa+uDrPyOzIcRIq1mZImZ8FrxVIZX
htNJtmaLBi9E0OAErFt4pHRz5XJ7Y/3Ccetp+bhTDNatjUHNoeyJhOyCHvX/4evtC/A
L4zoU2CWJ/04VpZchawUVpB5iA114BS6G/RW9bOKPWtLDtRogWVNd1qj+B/kLrKTjvM
QTEe3UjGZpmZn393ynTT6e3FefaD3OyMvGEoWtyKiMSl3NI18X51EAy5eSBvlX4qL45
i1/9d1XK7hmm52dshPmYg4FSTwtyRBimxNbZ8CdGCo5Q== postgres@pg1
```

Now, using the Postgres user, it is possible to connect the two servers together without providing a password.

Installing pgbackrest

Before installing, we have to know that each host has the same version of `pgbackrest` installed, so we have to install on the `pgdr` server and on the `pg1` server. On a Debian-like server as the root user, let's execute these commands on both servers:

```
# apt-get update
# apt-get install -y pgbackrest
```

If we use a RedHat-like server, we have to use the `yum` command instead of the `apt-get` command.

Configuring pgbackrest

Now let's look at how to configure the `pgbackrest` tool. `pgbackrest` needs the configuration of both servers; it needs the configuration of the repository server, which is where the data will be stored, and it needs the configuration of the PostgreSQL server so that it is able to send all the data to the repository server. So, we will address both of these configurations in turn:

- The repository configuration of the `pgdr` server
- The PostgreSQL configuration of the `pg1` server

The repository configuration

The configuration file can be found here:

```
/etc/pgbackrest/pgbackrest.conf
```

Using a different configuration file is possible but this must be specified consistently in each use of the program, so it is better to leave the default one. Each parameter specified in the configuration file can be overwritten by the relative parameter provided on the command line. Each parameter contained in a section is specified with a key-value pair. In the stanza configuration, the parameters of a cluster always start with `pgN-`, with `N` being a progressive number. The main (master) cluster is always number 1. The standby clusters are therefore numbered 2 in sequence. Similarly, in global parameters, repositories are numbered starting from 1 (repo1), but currently, multiple repositories are not supported. `pgbackrest` is symmetric; that is, every command can be executed on the backup machine or on the target machine. We will have a configuration file for the repository server and a configuration file for the PostgreSQL server and the two configuration files are different. `pgbackrest` by default has enabled the compression of WAL segments and basebackups with a compression factor of six. We can force the compression to a different level using the compress-level directive; for example, we can set the compression level to nine to have the maximum compression.

It is also possible to encrypt the repository managed by `pgbackrest`; this feature is useful for storing our backups on a low-cost cloud, for example.

Let's start now with a simple configuration; let's start with the global config section:

```
[global]
start-fast=y
archive-async=y
process-max=2
repo-path=/var/lib/pgbackrest
```

```
repo1-retention-full=2
repo1-retention-archive=5
repo1-retention-diff=3
log-level-console=info
log-level-file=info
```

We see the following options here:

- `start-fast=y`: Forces a checkpoint on the remote server, so that `pg_start_backup ()` starts as soon as possible.
- `archive-async=y`: Enables the asynchronous transfer of WAL for push/pull operations.
- `process-max=2`: Sets the maximum number of processes that the system can use for transfer/compression operations.
- `repo-path=/var/lib/pgbackrest`: Sets the path where the repository will be stored; the user running the `pgbackrest` command must have read/write permissions for this directory.
- `repo1-retention-full=2`: The number of full backups to keep. When a full backup expires, all differential and/or incremental backups associated with the full backup will also expire. When the option is not defined, the system issues a warning. If indefinite retention is desired, set the option to the maximum value (9,999,999).
- `repo1-retention-archive=5`: Represents the backup number of the WAL files to keep. The WAL segments required to make a backup consistent are always maintained until the backup expires, regardless of the configuration of this option. If this value is not set, the expiring archive will automatically expire at the `repo-retention-full` (or `repo-retention-diff`) value corresponding to the type of repo-retention archive if set to `full` (or `diff`). This will ensure that the WAL files are considered expired only for backups that have already expired.
- `repo1-retention-diff = 3`: The number of differential backups to keep. When a differential backup expires, all incremental backups associated with the differential backup will also expire. If not defined, all differential backups will be kept until the full backups on which they depend expire.
- `log-level-console=info/log-level-file=info log`: Settings for log management; set the terminal log level (`log-level-console`) and the logging level on the log file (`log-level-file`).

The configuration file shown here is just a simple example; if we want to add some more features, we just need to add them to the configuration file.

For example, if we want to modify the compression level and increase it to level nine, we can add these lines:

```
compress = y
compress-level = 9
compress-level-network = 9
```

In the same way, if we want to add the cipher feature, we can add these lines:

```
repo1-cipher-type = aes-256-cbc
repo1-cipher-pass = SuperSecret
```

After configuring the global section, we are ready to look at how to configure the stanza. `pgbackrest` introduces the idea of stanzas; in practice, we can associate each stanza with a cluster database. The following is an example of a room; it is only a coincidence that the name of the stanza, `[pg1]`, has the same name as the cluster. It is necessary to create a stanza for each remote PostgreSQL server on which we want to manage backups using `pgbackrest`. Each stanza must have a different name:

```
[pg1]
pg1-host = 192.168.12.35
pg1-host-user = postgres
pg1-path = /var/lib/postgresql/12/main
pg1-port = 5432
```

We see the following options here:

- `pg1-host`: This is the remote host of the PostgreSQL master server.
- `pg1-host-user = postgres`: When the `pg-host` parameter is set, this is the user that we want to use to access the remote PostgreSQL server. This user will also be the owner of the remote `pgbackrest` process and it starts the connection to the PostgreSQL server. This user should be the owner of the PostgreSQL database cluster. Usually, we can leave the default user, `postgres`, which is why it is usually the same user for whom we made the exchange of public keys.
- `pg1-path = /var/lib/postgresql/12/main`: The path on the PostgreSQL cluster where the data is stored. We can find it in the `data_directory` parameter inside the `postgresql.conf` file.
- `pg1-port = 5432`: The listen port of the remote PostgreSQL server.

The PostgreSQL server configuration

On the PostgreSQL server, `pg1`, we need to modify the `postgresql.conf` file and we need to set the `pgbackrest.conf` file as well.

The postgresql.conf file

For the `postgresql.conf` file, we have to set `wal_level` to `replica` or `logical`. It is important that the WAL level is not set to `minimal`. We also need to tell PostgreSQL the command that will send the WAL segment to the `pgbackrest` repository server.

Let's add these lines at the and of the `postgresql.conf` file:

```
#PGBACKREST
archive_mode = on
wal_level = logical
archive_command = 'pgbackrest --stanza=pg1 archive-push %p'
```

With the second line, we say to PostgreSQL that the WAL segments will be archived on the `pg1` stanza of the repository server using the `pgbackrest` command. After restarting PostgreSQL, these new lines will be available. As the root user, let's perform a restart of the PostgreSQL service:

```
# systemctl restart postgresql
```

The pgbackrest.conf file

Now, after modifying `postgresql.conf`, let's go to modify the `pgbackrest.conf` file of the PostgreSQL server. Let's remember that the PostgreSQL server has `ip=192.168.12.35`, and that the IP of the disaster recovery server is `192.168.12.35`. Let's now edit the `/etc/pgbackrest.conf` file; delete what is present and add these lines:

```
[global]
backup-host=192.168.12.37
backup-user=postgres
backup-ssh-port=22
log-level-console=info
log-level-file=info

[pg1]
pg1-path = /var/lib/postgresql/12/main
pg1-port = 5432
```

As for the repository configuration, the file is composed of sections: a global section and a section for each stanza.

For the global section, we have the following options:

- `backup-host`: The repository host
- `backup-user`: The user used for the backup
- `backup-ssh-port`: The `ssh` port
- `log-level-console=info - log-level-file=info`: As per the configuration seen in the previous section

For the stanza section, we have the following options:

- `pg1-path = /var/lib/postgresql/12/main`: The path on the PostgreSQL cluster where the data is stored. We can find it in the `data_directory` parameter inside the `postgresql.conf` file.
- `pg1-port = 5432`: The listen port of the remote PostgreSQL server.

Creating and managing continuous backups

Now that we have our system well configured, let's start to manage our backup.

Creating the stanza

The first thing we have to do is create the stanza on the repository server. To do this, as the root user, let's perform this command:

```
# sudo -iu postgres pgbackrest --stanza=pg1 stanza-create
2020-05-30 15:31:46.777 P00 INFO: stanza-create command begin 2.27: --log-level-console=info --log-level-file=info --pg1-host=192.168.12.35 --pg1-host-user=postgres --pg1-path=/var/lib/postgresql/12/main --pg1-port=5432 --repo1-path=/var/lib/pgbackrest --stanza=pg1
2020-05-30 15:31:48.151 P00 INFO: stanza-create command end: completed successfully (1375ms)
```

Now our stanza is created. If we go to `/var/lib/pgbackrest`, we can find the directory structure that will be used by the continuous backup system:

```
# ls -l
totale 8
drwxr-x--- 3 postgres postgres 4096 mag 30 15:31 archive
drwxr-x--- 3 postgres postgres 4096 mag 30 15:31 backup
```

Checking the stanza

After creating our stanza, let's check whether the system is ready to accept the continuous backup by performing this:

```
# sudo -iu postgres pgbackrest --stanza=pg1 check
2020-05-30 15:51:36.589 P00   INFO: check command begin 2.27: --log-level-
console=info --log-level-file=info --pg1-host=192.168.12.35 --pg1-host-
user=postgres --pg1-path=/var/lib/postgresql/12/main --pg1-port=5432 --
repo1-path=/var/lib/p
gbackrest --stanza=pg1
2020-05-30 15:51:46.614 P00   INFO: WAL segment 000000010000000000000000B
successfully archived to
'/var/lib/pgbackrest/archive/pg1/12-1/0000000100000000/00000001000000000000
000B-ef7b8a4cb2cad342a2c37131196d6ae0b3807950.gz'
2020-05-30 15:51:46.715 P00   INFO: check command end: completed
successfully (10126ms)
```

If everything is OK, we will receive a completed-successfully message; now we are ready to manage continuous backup.

Managing basebackups

As we previously mentioned, pgbackrest is able to handle full backups, differential backups, and incremental backups with a simple command-line statement.

To create a full basebackup, we can do this:

```
root@pgdr:/var/lib/pgbackrest# sudo -iu postgres pgbackrest --stanza=pg1 --
type=full backup
```

When we press the *Enter* key on the keyboard, if everything is OK, we get this message:

```
2020-05-30 15:58:09.374 P00 INFO: expire command end: completed
successfully (536ms)
```

Now, if we want information about our repository, we can use the info command as follows:

```
# sudo -iu postgres pgbackrest --stanza=pg1 info
stanza: pg1
    status: ok
    cipher: none

    db (current)
```

```
        wal archive min/max (12-1):
0000000100000000000000001/000000010000000000000000D

        full backup: 20200530-155639F
            timestamp start/stop: 2020-05-30 15:56:39 / 2020-05-30 15:58:07
            wal start/stop: 000000010000000000000000D /
0000000100000000000000000D
            database size: 33.5MB, backup size: 33.5MB
            repository size: 4MB, repository backup size: 4MB
```

The `info` command tells us about WAL segments, the full backup start time, the original database size, and the repository backup size.

In a similar way, starting with this full backup, we can make a differential backup:

```
root@pgdr:/var/lib/pgbackrest# sudo -iu postgres pgbackrest --stanza=pg1 --
type=diff backup

2020-05-30 16:06:49.684 P00 INFO: expire command end: completed
successfully (485ms)
```

We can also make an incremental backup:

```
# sudo -iu postgres pgbackrest --stanza=pg1 --type=incr backup

2020-05-30 16:07:32.044 P00   INFO: expire command end: completed
successfully (511ms)
```

Now an `info` command will track the three backups:

```
# sudo -iu postgres pgbackrest --stanza=pg1 info
stanza: pg1
    status: ok
    cipher: none

    db (current)
        wal archive min/max (12-1):
0000000100000000000000001/000000010000000000000011

        full backup: 20200530-155639F
            timestamp start/stop: 2020-05-30 15:56:39 / 2020-05-30 15:58:07
            wal start/stop: 000000010000000000000000D /
0000000100000000000000000D
            database size: 33.5MB, backup size: 33.5MB
            repository size: 4MB, repository backup size: 4MB

        diff backup: 20200530-155639F_20200530-160644D
            timestamp start/stop: 2020-05-30 16:06:44 / 2020-05-30 16:06:48
```

```
                  wal start/stop: 000000010000000000000000F /
000000010000000000000000F
                  database size: 33.5MB, backup size: 8.3KB
                  repository size: 4MB, repository backup size: 430B
                  backup reference list: 20200530-155639F

          incr backup: 20200530-155639F_20200530-160728I
                  timestamp start/stop: 2020-05-30 16:07:28 / 2020-05-30 16:07:30
                  wal start/stop: 0000000100000000000000011 /
0000000100000000000000011
                  database size: 33.5MB, backup size: 8.3KB
                  repository size: 4MB, repository backup size: 429B
                  backup reference list: 20200530-155639F
```

As we have set `repo1-retention-full=2` on the `pgbackrest.conf` file, `pgbackrest` after two backups will delete the first full backup and its linked differential or incremental backups. For example, here's the execution of two full backups:

```
# sudo -iu postgres pgbackrest --stanza=pg1 --type=full backup
2020-05-30 16:15:09.999 P00 INFO: expire command end: completed
successfully (695ms)
# sudo -iu postgres pgbackrest --stanza=pg1 --type=full backup
2020-05-30 16:16:45.930 P00  INFO: expire command end: completed
successfully (509ms)
```

We will then have the following outcome:

```
# sudo -iu postgres pgbackrest --stanza=pg1 info
stanza: pg1
    status: ok
    cipher: none

    db (current)
        wal archive min/max (12-1):
000000010000000000000001/000000010000000000000015

        full backup: 20200530-161343F
                timestamp start/stop: 2020-05-30 16:13:43 / 2020-05-30 16:15:08
                wal start/stop: 0000000100000000000000013 /
0000000100000000000000013
                database size: 33.5MB, backup size: 33.5MB
                repository size: 4MB, repository backup size: 4MB

        full backup: 20200530-161538F
                timestamp start/stop: 2020-05-30 16:15:38 / 2020-05-30 16:16:44
                wal start/stop: 0000000100000000000000015 /
0000000100000000000000015
                database size: 33.5MB, backup size: 33.5MB
```

```
           repository size: 4MB, repository backup size: 4MB
root@pgdr:/var/lib/pgbackrest#
```

As we can see, the system has automatically deleted the first full backup and its related incremental and differential backups.

Managing PITR

In this section, we will look at how to restore a PostgreSQL cluster after a disaster.

To build an example, let's see what time it is on the PostgreSQL server:

```
db1=# select now();
 now
-------------------------------
 2020-05-30 16:23:38.609572+02
(1 riga)
```

Now let's see the records that are present in the users table:

```
db1=# select count(*) from users;
 count
-------
 10000
(1 riga)
```

Let's suppose that a disaster has happened after this point in time; for example, suppose that we dropped a table after this time:

```
db1=# drop table users;
DROP TABLE
```

Now let's try to make a recovery at 2020-05-30 16:23:3, which is the time before the disaster happened. On the pg1 server, we need to stop the postgresql server:

```
# systemctl stop postgresql
```

Then we perform the pgbackrest restore command:

```
# sudo -u postgres pgbackrest --stanza=pg1 --delta --log-level-console=info
--type=time "--target=2020-05-30 16:23:38" restore

2020-05-30 16:40:02.588 P00   INFO: restore command end: completed
successfully (3406ms)
```

Now let's start the `postgresql` server:

```
# systemctl start postgresql
```

Then we check the `postgresql` log:

```
2020-05-30 16:53:38 CEST [1438]: user=,db=,app=,client= LOG:  restored log
file "000000020000000000000016" from archive
2020-05-30 16:53:38 CEST [1438]: user=,db=,app=,client= LOG:  recovery
stopping before commit of transaction 508, time 2020-05-30
16:52:48.399783+02
2020-05-30 16:53:38 CEST [1438]: user=,db=,app=,client= LOG:  recovery has
paused
2020-05-30 16:53:38 CEST [1438]: user=,db=,app=,client= HINT:  Execute
pg_wal_replay_resume() to continue.
```

As we can see, to end our PITR procedure, PostgreSQL suggests we execute `pg_wal_replay_resume()`. So, let's go into the PostgreSQL environment and perform the following:

```
postgres=# select pg_wal_replay_resume();
 pg_wal_replay_resume
----------------------

(1 riga)
```

Now if we go to check our database, `db1`, the `users` table is now present and the database is now in the state that it was in at `2020-05-30 16:23:38`:

```
 Schema |     Name     |      Type      |   Owner
--------+--------------+----------------+------------
 public | f_t2         | foregn table   | postgres
 public | t1           | table          | postgres
 public | users        | table          | postgres
 public | users_id_seq | sequence       | postgres
(4 rows)
```

Finally, we can execute this:

```
db1=# select count(*) from users;
 count
-------
 10000
(1 riga)
```

We have now restored the situation that was present before the disaster.

Summary

In this chapter, we have tried to show some extensions and some tools available for PostgreSQL. We chose not to give a rundown of everything that is available for PostgreSQL but instead to look specifically at some tools and extensions. We have talked in more detail about `pgbackrest`, which is a very useful tool for managing recovery and PITR. We also talked about how to do `like` searches using GIN indices and how to connect PostgreSQL to other data sources using foreign data wrappers. We saw a different way of using GIN indices in order to have a feature present on Oracle using the `btree_gin` extension.

In the next chapter, we will look at the next stage of the evolution of PostgreSQL: PostgreSQL 13.

References

- https://www.postgresql.org/docs/12/pgtrgm.html
- https://wiki.postgresql.org/wiki/Foreign_data_wrappers
- https://www.postgresql.org/docs/12/postgres-fdw.html
- https://en.wikipedia.org/wiki/Inverted_index
- https://www.postgresql.org/docs/12/btree-gin.html
- http://hlinnaka.iki.fi/2014/03/28/gin-as-a-substitute-for-bitmap-indexes/
- https://wiki.postgresql.org/wiki/Binary_Replication_Tools
- https://pgbackrest.org/
- https://github.com/darold/pgbadgerhttp://pgbadger.darold.net/samplev7.html
- http://pgbadger.darold.net/

20
Toward PostgreSQL 13

In this final chapter of the book, you will discover the new features that the upcoming PostgreSQL 13 release is going to introduce. You will also learn what the main differences between PostgreSQL 13 and the previous stable 12 version are and how to prepare for a version upgrade. Presenting every new feature in detail is out of the scope of the chapter, and a few changes have been already introduced in the previous chapters and will be reviewed here.

The new features have been organized into categories as follows:

- Replication
- Administration of the cluster
- The command-line interface (`psql`)
- Performance
- Backup and related tools

As you can imagine, not every feature fits into a single category, and not every category can summarize the new features, but what's presented in this chapter should help you get a clear idea of what is new in PostgreSQL 13.

The following topics are covered here:

- Introducing PostgreSQL 13 and its new features
- Upgrading to PostgreSQL 13

Introducing PostgreSQL 13's new features

In every release, the PostgreSQL Global Development Group introduces a set of new features that span from performance improvements to configuration management to adherence to the SQL standard and so on. PostgreSQL 13 follows its own path and introduces a rich set of new features; many of them are "under the hood" and so you will not perceive them during day-to-day database usage, but they are really important for the cluster to work more efficiently.

This section provides you with a categorization of the main features introduced with the upcoming PostgreSQL 13 release, which at the time of writing is still in beta 2 release, so it's *stable enough* for usage and testing but not yet production-ready. As you can imagine, it is not possible to explain (or even worth explaining) every single feature in detail, so you will just get an at-a-glance idea of the main changes. The changes are presented in no particular order.

Replication

There are several important changes to the replication aspects of the cluster. One important change is related to streaming replication: now stand-bys can apply configuration parameter changes without needing to be restarted, but rather by using a simple *reload* event.

Another important change is to do with the fact that, during crash recovery, if the `recovery_target` point in time specified is not reached, the server generates an error. Before this, if, for example, you specified a non-existent point in time (for instance, a point in time in the future), the server recovered all the WAL segments and promoted itself as usable; now it will generate an error so that you can be sure that the recovery target point in time is set up correctly.

It is now possible to promote a standby node even if it has paused; therefore, the promotion has a higher priority against the action that pauses the standby node (that is, a mode where the stand-by node waits to continue following the WAL segments).

Great care has been taken regarding replication slots: now WAL receiver processes that are not using a permanent slot can get access to a temporary slot to follow changes from the master (the `wal_receiver_create_temp_slot` tunable), and moreover there is a tunable named `max_wal_slot_keep_size` that indicates how many slots must be kept, invalidating those that exceed the mentioned size and therefore freeing resources.

Administration

In general, several administrative commands have been enhanced and enriched with new options and behaviors to simplify the tasks they are related to.

VACUUM now has a PARALLEL option that allows the administrator to specify the number of parallel worker processes that will work over indexes.

There is now an ALTER VIEW version that accepts RENAME to change the name of a column in a view, something that was not possible before.

The DROP DATABASE statement now has a WITH FORCE option that immediately disconnects users using a database that is going to be dropped, making sure, that the database drop immediately succeeds. In previous PostgreSQL versions, the command fails to drop a database if there are users connected to the database.

There is an OVERRIDING USER VALUE clause for the INSERT statement that allows the overriding of all nonidentity and other columns in a table, in a way similar to the pre-existing OVERRIDING SYSTEM VALUE (which acts on identity columns only).

The TRUNCATE statement is now faster on large tables.

The EXPLAIN command has now a WAL option that allows the inspection of WAL usage, as explained in Chapter 13, *Indexes and Performance Optimization*.

Several system views have been introduced and improved:

- pg_stat_progress_basebackup has been added to report information about a streaming base backup.
- pg_stat_progress_analyze has been added to report progress about ANALYZE execution.
- pg_stat_replication has been improved to report more information about logical replication.
- pg_stat_activity has been improved to report the leading process in the case of parallel execution.

psql

The command-line client `psql` has been improved and now includes a more verbose description of tables and data structures, as well as some other changes to the line prompt. For example, it now automatically marks the presence of an unsaved transaction:

```
$ psql -U postgres template1
psql (13beta2)
Type "help" for help.

template1=# BEGIN;
BEGIN
template1=*# SELECT current_date;
 current_date
--------------
 2020-07-25
(1 row)

template1=*# ROLLBACK;
ROLLBACK
template1=#
```

As you can see, there is a `*` mark after `BEGIN`, making it clear that the user is within a new transaction. If the transaction aborts, the symbol is changed into a `!`. Using the special `\dt+` command, it is even possible to see the storage used by a table, whether it is permanent (the default), unlogged, or temporary:

```
forumdb=> \dt+
                            List of relations
 Schema |    Name    | Type  |  Owner   | Persistence |    Size     |
Description
--------+------------+-------+----------+-------------+------------+-------
------
 public | categories | table | postgres | unlogged    | 16 kB       |
 public | users      | table | postgres | permanent   | 8192 bytes  |
(2 rows)
```

Other minor changes to similar introspection commands have been introduced.

Performance

Pl/pgSQL has been improved and can now execute simple and immutable expressions in a faster way.

There is now a lookup mechanism to speed-up the conversion between numbers and text.

The optimizer has been improved in order to get better row estimations in particular cases, and moreover it is now possible to set the statistics target even for user-defined extended statistics. What this means is that the optimizer can be instrumented to better understand the underlying data, thus providing a better execution plan.

Sortings have been improved too, and now it is possible to specify `enable_incrementalsort` (already activated by default) so that the system can exploit an already-sorted result: if the result is sorted by a few leading keys, there is the possibility of continuing the sorting while concentrating only on the remaining trailing keys for the final result, instead of doing the whole sort over again.

The new configuration parameter `wal_skip_threshold` allows the specification of a limit for the size of data, beyond which `COMMIT` will not force a WAL full-page write but rather sync the data files on disk directly. If the WAL level is *minimal* (that is, no replication), and at the `COMMIT` event the amount of data is greater than `wal_skip_threshold`, then the data pages are synced on storage directly without the data page being included in the WAL logs as a full-page write. This can help to fine-tune the concurrency about transactions.

Backup tools

The `pg_verifybackup` tool has already been introduced in `Chapter 15`, *Backup and Restore*, and has been added to PostgreSQL 13 to allow you to rely on your backups to not have been messed up by some disaster.

The other backup-related tools have been changed too. For example, `pg_basebackup` now provides an estimation of the work to be performed, and this is used to feed the special `pg_stat_progress_basebackup` view described earlier.

The `pg_rewind` tool, even if not strictly speaking a backup tool, has been improved to use the stand-by recovery command to fetch needed WAL segments. Moreover, it now supports the configuration of the stand-bys in a similar way to what `pg_basebackup` does (that is, by writing the `recovery.signal` file).

Now that you know what the main new features of PostgreSQL 13 are, it is time to discuss how to get PostgreSQL 13 running on top of your existing cluster, which means how to upgrade to version 13. The next section briefly discusses the upgrading process.

Upgrading to PostgreSQL 13

When upgrading from a previous PostgreSQL version, there are different approaches that you can follow depending on the specific context, the amount of data you need to migrate, and the version you are coming from.

One approach that always works is to *dump* and *restore* the databases: you execute pg_dumpall against your previous cluster and pg_restore against the PostgreSQL 13 one. The main advantages of this approach are that it is simple and works with any version of PostgreSQL you are upgrading from and to. The main drawbacks are that it requires an *off-line* migration, meaning you are going to have a period of time where the two databases are not usable; plus, it can require a lot of time and space depending on how much data you need to migrate.

Another approach is to use pg_upgrade, a tool designed to transfer and migrate data from one cluster to another one side by side. The idea is that pg_upgrade will have access to both the clusters at the same time and will perform all the required operations to "move" data from one cluster to the other one, under some circumstances even without copying the data but linking it to save storage space. Describing pg_upgrade in detail is out of the scope of this book.

Lastly, if you are upgrading from a quite recent version of PostgreSQL (at least 10), you can set up a logical replication between your current cluster and the PostgreSQL 13 one. Once the replication has been completed, you can perform the switch over and leave PostgreSQL 13 as your new cluster.

Summary

PostgreSQL 13 presents a lot of changes: many of them will not be immediately visible to the user, while others will be available to administrators, and others still will be integrated into the command-line tools. Describing all the new features is almost impossible, because every new release includes a lot of human work, and that work goes into making PostgreSQL more stable, usable, and efficient.

Now, thanks to the knowledge you have gained throughout this whole book, you are able to install PostgreSQL 13 or even a higher version and go find the features you like the most and fit your needs best. Use this book as a teammate during your journey of exploring PostgreSQL, and feel free to jump back and forth between chapters depending on what aspect of PostgreSQL 13 you are learning about or faced with.

References

- PostgreSQL 13 beta 2 release notes: `https://www.postgresql.org/docs/13/release-13.html`
- The `pg_upgrade` migration tool – official documentation available at `https://www.postgresql.org/docs/13/pgupgrade.html`

Other Books You May Enjoy

If you enjoyed this book, you may be interested in these other books by Packt:

Mastering PostgreSQL 12

Hans-Jürgen Schönig

ISBN: 978-1-83898-882-1

- Understand the advanced SQL functions in PostgreSQL 12
- Use indexing features in PostgreSQL to fine-tune the performance of queries
- Work with stored procedures and manage backup and recovery
- Master replication and failover techniques to reduce data loss
- Replicate PostgreSQL database systems to create backups and to scale your database
- Manage and improve the security of your server to protect your data
- Troubleshoot your PostgreSQL instance for solutions to common and not-so-common problems

PostgreSQL 12 High Availability Cookbook - Third Edition
Shaun Thomas

ISBN: 978-1-83898-485-4

- Understand how to protect data with PostgreSQL replication tools
- Focus on hardware planning to ensure that your database runs efficiently
- Reduce database resource contention with connection pooling
- Monitor and visualize cluster activity with Nagios and the TIG (Telegraf, InfluxDB, Grafana) stack
- Construct a robust software stack that can detect and avert outages
- Use multi-master to achieve an enduring PostgreSQL cluster

Leave a review - let other readers know what you think

Please share your thoughts on this book with others by leaving a review on the site that you bought it from. If you purchased the book from Amazon, please leave us an honest review on this book's Amazon page. This is vital so that other potential readers can see and use your unbiased opinion to make purchasing decisions, we can understand what our customers think about our products, and our authors can see your feedback on the title that they have worked with Packt to create. It will only take a few minutes of your time, but is valuable to other potential customers, our authors, and Packt. Thank you!

Index

about 18, 20, 40, 323, 353, 354, 355, 356,
  497, 534
as rescue method, in event of crash 356
checkpoints 357, 358

X

XID wraparound problem 332, 333

www.ingramcontent.com/pod-product-compliance
Lightning Source LLC
Chambersburg PA
CBHW060920060326
40690CB00041B/2725